Ion Mobility Spectrometry- Mass Spectrometry

Theory and Applications

Ion Mobility Spectrometry-Mass Spectrometry

Theory and Applications

Edited by

Charles L. Wilkins
Sarah Trimpin

CRC Press
Taylor & Francis Group
Boca Raton London New York

CRC Press is an imprint of the
Taylor & Francis Group, an **informa** business

Ion mob
Charle
    p
Includ
ISBN
1. Ion
Lee), 193

QD96.P
543'.65--

Contents

Fundamentals

Instrumentation

Applications

Preface

Rapid advances in ion mobility spectrometry–mass spectrometry (IMS-MS) are beginning to have significant impact on biological and materials research and in analytical laboratories worldwide. Although the history of ion mobility spectrometry goes back at least 40 years, when it was more commonly known as plasma chromatography, the more recent hyphenation of ion mobility with mass spectrometers has greatly expanded its scope. Thus, IMS is a gas-phase separation method that is applicable for a wide variety of substances. For example, many airport explosives detectors are currently based upon IMS. However, it can also be used for biomedical applications, such as understanding the factors associated with Alzheimer's disease or cancer, as will be seen in several of the chapters.

Introduction of more or less turnkey commercial IMS-MS instruments are just now making this technology generally available. The analytical power afforded by IMS-MS instruments is certain to drive the technology from research to analytical laboratories. Thus, this book appears at a critical time and presents contributions from developers, as well as more recent users of this technology. It is a goal of this work to provide key information in a single location to help readers appreciate the value of having molecular size and shape information combined with the well-known analytical advantages of high-performance mass spectrometry. Armed with this understanding, it is expected that some of the readers will develop sufficient appreciation of the possible analytical uses of IMS-MS that they will become interested in further exploring the power of the method.

To this end, both fundamentals and applications are presented. Accordingly, the book begins with an overview chapter and fundamentals (Chapters 1 to 3) followed by sections emphasizing instrumentation (Chapters 4 to 7) and ionization sources (Chapters 8 and 9). In the subsequent applications (Chapters 10 to 16), homebuilt and commercial instrumentation using electrospray ionization and matrix-assisted laser desorption/ionization methods are employed to solve biological and synthetic motivated questions. In this way, it is the intent of the editors to cover the current status of IMS-MS in such a way as to make it convenient for those readers unacquainted with this technique to understand its fundamental theory and practical applications. As a consequence, it is expected that this volume could serve as a useful specialized textbook for an advanced course on IMS-MS.

Sarah Trimpin
Detroit, Michigan

Charles L. Wilkins
Fayetteville, Arkansas

Publisher's Note: Please see CD, which can be found at www.routledge.com/9781439813249, for color figures.

The Editors

Charles l. Wilkins is currently a Distinguished Professor in the Department of Chemistry and Biochemistry at the University of Arkansas (Fayetteville). His interests include mass spectrometry of polymer and copolymer materials, Fourier transform mass spectrometry, and the development of new methods to improve the utility of analytical mass spectrometry. Past research has dealt with applications of laboratory computers in chemistry, graph theoretic analysis of chemical problems, and research in chemometrics. Investigations of hyphenated analytical systems such as gas chromatography–infrared mass spectrometry and HPLC-NMR have also been of interest. He is the author of more than 235 scientific papers and 21 book chapters, in addition to editing eight books on a variety of chemistry topics.

He has received a number of honors recognizing his research contributions. Among them are the Tolman Medal of the Southern California American Chemical Society, the New York Section of the Society for Applied Spectroscopy Gold Medal, the American Chemical Society Franklin and Field Award for Outstanding Achievement in Mass Spectrometry, the Eastern Analytical Symposium Award for Outstanding Achievements in the Field of Analytical Chemistry, the Pittsburgh Analytical Award, and the University of Oregon Alumni Award for Outstanding Achievement in Pure Chemistry. He is a Fellow of the American Association for the Advancement of Science, a Fellow of the Society for Applied Spectroscopy, and a Fellow of the American Chemical Society. He is a lifetime Honorary Member of the Society of Applied Spectroscopy.

Professor Wilkins has served the chemistry profession through membership on numerous editorial advisory boards, including those of *Analytical Chemistry*, the *Journal of the American Society for Mass Spectometry*, *Applied Spectroscopy Reviews*, and *Mass Spectrometry Reviews*, among others. He currently serves as a contributing editor for *Trends in Analytical Chemistry*, a position that he has held for almost 20 years.

Sarah Trimpin is an assistant professor at Wayne State University with interest in improving and applying mass spectrometry to difficult problems involving both complexity and insolubility. She obtained the PhD equivalent from the Max-Planck-Institute for Polymer Research, Mainz, Germany, where she pioneered the development of the solvent-free matrix-assisted laser desorption (MALDI) method demonstrating its potential with insoluble materials. After completing a postdoctoral joint position between Oregon State University, Corvallis, and the Oregon Health & Science University, Portland, Oregon, she joined David Clemmer's laboratory at Indiana University, Bloomington, IN, as a senior research associate to study ion mobility spectroscopy–mass spectrometry (IMS-MS) instrumentation and methods. Combining solvent-free MALDI and IMS-MS has led to a total solvent-free analysis approach to analyze solubility-restricted materials. She recently discovered a new ionization method she named laserspray ionization, which combines the attributes of MALDI and electrospray ionization. The long-term goals of her research are to develop methods and instrumentation for the structural characterization and imaging of the soluble *and* insoluble components in single cells.

Dr. Trimpin has over 40 publications, including four book chapters, four reviews, and one perspective article, and has given numerous invited lectures at national and international meetings. She has received a number of honors including the German Society for Mass Spectrometry Wolfgang-Paul-Studienpreis and the Wolfgang-Paul-Promotionspreis. She was highlighted as one of *Genome Technology* magazine's Rising PIs and recently received the NSF CAREER award, the American Society for Mass Spectrometry Research Award, and the DuPont Company Young Investigator Award.

Contributors

Peter B. Armentrout
Department of Chemistry
University of Utah
Salt Lake City, Utah

Perdita E. Barran
School of Chemistry
University of Edinburgh
Edinburgh, United Kingdom

Gökhan Baykut
Bruker Daltonik GmbH
Bremen, Germany

Michael T. Bowers
Department of Chemistry and
    Biochemistry
University of California Santa Barbara
Santa Barbara, California

Richard M. Caprioli
Mass Spectrometry Research Center
Departments of Chemistry and
    Biochemistry
Vanderbilt University School of Medicine
Nashville, Tennessee

Emmanuelle Claude
Waters Corporation
Manchester, United Kingdom

David E. Clemmer
Department of Chemistry
Indiana University
Bloomington, Indiana

Claudia Cozma
Laboratory of Analytical Chemistry
    and Biopolymer Structure Analysis
Department of Chemistry
University of Konstanz
Konstanz, Germany

Juan Fernandez de la Mora
Mechanical Engineering Department
Yale University
New Haven, Connecticut

Michael Desor
Waters Corporation-MS Technologies
    Centre
Manchester, United Kingdom
and
Waters GmbH
Eschborn, Germany

Peter A. Faull
LGC Ltd.
Teddington
Middlesex, United Kingdom

Facundo M. Fernández
School of Chemistry and Biochemistry
Georgia Institute of Technology
Atlanta, Georgia

Jochen Franzen
Bruker Daltonik GmbH
Bremen, Germany

Jennifer Gidden
Arkansas Statewide Mass Spectrometry
    Facility
University of Arkansas
Fayetteville, Arkansas

Kevin Giles
University of the Sciences in
    Philadelphia
Philadelphia, Pennsylvania

Martin Green
Waters Corporation
Manchester, United Kingdom

Glenn A. Harris
School of Chemistry and Biochemistry
Georgia Institute of Technology
Atlanta, Georgia

Herbert H. Hill, Jr.
Department of Chemistry
Washington State University
Pullman, Washington

Ellen D. Inutan
Department of Chemistry
Wayne State University
Detroit, Michigan

Marius-Ionuţ Iuraşcu
Laboratory of Analytical Chemistry
    and Biopolymer Structure Analysis
Department of Chemistry
University of Konstanz
Konstanz, Germany

Shelley N. Jackson
NIDA IRP, NIH
Baltimore, Maryland

Kimberly Kaplan
Department of Chemistry
Washington State University
Pullman, Washington

Mark Kwasnik
School of Chemistry and Biochemistry
Georgia Institute of Technology
Atlanta, Georgia

James Langridge
Waters Corporation-MS Technologies
    Centre
Manchester, United Kingdom
and
Waters GmbH
Eschborn, Germany

Barbara S. Larsen
DuPont Corporate Center for Analytical
    Sciences
Wilmington, Delaware

Hilary Major
Waters Corporation
Manchester, United Kingdom

Jody C. May
Laboratory for Biological Mass
    Spectrometry
Department of Chemistry
Texas A&M University
College Station, Texas

Bryan J. McCullough
LGC Ltd.
Teddington
Middlesex, United Kingdom

Charles N. McEwen
University of the Sciences in
    Philadelphia
Philadelphia, Pennsylvania

John A. McLean
Department of Chemistry
Vanderbilt Institute of Chemical
    Biology
Vanderbilt Institute for Integrative
    Biosystems Research and Education
Vanderbilt University
Nashville, Tennessee

Jasna Peter-Katalinić
Institute for Medical Physics and
    Biophysics
Westfalian Wilhelms University
    Muenster
Muenster, Germany
and
Department of Biotechnology
University of Rijeka
Rijeka, Croatia

Michael Przybylski
Laboratory of Analytical Chemistry
    and Biopolymer Structure Analysis
Department of Chemistry
University of Konstanz
Konstanz, Germany

Oliver Raether
Bruker Daltonik GmbH
Bremen, Germany

Whitney B. Ridenour
Mass Spectrometry Research Center
Departments of Chemistry and
    Biochemistry
Vanderbilt University School of
    Medicine
Nashville, Tennessee

David H. Russell
Laboratory for Biological Mass
    Spectrometry
Department of Chemistry
Texas A&M University
College Station, Texas

J. Albert Schultz
Ionwerks Inc.
Houston, Texas

Nick Tomczyk
Waters Corporation-MS Technologies
    Centre
Manchester, United Kingdom
and
Waters GmbH
Eschborn, Germany

Sarah Trimpin
Department of Chemistry
Wayne State University
Detroit, Michigan

Sergey Y. Vakhrushev
Copenhagen Center for Glycomics
Department of Cellular and Molecular
    Medicine
Panum Institute
University of Copenhagen
Copenhagen, Denmark

Oliver von Halem
Bruker Daltonik GmbH
Bremen, Germany

Amina S. Woods
NIDA IRP, NIH
Baltimore, Maryland

Thomas Wyttenbach
Department of Chemistry and
    Biochemistry
University of California Santa Barbara
Santa Barbara, California

Fundamentals

1 Developments in Ion Mobility

Theory, Instrumentation, and Applications

Thomas Wyttenbach, Jennifer Gidden, and Michael T. Bowers

CONTENTS

1.1 INTRODUCTION

In this chapter we focus on ion mobility spectrometry (IMS) employed to obtain structural information of polyatomic ions. In these applications—reviewed by Clemmer and Jarrold,[1] Wyttenbach and Bowers,[2] Creaser et al.,[3] Weis,[4] and

Bohrer et al.[5]—IMS always occurs in combination with mass spectrometry (MS). We divide the chapter into three sections covering theory, instrumental aspects, and research applications from our lab. In the theoretical section we cover some of the basic theory of IMS, how ions exposed to an electric field move in a buffer gas, and how this motion relates to the structure of the ion. In the instrumentation section we present basic hybrid IMS-MS setups and their components and discuss issues to consider in the design of such an instrument. And finally, in the applications section we present a few instructive application examples from our lab to illustrate the concepts outlined in the theory and instrument sections and to demonstrate the potential of the method to solve biochemically relevant problems.

1.2 THEORY

1.2.1 Ion Mobility

Ions exposed to an electric field experience a force and are accelerated along the field lines. Upon addition of a buffer gas, the motion of the ions becomes more complicated as collisions with the gas scatter the ions in random directions as it diffuses. However, if an ion cloud is given enough time to reach equilibrium and the electric field is uniform throughout, the ion cloud will travel with constant velocity parallel to the field lines and simultaneously grow in size due to diffusion. This constant equilibrium velocity is the result of forward acceleration by the field and decelerating friction by collisions. Following Mason and McDaniel,[6] for weak electric fields of magnitude E, the drift velocity v is directly proportional to E with the proportionality constant K called ion mobility

$$v = K E \tag{1.1}$$

Since v is inversely proportional to the buffer gas number density N, the mobility K is also inversely proportional to N. Here N (in units of molecules per volume) is used as the relevant quantity to express pressure because N is, in contrast to pressure p, decoupled from the temperature T. Because K depends on N it is practical to convert K into the pressure-independent quantity $K_o \propto NK$, where K_o is termed the reduced mobility

$$K_o = \frac{p}{p_o} \frac{T_o}{T} K \tag{1.2}$$

with the constants $p_o = 760$ Torr and $T_o = 273.15$ K.

A field is considered weak if the average ion energy acquired from the field is small compared to the thermal energy of the buffer gas molecules. This ion field energy is proportional to v^2 or $(KE)^2$. However, for a given ion with given $K_o \propto NK$ it is the ratio E/N which determines whether a field is weak or strong, and collisional heating due to the field is given by Equation (1.3):

$$T_{eff} - T = (M/3k_B) (NK)^2 (E/N)^2 \tag{1.3}$$

Here T_{eff} is the effective temperature, M the mass of the buffer gas particle, and k_B the Boltzmann constant.

Following Equation (1.1), a measurement of K involves measuring the drift time t of a pulse of ions traveling in a weak field E over a given drift length L. The spread, Δx, of a cloud of identical ions due to diffusion, the random part of motion, is given by

$$\Delta x = \sqrt{\frac{4k_B TL}{\pi Ee}} = \sqrt{\frac{4k_B TL^2}{\pi Ve}} \tag{1.4}$$

where e is the ion charge and V the voltage across the drift region.

Because the resulting acceleration of the ion by E is proportional to e, the mobility is also proportional to e. In addition, because the deceleration of the ion by friction is proportional to the buffer gas number density N and because large ions experience more friction than small ions, the mobility is inversely proportional to both N and the collision cross section σ of the ion. Using momentum transfer theory, a statistical approach to balance ion energy and momentum gained in the electric field and lost in buffer gas collisions, a quantitative relationship between K and the quantities e, N, and σ can be derived,

$$K = \frac{3e}{16N} \sqrt{\frac{2\pi}{\mu k_B T}} \frac{1}{\sigma} \tag{1.5}$$

where $\mu = Mm/(M + m)$ is the reduced mass of buffer gas (with mass M) and ion (with mass m).

1.2.2 Collision Cross Section

The collision cross section is a measure of the size of the ion. However, size in this context is determined by a number of factors, all ultimately determined by the ion–buffer gas interaction potential extending in all three dimensions of space and the energy distribution of the particles. Depending on the system and conditions, either the repulsive wall of atoms on the surface of a polyatomic ion may constitute the relevant part of the interaction potential, or the long-range attractive terms may also be important. Hence, ion–neutral collision theory dictates that not only does a large object extending in space (such as a protein molecule) have an appreciable collision cross section, but also an infinitely small point charge.

If the objective is to obtain structural information from cross section measurements, hard sphere-type collisions are desirable and contributions to σ by long-range interactions should be minimized. This is achieved by avoiding low temperatures and by using a buffer gas such as helium, which does not have a dipole moment and has a low polarizability. In addition, ion mobility is in a better position to deliver size and shape information for large polyatomic ions composed of many atoms (e.g., proteins) compared to small ions composed of few atoms where G depends on the details of the interaction potential.

Evaluating the cross section of a model structure of a polyatomic ion appears to be a fairly simple matter, at least to a first approximation. Placing spheres at the positions of each atom and projecting a shadow of the molecule onto a screen yields a cross section value corresponding to the collision area, and averaging the projection cross section over all possible molecule orientations yields an averaged cross section. Using this type of model to test candidate structures of molecules with unknown structure only works if atomic radii are known, and this is the tricky part. Researchers working in the field noticed early on that atomic radii are not constant from system to system, from molecule to molecule, or even from conformation to conformation of the same molecule in extreme cases.[7–9] Thus it is found that the fullerene carbon atom radius measured at 300 K is not the same as that at 400 K. Similarly a carbon atom in fullerene C_{60} does not have the same radius as a carbon atom in glycine, and the glycine atomic radii in the free amino acid are not the same as those of glycine embedded in a protein. Therefore, using the simple projection cross section with fixed atomic radii can lead to errors when comparing its predictions to ion mobility data.

There are at least two important reasons for the failure of the simple projection approximation. First, the long-range potential in the ion–neutral interaction is not negligible in the collision process at room temperature or below, since the well depth ε of the interaction potential is generally not small compared to the thermal energy k_BT at 300 K even if helium is used as a buffer gas. Simply increasing T to decrease the ε/k_BT ratio is usually not a good solution for large polyatomic ions because the increased thermal motion may lead to undesirable effects such as change of structure (unfolding of polymers/denaturing of proteins) and thermal decomposition. Hence a model including adjustments of atomic radii according to temperature and ion size and type is needed to minimize these problems.

However, a second problem persists for any projection model: The experimental quantity σ is a collision cross section, not a projection cross section. Therefore, a momentum transfer cross section must be evaluated so that a more successful comparison of a model structure with experiment can be made. It is intuitively clear that momentum transfer depends on the scattering angle of the colliding particles, a parameter that is not considered at all in the projection approximation. And it is also obvious that evaluating scattering angles can be a tricky task especially for particles with concave surfaces where multiple bounces off the surface are possible within one collision trajectory.[10] Hence more sophisticated cross section models include propagation of trajectories, derivation of the statistics of scattering angles, and ultimately evaluation of a set of irreducible collision integrals $\overline{\Omega}^{(\ell,s)}$.[6,7] Kinetic theory readily establishes a connection between the most relevant term, $\overline{\Omega}^{(1,1)}$, and a first order approximation $[K]_1$ of the mobility

$$[K]_1 = \frac{3e}{16N}\sqrt{\frac{2\pi}{\mu k_BT_{eff}}}\frac{1}{\overline{\Omega}^{(1,1)}(T_{eff})} \tag{1.6}$$

For the applications discussed here (low field limit) $[K]_1$ is a very good approxima-tion for K and σ is essentially given by the quantity $\overline{\Omega}^{(1,1)}$. Higher order terms with indices $\ell > 1$ and $s > 1$ would be required to improve the accuracy for high fields.

Calculation of $\overline{\Omega}^{(\ell,s)}$ for a given polyatomic candidate structure requires applica-tion of numerical integration methods and evaluation of trajectories in a reasonable ion–neutral interaction potential. Generally a 12-6-4 type of potential (Lennard-Jones and ion–induced dipole interaction) is used in these calculations.[7] However, these trajectory calculations are very time-consuming and therefore a hard sphere-based potential is sometimes used, which cuts the computation time considerably.[10] In particular for large molecular structures (e.g., proteins), $\overline{\Omega}^{(1,1)}$ is given predomi-nantly by the molecular structure and the exact nature of the potential (or the exact size of each atom in the structure) has less effect on $\overline{\Omega}^{(1,1)}$.

1.2.3 RESOLUTION

The resolution of an IMS device depends in practice not only on the spread of ions Δt due to diffusion in the drift region relative to the drift time t, but also on the drift tube-independent spread of the ion cloud, Δt_o, brought about by the initial width of the ion pulse entering the drift region and the spread of ions after the drift region (e.g., in the ion funnel). Following Equation (1.4), the resolution $t/\Delta t$ is affected only by the two parameters: temperature T and drift voltage V.

$$\frac{t}{\Delta t} = \frac{L}{\Delta x} = \sqrt{\frac{\pi V e}{4 k_B T}} \tag{1.7}$$

Reducing T decreases diffusion and thus Δt. Increasing the voltage reduces both t and Δt, with the ratio $t/\Delta t$ being proportional to $\sqrt{V/T}$. However, whereas increasing L is a reasonable approach to improve resolution, increasing E requires simultaneous increase in N in order to keep E/N constant and to stay in the low-field regime desirable for ion mobility experiments. For high-resolution devices the pulse width Δt_o may be an additional limitation to resolution. Once Δt reaches a small value and the resolution is determined by Δt_o, further optimization of the parameters E, N, and T does not further improve the resolution. Hence the design of an ion mobility instrument requires a careful balance of ion energy, technical feasibility, and all the quantities determining the resolution. High voltage sup-plies, insulators, and discharge problems set limits to V; discharge and pumping requirements may set limits to N; space requirements set limits to L; choice of materials, insulation, discharge due to water condensation, and thermal equilibra-tion set limits to T; and speed of switches, ion space charge, and signal intensity set limits to Δt_o.

Therefore, an instrument with given fixed parameters L, Δt_o, and T and with maximum allowable pressure N may operate with maximum resolution at a field value E_{max} smaller than the instrument limit and smaller than allowed due to E/N limitations. Hence whereas the resolution is optimum for $E = E_{max}$, it drops both for

$E < E_{max}$ due to smaller $t/\Delta t \propto \sqrt{EL/T}$ values [Equation (1.7)] and for $E > E_{max}$ due to smaller $t/\Delta t_0 \propto t \propto 1/E$ values [Equation (1.1)].

$$t = \frac{L}{v} = \frac{L}{KE} \tag{1.8}$$

1.2.4 REACTIVE AND DYNAMIC IONS

In the discussion above σ was considered constant with time. However, there are a number of situations where this is not true. In one case, ions may react as they drift, generating product ions with different cross sections than the original ions. Possible reactions occurring in the drift cell include thermal unimolecular decay, dissociation of non-covalently bound complexes, and isomerization. The resulting product ions may not necessarily have a different mass-to-charge ratio (m/z) value than the reactant ions. For example, m/z does not change if a doubly charged dimer dissociates into two singly charged monomers or if isomerization occurs.

The reaction rate relative to the drift time determines whether the arrival time distribution (ATD) is more representative of the mobility of the reactant or of the product. A distribution of ions reacting at various places throughout the drift region gives rise to a broad ATD, covering the range from reactant to product ion mobility. However, in certain cases (e.g., isomerization) back reaction may be possible. If the forward and reverse reactions are rapid, an equilibrium between reactant and product is established and all individual ions spend the same fraction of time as reactant and as product. In this case a narrow ATD is observed at a position corresponding to the equilibrium and with a width given by the instrument resolution.

Another factor that can influence σ is molecular dynamics. Many floppy molecules are extremely flexible and sample a broad range of conformations, each with a slightly different cross section. These intramolecular motions occur on all levels of time frame—from femtosecond to second. Dynamics taking place on the submicrosecond scale are not resolved by IMS but some low-frequency motions may be frozen out in the IMS experiment. However, in many cases local minima separated by substantial barriers simply give rise to broadening of the ATD rather than yielding separate peaks that correspond to a given conformation. The individual conformations may not be resolved due to the presence of many conformations with barely different cross sections and due to limited instrument resolution.

In the examples given below in the application section, ATDs of reactive and dynamic ions are shown. The VEALYL example demonstrates dissociation in the drift tube of a doubly charged dimer into two singly charged monomers. The dCG dinucleotide example shows interconversion of two distinctly different conformations as a function of temperature. The rather complex theory to calculate ATDs of two swarms of ions A and B undergoing the reactions A→B and B→A inside the drift tube has been worked out in a paper by Gatland[11] where a Green's function solution to the transport equations is presented. In this approach the histories of the ions A and B are evaluated as a function of drift time t. For instance, an ion A exiting the drift tube may have started out as A produced in the source, or it may have started as B and converted to A (B→A), or it may have started as A, converted to B, and converted back to A (A→B→A), etc. The mathematical description of this relatively simple scheme becomes fairly complex because the probability of each history is different for each value of t and we refer the interested reader to Gatland's publication.[11]

1.2.5 REACTIVE BUFFER GASES

As mentioned above cross section measurements designed to obtain shape information require an inert buffer gas. However, under certain circumstances it may be desirable to use a reactive buffer gas. For instance, the two steroisomers of a chiral ion may interact differently with a chiral buffer gas and may therefore be distinguished by their ion mobility in the chiral buffer.[12]

For some buffer gases its interaction energy with the ion is a highly relevant quantity of great interest. A good example is the analyte–water interaction energy corresponding to the hydration energy of the analyte. If the formation and dissociation processes of ion–water complexes, $M \cdot (H_2O)_n$, are rapid compared to the ion drift time in water, an equilibrium distribution of the complexes is established, which can be analyzed by mass spectrometry after ions exit the drift tube. The equilibrium constant K_n^{eq}, given by the ratio of ion intensities in the mass spectrum, $[M \cdot (H_2O)_n]/[M \cdot (H_2O)_{n-1}]$ and by the water pressure $p(H_2O)$ relative to the standard pressure $p^\circ = 760$ Torr,

$$M \cdot (H_2O)_{n-1} + H_2O \rightleftarrows M \cdot (H_2O)_n \tag{1.9}$$

$$K_n^{eq} = \frac{[M \cdot (H_2O)_n]}{[M \cdot (H_2O)_{n-1}]} \frac{p^\circ}{p(H_2O)} \tag{1.10}$$

$$\Delta G_n^\circ = -RT \ln K_n^{eq} \tag{1.11}$$

is related to the free energy ΔG_n° of hydration. In order to sort out energetic and entropic contributions to ΔG_n°, the equilibrium measurement has to be carried out at different temperatures. A subsequent van't Hoff analysis yields ΔH_n° and ΔS_n° of hydration.

$$\ln K_n^{eq} = -\frac{\Delta H_n^\circ}{RT} + \frac{\Delta S_n^\circ}{R} \tag{1.12}$$

Hence obtaining thermodynamic information in equilibrium measurements with water or any other ligating buffer gas requires instrumentation very similar or identical to most IMS-MS setups. Specific instrumental requirements are that mass selection occurs after the drift cell and that the drift cell temperature can be varied.

1.3   INSTRUMENTATION

The need to couple IMS and MS can either be driven by the desire to identify IMS features by mass analysis or by the wish to shape-analyze mass-selected ions. Although the two motivations approach the problem from opposite directions, they lead to instrument designs with basically identical capabilities as far as shape-mass analysis is concerned. The first perspective, upgrading an ion mobility spectrometer by adding an MS detector, logically leads to an IMS-MS configuration. The second approach may lead to a modified tandem mass spectrometer with the collision cell replaced by a drift cell.

The most general setup is schematically shown in Figure 1.1. One or several of the components shown in the figure may be omitted in individual designs. Nevertheless, in all cases ions are produced in a source, analyzed by IMS and MS, and finally detected. In addition, the IMS unit typically requires an entrance section to condition the ions prior to IMS analysis and an exit section to collect the ions afterwards. In the following paragraphs each of the individual components of a typical IMS-MS instrument are discussed.

1.3.1   Ion Source

Electrospray ionization (ESI) has emerged as a powerful technique to create ions in many IMS-MS applications, although other ionization methods including matrix-assisted laser desorption/ionization (MALDI) are also common for certain applications. In fact, any ionization method used in IMS or MS will work for the IMS-MS

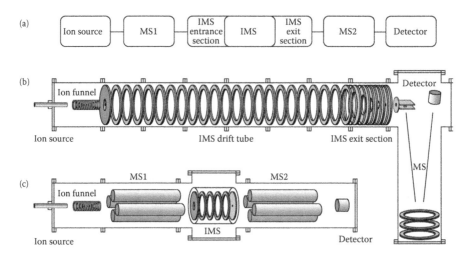

FIGURE 1.1   (a) Schematic setup of an IMS-MS device. The MS1 mass selection step before IMS analysis is omitted in many designs, in particular in combination with high-pressure (≥10 Torr) IMS devices and/or atmospheric pressure ion sources such as ESI. In alternative designs MS2 may be omitted. (b) Schematic drawing of a high-resolution IMS device coupled to a TOF mass spectrometer. (c) Schematic drawing of a mass spectrometer equipped with an IMS drift cell in a quad-EMS-quad arrangement.

combination as well. A review of ion sources is beyond the scope of this chapter and can be found elsewhere[13] and in any book on MS.[14]

1.3.2 IMS ENTRANCE SECTION

IMS analysis requires a narrow pulse of ions entering the drift region. Whereas certain ion sources prepare ion pulses naturally (e.g., MALDI employing a pulsed laser), many other sources produce a continuous ion beam (e.g., ESI). In this case the continuous ion beam has to be converted into a pulsed beam. To increase the duty cycle, this is done most efficiently in an ion trap where ions are accumulated for a period of time and periodically pulsed out. Whereas any type of ion trap works in theory,[15] in practice most designs are based on ion funnels.[16] Ion funnels[17] are a natural choice because they are most often part of the front end of an MS or IMS device anyway, taking care of efficient ion transmission from the source into the analyzer, often through a number of differential pumping stages. Smart ion funnel designs allow storage of a large number of ions[18] and include mechanisms to empty the storage volume efficiently. Such designs include hourglass-shaped funnels with small DC ramps for storage and large ramps for emptying.[19]

Whereas ion pulsing may occur anywhere between the source and the IMS section, ion injection into the drift region is handled directly at the IMS entrance interface. Depending on source design and instrument configuration a vacuum may be required prior to IMS analysis. In this case ions have to be transferred from vacuum into the buffer gas, a process that becomes increasingly more difficult with increasing buffer gas pressure and ion cross section. For high pressures, injection may become nearly impossible. However, the ion injection energy can easily be adjusted and provides a convenient tool to control the ion–neutral collision energy for collisions taking place at the entrance to the drift region before thermal equilibration has occurred. Gentle injection conditions provide just enough energy for ions to reach the drift region against the headwind. More energetic injection processes can be used to induce structural changes or fragmentation in front of the drift region. Both processes may yield important additional information about the ions to be analyzed given that the sequence of high-energy collisions followed by thermal equilibration corresponds to an annealing cycle and may lead to structural annealing. The drawback of an injection setup is the possibility of discrimination effects as gentle injection conditions tend to also yield lower transmission for large ions.

IMS units operating at pressures significantly higher than 1 Torr make ion injection from a vacuum impractical and are generally not compatible with mass selection prior to IMS analysis. In these cases ions are either generated inside the drift chamber or are transferred from the ion source via ion guides and ion funnels. For an ESI source operating under ambient conditions coupled to an IMS unit operating with helium, the IMS interface requires a section with helium counter flow to prevent air from entering the drift region.[19,20] Keeping air out of the drift cell even in trace amounts is very important, since cross sections of ions in air are much larger than in helium. The counter flow of buffer gas is generally achieved by keeping the drift region at a slightly higher pressure than the chamber preceding it. This can technically be accomplished by simultaneously pumping the IMS entrance interface and pressurizing it with helium employing a valve controlled by a feedback circuit involving a differential Baratron pressure gauge, which compares pressures in the entrance and drift section.[19]

A final function of the IMS entrance section is to restrict the flow of buffer gas out of the drift cell. This is accomplished by using a small entrance aperture. If ions enter the drift cell from vacuum then electrostatic lenses can be used to steer the ions and focus them onto the aperture. Otherwise, ion funnels are used to guide and focus them.

In summary, the IMS entrance section is used to convert a continuous ion beam into ion pulses (if necessary), admit ion pulses into the drift region, keep unwanted neutrals out, and restrict the flow of buffer gas out of the drift region.

1.3.3 IMS DRIFT TUBE

Drift tubes come in all shapes and sizes. The drift length may be as short as a few centimeters or as long as several meters. Small drift cells may be fabricated of a cubic piece of copper and inserted into the vacuum chamber of a mass spectrometer. Large drift tubes may be fabricated of cylindrical sections of pipe with a mass spectrometer bolted to the last IMS section. Independent of size, the interior usually looks the same; stacks of equally spaced guard rings connected by a resistor chain provide a uniform electric field. To prevent ion loss, the inside diameter (i.d.) of the guard rings has to be larger than the spread of ions due to diffusion [Equation (1.4)]. The drift voltage supplied to the first and last guard ring is proportional to the drift length for equal pressure and field strength. Hence whereas 50 V may be sufficient for 2 cm of drift length, 5 kV are required for 2 m, and whereas 2 cm i.d. may be sufficient for 2 cm drift length, 20 cm are required for 2 m ($\Delta x \propto \sqrt{L}$).

The IMS resolution does not directly depend on the drift length. Rather, it depends on the square root of the drift voltage. However, longer drift tubes allow larger voltages to be applied across the cell (at constant field), thereby increasing the resolution. Increasing the pressure at constant drift length, on the other hand, will have the same effect. Higher pressure allows application of a higher voltage while keeping the pressure-to-field ratio constant. Drift tubes work equally well at high and low pressure, although voltages >100 V may lead to discharges in a certain pressure-distance range (4 Torr·cm for helium, see Paschen curves).[21,22] However, whereas ions have no difficulty drifting in high pressure, interfacing a high-vacuum chamber at the front and back ends of the drift tube is the challenging part.

In practice, design of a 2-m drift tube operating at 10 Torr and 5 kV and scaled-down versions thereof are manageable. A pressure of ≤10 Torr allows reasonably sized apertures and pumps, although much higher pressures have been used in individual designs.[20] Handling a voltage of 5 kV is reasonable. Significantly higher voltages require significantly more expensive and clunky high-voltage components.

IMS resolution can also be influenced by temperature, although drift tubes with adjustable buffer gas temperature are relatively difficult to design. Careful choice of materials compatible with the temperature range desired is necessary. In addition, care must be taken that the material is allowed to expand and contract with changing temperature without deformation of surfaces involved in making a vacuum seal and without cracking brittle materials such as ceramics and glass. Adjusting the tempera-ture in larger drift tubes creates additional challenges including thermal equilibra-tion, thermal insulation, and water condensation on cold metal parts exposed to air and carrying high voltage.

Measurement of an absolute ion mobility requires knowledge of the drift time t, drift length L, drift voltage V, buffer gas pressure p, and temperature T. All of these parameters can easily be measured with good precision. However, any uncertainty in the parameters t, L, $E = V/L$, N, and the square root of T [Equations (1.1) and (1.5)] translates directly into an uncertainty in the mobility measurement.

With quick ion counting technology available for time-of-flight (TOF) MS and other applications accurate measurement of drift times in the 0.1- to 100-ms time range is not a challenge. A multichannel scaler or similar device is often used to bin arriving ions by time with a resolution in the nanosecond range if desired. Determining the effective drift length accurately may be somewhat of a challenge in short drift cells due to ion injection effects and other perturbations near the entrance and exit holes. Note that accurate knowledge of L is essential as K depends on L^2, according to $K = v/E = (L/t)/(V/L)$. Whereas measuring a voltage is straightforward, accurate measurement of a pressure can be tricky especially if there is a temperature gradient between the drift tube (at high voltage) and the pressure gauge (at ground). However, a factory-calibrated Baratron gauge connected to the drift tube via a rea-sonably dimensioned tube ($\geq$1-cm i.d.) is an appropriate setup for absolute pressure measurements in ion mobility applications. A number of thermocouple tempera-ture sensors or platinum resistance thermometers mounted in various places in and around the drift tube provide a good readout for the temperature and give a measure for potential temperature gradients inside the drift region.

1.3.4 IMS Exit Section

In traditional IMS, ions are collected in a Faraday cup detector as they reach the end of the drift region. In an IMS-MS setup a small aperture at the end of the drift region allows a small fraction of ions to exit the buffer gas chamber and enter a mass spectrometer. In this setup most ions are lost even for short drift lengths with a small spread of the ion cloud and relatively large apertures. As an example, for a 1-cm ion cloud only ~15% of the ions make it through a 1-mm diameter orifice. However, even if pumping capacity allows a 1-mm orifice, it has to be kept in mind that a hole of this size produces a considerable pressure gradient within at least 1 mm inside the IMS chamber towards the hole. Additionally, there is a significant flow of gas out of the cell pushing ions out. In a design that does not address these issues the actual drift length may be considerably shorter than the drift cell length. Hence, from this point of view exit (and entrance) apertures have to be kept as small as possible, typically $\leq$0.5-mm diameter.

If sensitivity is not an issue, the setup with a simple small exit aperture is the most straightforward approach both with respect to design and data analysis. Arrival time and spread of the ion cloud are determined entirely by the drift velocity and diffusion, respectively, in the IMS cell without perturbation by subsequent refocusing devices.

However, for reasonable dimensions (drift length, ion cloud spread, exit hole) ion transmission is <<15% and often <<1% in many IMS designs. Therefore, if sensitivity is an issue, refocusing at the end of the drift region is of paramount importance as it improves the signal for large drift lengths by orders of magnitude. Whereas electrostatic focusing was attempted in early designs,[23] today the ion funnel is the focusing device of choice.[18,19] Hence the IMS exit interface consists of a high-transmission metal screen defining the end of the drift field, an ion funnel focusing the ion cloud down to a small diameter, and an exit aperture into the vacuum of the mass spectrometer at the exit of the funnel. Certain designs may require an additional differential pumping stage between buffer gas chamber and MS chamber. The drawback of a focusing device is that the IMS resolution may suffer due to a spread in time of the different ion trajectories inside the ion funnel. Nevertheless, for most IMS designs inclusion of an exit ion funnel is not just an option, it is required to make it work.

1.3.5 ALTERNATIVES TO IMS DRIFT TUBES

Whereas high-field asymmetric waveform ion mobility spectrometry (FAIMS) is able to separate ions in a manner similar to IMS, the method cannot be considered an alternative to IMS in the context (elucidation of ion structure) discussed here, since electric fields employed in FAIMS are way above the low-field limit,[24–26] However, two other newer developments, traveling-wave IMS and overtone ion mobility spectrometry (OMS), are worth mentioning here briefly.

Traveling-wave IMS became popular with the recent introduction of a commercial mass spectrometer with ion mobility separation capability.[27] This device employs traveling-wave technology in a cell composed of a stack of lenses. The cell acts as a radio frequency (RF) ion guide to trap ions radially, similarly to an ion funnel but with constant lens i.d. and without DC component, with superimposed voltage waves traveling from lens to lens towards the exit of the lens stack. This setup pushes ions through the lens stack in the presence of a buffer gas. Ions catch a wave, ride on it for a while, and eventually fall behind it until they catch the next wave. Ions with a small collision cross section experience less friction by the buffer gas and fall back behind the wave less frequently than large ions. This complicated ion motion executed within the low-field limit leads to a separation of ions based on their ion mobility. Whereas this setup provides a great tool for analytical applications with outstanding sensitivity, it is not designed to measure ion mobilities or cross sections quantitatively. Although recent advances in theory and calibration methods may make it possible to use traveling-wave IMS for cross section measurements, the method employs an intrinsically complex (and not necessarily linear) relationship between ion travel time and the quantity K defined in Equation (1.1) and is therefore not suitable to measure K on first-principles grounds. Theory indicates that the transit ion drift velocity is proportional to the square of K for certain ions, wave speeds, and wave forms, and deviations from the quadratic relationship occur for other ions, wave speeds, and wave forms.[28]

OMS is conceptually identical to traditional IMS but with the drift field broken up into a number of sections.[29,30] There is a weak uniform electric field within each section and a potential wall at the end of each section, lending the overall field a sawtooth-like form. Ions drift within a section as they would in regular IMS, but they get lost once they reach the potential wall. However, every once in a while at a given frequency the sawtooth potential is shifted by a certain length along the drift direction. Thus, ions moving in resonance with the frequency of the shift events never reach any potential wall and never get lost. All other ions will reach the wall eventually and will be filtered out. Hence, OMS provides an ion mobility filter that can be scanned over a mobility range by scanning the frequency. An advantage of OMS over IMS is that high resolution can be obtained for ion signals observed at a frequency corresponding to an overtone of the fundamental resonance frequency. Of course sensitivity rapidly becomes an issue since only a fraction of the ions have the proper frequency to transit the device.

1.3.6 MASS SPECTROMETERS

The IMS-MS combination works well in a number of different configurations. In particular, shape-analysis of mass-selected ions works both in an MS-IMS or IMS-MS configuration and for many different types of mass spectrometers. Quadrupole (quad) mass filters are frequently used and work well for this application. They can be positioned in front or after the IMS tube or in both positions. However, if ionization does not require a vacuum (e.g., ESI) and if tandem MS is not required, the IMS-quad sequence may be more meaningful than quad-IMS from a pressure profile point of view. In the ESI-IMS-quad setup the higher-pressure components are grouped together at the front end of the instrument and the vacuum components, MS and detector, at the back end. A quad-IMS-quad configuration, analogous to a triple quad arrangement in mass spectrometry, is very versatile and allows analysis of chemistry occurring in the IMS cell (fragmentation, clustering), an extremely powerful feature that should not be overlooked in design decisions.[31]

The choice of mass spectrometer type obviously depends on the requirements with respect to mass resolution and other performance parameters. Vacuum requirements for various MS types may be another factor to consider. Since IMS requires a buffer gas, which potentially leaks into the mass spectrometer, not all types of mass spectrometers are easily coupled with IMS to the same degree. TOF mass spectrometers, although they require a rather good vacuum, have emerged as a powerful component in IMS-MS combinations.[19,32] Some of the advantages of TOF MS are the theoretically unlimited mass range, the high sensitivity compared to scanning MS filters where unselected ions are lost, and the high mass resolution in a reflectron arrangement. Additionally, in high-resolution IMS tubes ion clouds are typically spread out in time over many milliseconds, which allows TOF mass analysis on the fly as ions exit the IMS tube. Hence, as ions may be pulsed into the IMS tube at 10 Hz, they may subsequently be pulsed into the TOF analyzer at 1000 Hz. In optimized designs, this type of data collection yields simultaneous mass and shape information for every ion entering the device essentially without any loss of ions.

With the emergence of highly optimized differentially pumped quad-TOF (qTOF) combinations with essentially inseparable components including integrated TOF pulsing elements, IMS-qTOF combinations may in some respect present a less difficult design than coupling TOF straight to IMS, with the added benefit of an MS-MS functionality following IMS.[19]

1.4 APPLICATIONS

Whereas general ion mobility applications range from executing routine analytical tasks to solving specific problems of interest in basic research, the focus of the combined IMS-MS approach was historically on basic research involving ion–neutral interactions[6,33] and in more recent years on structural analysis of polyatomic and macromolecular ions.[2] However, with commercial IMS-MS instruments emerging,[27] applications expand more and more into the field of analytical chemistry where the method is used as a tool to separate complex mixtures of ions and simplify mass spectra. Here, we do not elaborate on this development and focus on examples of IMS-MS applications carried out in our labs, aiming at elucidating structural and energetic aspects of biochemically relevant systems. The examples are specifically chosen to illustrate the concepts outlined above.

1.4.1 PEPTIDES, PROTEINS, AND AGGREGATION MECHANISMS

Peptides and proteins are highly flexible polymers of a set of building blocks—the 20 naturally occurring amino acids. Each amino acid has its own unique properties. With respect to solubility in water, the properties of the amino acid side chains cover the entire spectrum from hydrophobic to hydrophilic, from nonpolar to polar to ionic. In an aqueous environment the combination of flexibility and diversity in side chain hydrophilicity leads to protein monomer conformations with hydrophilic residues exposed to the protein surface and hydrophobic residues buried in the protein interior. However, oligomerization is often an effective alternative in optimizing intermolecular interactions. For many proteins the aggregation process leads to fibrils, large ordered homopolymeric assemblies extending predominantly in one dimension.

In living organisms aggregation of one particular protein is often associated with a particular disease.[34,35] For instance, aggregation of the amyloid β-protein (Aβ) is associated with Alzheimer's disease. The monomer structure of small proteins or peptides such as Aβ is generally best described as a random coil without a well-defined global minimum. In the fibril, however, each monomer unit has a well-defined ordered structure, which can, in some cases, be analyzed by x-ray and nuclear magnetic resonance (NMR) methods. For large aggregating proteins the reaction from monomer to fibril involves a structural transition from a well-defined folded monomer state (often high in α-helix content) to a very different well-defined conformation in the fibril state (often high in β-sheet content). Therefore, diseases connected to protein aggregation are often termed protein misfolding diseases. However, the mechanism for the structural transition is not understood for any protein or peptide. It is not known whether the monomer misfolds, thereby causing aggregation, or whether refolding occurs in the oligomeric state.

Investigation of small oligomers of disease-causing proteins is also relevant because it has been shown for an increasing number of systems that it is these small species that are neurotoxic rather than the macroscopic fibrils, which appear to be simply the final product of aggregation. We are currently investigating a number of aggregating model peptides and disease-related proteins including Aβ and the tau protein (Alzheimer's), α-synuclein (Parkinson's), prion protein (spongiform encephalopathies), and amylin (type II diabetes). Here we focus on methods and conceptual aspects of the problem and report very few specific results; this is not a review or summary of our work in the field. For an overview of results and conclusions we refer to our review articles and original research papers in the literature. [36–43] In addition, several studies, including aggregation of the VEALYL peptide introduced below, are not concluded and are presently in progress.

As mentioned above, understanding the first steps of aggregation, including structural aspects of the monomers involved, is crucial for understanding the mechanism of misfolding diseases and for designing inhibitors. Ion mobility coupled to mass spectrometry is in an excellent position to address structural issues of peptides and proteins and their oligomers. In contrast to bulk measurements where a mixture of monomers and the various oligomeric species has to be analyzed, in the IMS-MS method each oligomeric species can be isolated from the mixture by MS and shape-analyzed by IMS.

The example of the simple pentaglycine (GGGGG) peptide illustrates how shape information is obtained for a monomeric species. [44] Parking the mass filter at $m/z = 304$, the mass of the protonated $(GGGGG)H^+$ monomer, the ion mobility spectrum shown as a solid line in Figure 1.2a is recorded using helium. Experimental parameters such as temperature T, drift voltage V, and pressure p used to obtain this $(V = 96 \text{ V})$ and five additional ATDs are given in Table 1.1 along with the average ion arrival times t_A deduced from the ATDs. The t_A values show the expected linear dependence on p/V indicated in Equation (1.13), which combines Equations (1.1) and (1.5).

$$t_A = t_o + t = t_o + \sigma \frac{16L^2}{3e} \sqrt{\frac{\mu k_B T}{2\pi}} \times \left(\frac{N}{EL} \right) = t_o + slope \times \frac{p}{V} \qquad (1.13)$$

The intercept t_o of the t_A versus p/V data corresponds to the ion travel time outside the drift region and the slope is proportional to the cross section with the proportionality constant given by parameters such as temperature, ion charge, drift length, and reduced mass.

From the data in Table 1.1 and the appropriate instrumental and other experimental parameters a $(GGGGG)H^+$ cross section of 98 Å^2 is deduced. Parking the mass spectrometer at $m/z = 326$, on the other hand, yields a substantially larger cross section of 108 Å^2 for the sodiated species $(GGGGG)Na^+$ (dashed line in Figure 1.2a). Evaluation of molecular-dynamics-based candidate structures provides an explanation for the 10% increase in cross section. [44] $(GGGGG)H^+$ forms a compact globular structure with the backbone forming a 13-membered quasi-planar ring closed by a COOH···O=C hydrogen bond (highlighted by the red arrow in Figure 1.2a) and with the ammonium group sitting above the ring tied to the ring by two hydrogen bonds with two backbone carbonyl oxygen atoms. In contrast, the backbone in $(GGGGG)Na^+$ forms a very open and extended structure with the sodium ion interacting with all five carbonyl oxygen atoms. Theoretical cross sections for these structures are 98 and 111 Å^2 for $(GGGGG)H^+$ and $(GGGGG)Na^+$, respectively, in good agreement with experiment.

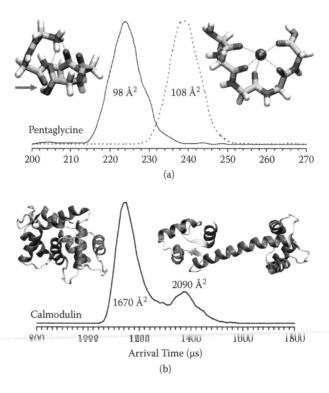

FIGURE 1.2 Ion arrival time distributions (ATDs) of (a) protonated (solid line) and sodiated (dashed line) pentaglycine GGGGG and (b) calmodulin (CaM) charge state +8. Cross sections deduced from the ATDs of the various features are indicated. Protonated GGGGG drifts through the cell much more quickly than sodiated GGGGG. Possible molecular-dynamics-based structures are indicated for (GGGGG)H$^+$ and (GGGGG)Na$^+$ with calculated cross sections of 98 and 111 Å^2, respectively. The red arrow points to a hydrogen bond between the C-terminus and a backbone C=O group. The CaM ATD shows two peaks corresponding to a compact and an extended conformation. The two x-ray structures 1PRW and 3CLN[45,46] are indicated with calculated cross sections of 1680 and 2060 Å^2, respectively. The GGGGG and CaM ATDs were obtained on similar instruments (4- to 5-cm drift length) described by von Helden et al.[56] and Wyttenbach, et al.[16] and under similar conditions (2- to 5-Torr He, 40- to 100-V drift voltage).

Hence the pentaglycine backbone folds into two distinctly different conformations depending on the type of ionic group attached to it. For some biopolymers two distinct features are observed simultaneously in one ATD (at one *m/z* value) indicating the presence of two types of folding for the same species. The two types of structures could correspond, for example, to a correctly folded and a misfolded protein. An example for an ATD with two features is given in the following section for a dinucleotide. Another example is shown in Figure 1.2b for the small protein calmodulin (CaM). Based on x-ray crystallography work it is known that CaM crystallizes in two distinctly different conformations, a compact globular structure and an extended dumbbell-like structure.[45,46] The theoretical cross sections for the two

TABLE 1.1

Experimental Pressure p and Drift Voltage V Values Used to Obtain (GGGGG)H+ Ion Arrival Time Distributions and Average Arrival Times t_A

p (Torr)	V (V)	p/V (Torr/V)	t_A (µs)
2.522	95.76	0.02634	222
2.519	57.42	0.04387	282
2.523	35.21	0.07166	382
2.533	28.22	0.08976	442
2.546	23.49	0.10839	508
2.567	23.44	0.10951	513

t_A versus p/V Data

Slope	3494 µs V/Torr
Intercept	130 µs
Correlation coefficient R^2	0.99990

types of x-ray structures, 1680 and 2060 Å^2, compare favorably with the IMS experimental (charge state 8) values of 1670 and 2090 Å^2, respectively.[47]

Several features are observed in the ATD of the protonated hexapeptide VEALYL, $m/z = 708$ (Figure 1.3). This peptide is known to aggregate and form fibrils.[48] The mass spectrum (Figure 1.4) reveals intense peaks for the trimer (charge state $z = 2$), tetramer ($z = 3$), and pentamer ($z = 3$) and smaller peaks that indicate the presence of oligomers up to the 30-mer and beyond.[49] Hence the intense features in the $m/z = 708$ ATD are assigned to a singly protonated monomer and doubly protonated dimer with cross sections of 191 and 310 Å^2, respectively, and some of the weaker features are likely multiply protonated oligomers. Interesting in the ATD of Figure 1.3 is the "fill-in" between the monomer and dimer peaks. The shape of this ATD is exactly what we expect for a reactive ion A converting into B. Solving the ion transport equation for the ion flux exiting the drift tube[11] using parameters corresponding to the drift tube geometry used in the experiment yields the theoretical ATD components shown in Figure 1.3a. The components are nonreactive species A and B present at the entrance to the drift tube, and reactive species A converting into B inside the drift tube. The unknown A→2B rate constant (dimer → 2 monomers) is an adjustable parameter fit to the experimental data. For the example shown in Figure 1.3a a good fit requires at least two types of dimers with slightly different cross sections by 1.2% (broadened dimer peak) and different rate constants of 10 and 32 s^{-1}. The fit indicates a population ratio of the more reactive to less reactive species of 3:4.

A systematic study of the ion mobility of the VEALYL monomer and increasingly larger oligomers has the potential of revealing a structural transition from a disordered globular random coil monomer and globular disordered small oligomers to highly ordered oligomers extending in one dimension. Understanding aggregation of the VEALYL or similar model system is an important first step in understanding

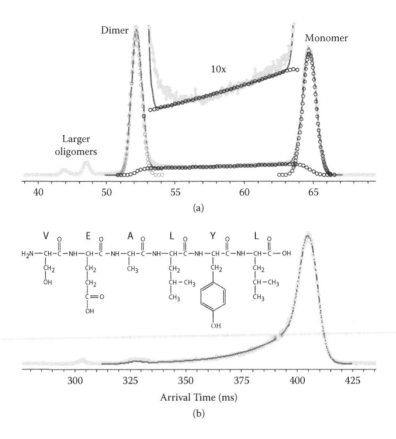

FIGURE 1.3 Positive ion mode ATDs of the hexapeptide VEALYL mass-selected at $m/z = 708$ recorded at (a) high and (b) low drift voltage (data shown in gray). Several peaks are apparent and assigned to monomer and oligomers of $(VEALYL)H^+$. The fill-in signal between monomer and dimer peaks corresponds to dimer reacting to monomer inside the drift tube, resulting in a drift time between monomer and dimer. Theoretically expected monomer/dimer ATDs are shown as a red solid line with the individual monomer and dimer components shown as strings of circles (green, unreactive dimer; blue, unreactive monomer; black, dimer reacting to monomer). At high drift voltage, in (a), ions drift more rapidly giving dimers less time to dissociate to monomers. At low drift voltage, in (b), almost all dimers have enough time to react to monomers. Both ATDs were obtained on a high-resolution instrument (2-m drift length, 12-Torr He, and 500- to 4000-V drift voltage) described by Kemper et al.[19]

aggregation of more difficult to handle systems such as the Aβ protein and other aggregating, ESI tip-clogging proteins.

1.4.2 Structure and Energetics of Dinucleotides

ATDs of mass-selected biopolymers come in all forms and shapes. However, on low-resolution instruments the ATD is characterized in many cases by one single peak with a width given more or less by the instrument resolution and an average drift

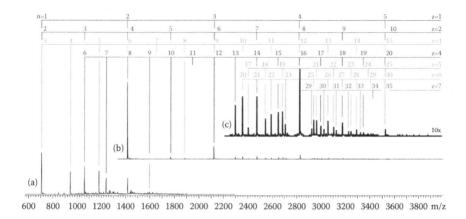

FIGURE 1.4 Mass spectra of the hexapeptide VEALYL recorded on (a) an IMS-quad[19] and (b,c) a Synapt qTOF instrument.[27] Each peak corresponds to a certain charge state ($z = 1–7$) and oligomer size (n) ranging from monomer to 32-mer.

time given by the (average) cross section of the mass-selected ions (Figure 1.2a). For rigid molecules the measured cross section is the orientation-averaged cross section of the rigid object. For floppy molecules the experimental value corresponds typically to the average of the lowest-energy family of structures. If multiple features are present in the ATD the various features can typically be assigned to different families of largely different structures separated by large barriers that prevent rapid interconversion between families.

Interpretation of a simple ATD with one peak may be somewhat puzzling when theory predicts two types of nearly isoenergetic structures, with largely different cross sections that should easily be resolved in the experiment, and neither cross section matches experiment. This situation occurs when interconversion between the two families of structures is rapid compared to the IMS experiment. Repeating the experiment at reduced temperature may slow down the reaction and may afford the opportunity to extract kinetic and energetic information along with the structural data. For example, Figure 1.5 shows ATDs of the MALDI-generated deprotonated dinucleotide dCG⁻ at different temperatures.[50] At room temperature, the ATD shows a single peak, but as the temperature is lowered to 80 K a second peak appears in the spectra. Because the ions were mass selected before injection into the drift cell, the two peaks in the ATDs are, undoubtedly, isomers of the dinucleotide that have significantly different structures. In fact, theoretical modeling predicts at least three distinct families of structures for a given dinucleotide, each with a different cross section. The most compact isomer is a "stacked" form in which the two nucleobases are stacked on top of each other. The largest isomer is an "open" form in which the two nucleobases are completely separated from each other and the dinucleotide is almost linear. The final isomer, with a cross section in between the stacked and open forms, is an "H-bonded" form where the two nucleobases are parallel to each other in the same plane and are hydrogen bonded to each other. Each of these isomers is shown in Figure 1.6 for dCC⁻.

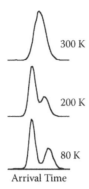

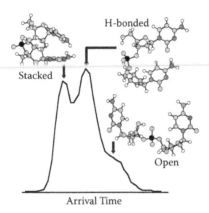

FIGURE 1.5 ATDs for the deprotonated dinucleotide dCG at different temperatures. (Reproduced from Gidden, J., and Bowers, M. T., *Eur Phys J D*, 20, 409–419, 2002. With permission of *The European Physical Journal.* Copyright SIF and Springer-Verlag Berlin Heidelberg.)

FIGURE 1.6 Three families of conformers for deprotonated dinucleotides shown for the example of dCC$^-$, with significantly different cross sections, predicted by theoretical modeling and resolved by IMS at 80 K. The "stacked" form has the smallest cross section while the "open" form has the largest. For some dinucleotides only two of the three types of conformers are observed. (Reproduced in part from Gidden, J., and Bowers, M. T., *Eur Phys J D*, 20, 409–419, 2002. With permission of *The European Physical Journal.* Copyright SIF and Springer-Verlag Berlin Heidelberg.)

Isomers are generally separated from each other on an energetic landscape by isomerization barriers and, to illustrate, a simple reaction schematic diagram for a two-conformer system (A and B) is shown in Figure 1.7. In this case, conformer B has a larger internal energy than conformer A by an amount given by ΔE, the height of the isomerization barrier is E_a, and k_{AB} is the rate constant for the conversion of A to B while k_{BA} is the rate constant for the conversion of B to A.

At high temperatures where the average energy of the system is also high, the two conformers can easily overcome the barrier and rapidly interconvert. If this

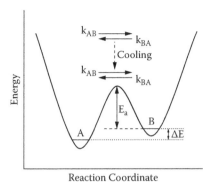

Reaction Coordinate

FIGURE 1.7 Reaction coordinate diagram for a hypothetical two-isomer system. (Reproduced from Gidden, J., and Bowers, M. T., *Eur Phys J D*, 20, 409–419, 2002. With permission of *The European Physical Journal*. Copyright SIF and Springer-Verlag Berlin Heidelberg.)

interconversion happens in the drift cell, the resulting ATD will show a single, time-averaged peak with a corresponding cross section that falls between the cross sections of the two, pure isomers, as the ion packet spends part of the time in form A and part of the time in form B. This is observed in the 300 K ATD in Figure 1.5 for dCG⁻. As the temperature is lowered, the average energy in the system decreases and approaches that of the isomerization barrier. As such, the isomerization process slows down and the two isomers begin to separate in the drift cell due to their different drift velocities. Eventually, the energy in the system decreases to a point lower than the isomerization barrier and the two isomers essentially "freeze" in their respective wells and appear as two distinct peaks in the ATD (with a separation dependent on the difference in cross section of the two isomers and the ability to observe the separation dependent on the resolution of the instrument). This is shown in Figure 1.5 for the lower-temperature ATDs of dCG⁻, which display two clearly resolved peaks. Comparison with model structures (see Figure 1.6) indicates that the two ATD features correspond to the stacked and open families of structures. The H-bonded type of structure is not observed for dCG⁻.[50]

Kinetic and energetic information about the isomerization process can be extracted by modeling the varying shapes of the ATDs at different temperatures with a theoretical simulation based on transport theory.[11] For nonreacting, multiple ions (with different mobilities) in a drift cell, the "total" ATD is simply the sum of the individual ATDs of each ion. If the ions isomerize while they drift, the shape of the ATD also becomes dependent on the frequency of those conversions. In the theoretical simulations, the experimental ATD data is entered as input along with various control variables such as temperature, pressure, cell length, etc. The rate constants for the conversion of one ion species into the other (i.e., k_{AB} and k_{BA}) are then adjusted until the theoretical ATD matches that of the experimental ATD. By matching theoretical and experimental ATDs over a given temperature range, the rate constants can be determined as a function of temperature and, assuming a first-order reaction, an

Arrhenius plot of $\ln(k)$ versus $1/T$ yields a straight line with a slope proportional to the isomerization barrier, E_a.

Estimates of the relative energy difference between the two isomers, ΔE, can be obtained from the relative intensities of the ATD peaks at a temperature where the isomerization has effectively stopped. The equilibrium constant at this energy is given by[51]

$$K(E) = \frac{[B]_E}{[A]_E} = \frac{\rho_B(E - \Delta E)}{\rho_A(E)} \tag{1.14}$$

where ρ_x is the density of states of isomer x and E and $E-\Delta E$ are the internal energies of isomers A and B, respectively. The ratio, [B]/[A], is determined from the peak intensities in the ATD and, using semiempirical vibrational frequencies, ρ_B and ρ_A are calculated at various values of ΔE until the ratio matches that of the experimental value.[52] Results obtained for a number of dinucleotides are summarized in Table 1.2.

1.4.3 HYDRATION OF BIOMOLECULES

A criticism IMS routinely faces as a structural analysis tool in the field of biochemistry is the solvent-free environment of the method. The biochemistry community

---

TABLE 1.2

Intensity Ratios of the Three Conformers "Closed," "H-Bonded," and "Open"; Energy Differences ΔE between Two of the Conformers; and Energy Barriers E_a for the Transition between Two States Determined from IMS Data of the Dinucleotides Indicated

Dinucleotide	Intensity Ratio[a]	ΔE (kcal/mol)	E_a (kcal/mol)	Direction of Isomerization
dAA	3:0:1	—	2.6	
dCA	15:0:1	—	2.3	
dTA	9:0:1	—	3.2	Open → stacked
dAG	18:0:1	—	0.8	
dCG	5:0:2	—	4.1	
dAT	1:0:19	—	2.1	
dCT	1:0:80	—	—	
dGT	1:0:6	0.3	2.8	Stacked → open
dTT	1:0:50	—	—	
dTG	1:0:12	2.4	3.9	
dGA	3:2:0	—	1.5	Stacked → H-bonded
dGG	1:3:0	—	1.8	
dCC	1:2:1	—	12.9	Stacked → H-bonded
			0.8	Open → H-bonded

[a] Closed : H-bonded : open ratios determined from ATDs measured at 80 K.

often considers IMS molecular shape information suspect or even irrelevant with respect to biochemistry. Skepticism persists even though numerous examples document that solution structure is largely maintained after desolvation in ESI-IMS-MS experiments (see, for example, Wyttenbach et al.[47] and references therein) and despite the widely acknowledged power of access to intrinsic properties of a molecule (e.g., intrinsic helix propensity of a peptide[53]). However, IMS instrumentation can be used to bridge the gap between solution and solvent-free environment by using volatile solvent molecules as the buffer gas in the IMS drift cell.

Water is one of the most important solvents in biochemical systems. Hence, the energetics of biomolecule hydration provide important information complementary to a cross section–based IMS analysis. Figure 1.8 shows data obtained for the example of deprotonated dinucleotide dCC exposed to water vapor in the drift cell.[54] In these experiments the drift voltage is set to a value small enough to ensure there is insignificant collisional heating [Equation (1.3)] and enough time to reach equilibrium. Since the mass spectra shown in Figure 1.8a and b are found to be independent of drift voltage in the range of low voltages used in the experiments ($E \leq 10$ V/cm, $t \cong 1$ ms), they represent equilibrium distributions of the $M \cdot (H_2O)_n$ species under the conditions indicated in the figure.

The dCC$^-$ mass spectra indicate that several water molecules add to the ion with the number of solvating water molecules dependent on temperature. Less water binds at higher temperature, up to $n = 2$ or 3 H_2O molecules at 311 K compared to 8 or 9 at 251 K. Following Equation (1.10), the hydration equilibrium constants K_n^{eq} are readily obtained from the relative intensities of the $M \cdot (H_2O)_{n-1}$ and $M \cdot (H_2O)_n$ peaks and the known water pressure. Following the change in K_n^{eq} as a function of temperature T yields the van't Hoff plot shown in Figure 1.8d. Slope and intercept [Equation (1.12)] yield ΔH_n° and ΔS_n° values of -8.1 kcal/mol and -17 cal/(mol K), respectively, for the $n = 3$ example shown.

As for ion mobility studies, the experimental values can be compared to theoretical values obtained for candidate geometries. For nucleotides it is found theoretically that the first water molecule binds—not surprisingly—to the charge-carrying phosphate group with a binding energy in agreement with experiment.[55] Some of the values are shown in Table 1.3 for mononucleotides. This agreement between theory and experiment is important to validate theoretical structures. Molecular mechanics-based theory further indicates that the subsequent water molecules bind about equally well to the phosphate group, the nucleobases, and to already bound water molecules. Experimentally it is found that the $n = 2$–7 water binding energies are smaller by 1–2 kcal/mol than $|\Delta H_1^\circ|$ and nearly constant in the $n = 2$–7 range. Constant values for a range of water molecules suggest several equally good hydration sites are available after the first site is occupied, which is in agreement with theory.[54,55]

The experimentally measured water binding energies of nucleotides provide an important set of data required, for instance, to work out the energetics of DNA duplex formation. The Watson-Crick base pair interaction, in solution, is in constant competition with hydration of the individual bases. Thus, the duplex stability depends on the base–base interaction energy relative to the base–water and water–water interaction. Knowing all of the energetic contributions to a complex system such as solvated

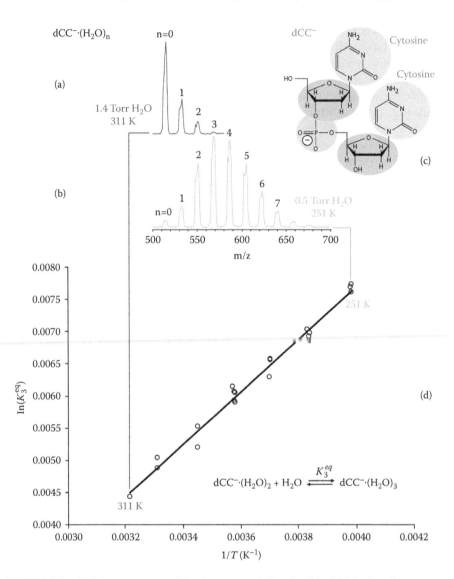

FIGURE 1.8 (a) Mass spectrum of the deprotonated dinucleotide dCC hydrated by water vapor under equilibrium conditions at 311 K and (b) 251 K. (c) Schematic structure of dCC highlighting the anionic phosphate group (blue), the deoxyribose sugar rings (red), and the cytosine nucleobases (green). (d) van't Hoff plot for the equilibrium indicated (addition of third water molecule).

DNA is essential to a deep understanding of structural and other properties of a biochemical system.

In summary, the synergy of experiment and theory extends from ion mobility studies to equilibrium studies covered here and provides important insight into energetic and structural aspects of fully desolvated, partially solvated, and—together with other methods—fully solvated biomolecules.

TABLE 1.3
Experimental and Calculated (B3LYP/6-311++G**) Water Binding Energies for the Four Deprotonated Deoxyribonucleoside 5′-Monophosphates Indicated

	$-\Delta H_1^\circ$ (kcal/mol)	
Mononucleotide	Exp	Calc
dCMP	11.5	11.6
dTMP	10.1	10.8
dAMP	10.3	11.4
dGMP	10.9	10.3

ACKNOWLEDGMENTS

The work described here has been supported for many years by the National Science Foundation and is currently supported by NSF under grant number CHE-0909743 and the National Institutes of Health under grant IPOIAG027818-D10003. We are very grateful for this support.

REFERENCES

1. Clemmer, D. E.; Jarrold, M. F., "Ion mobility measurements and their applications to clusters and biomolecules", *J Mass Spectrom* **1997**, *32*, 577–592.
2. Wyttenbach, T.; Bowers, M. T., "Gas-phase conformations: The ion mobility/ion chromatography method", *Top Curr Chem* **2003**, *225*, 207–232.
3. Creaser, C. S.; Griffiths, J. R.; Bramwell, C. J.; Noreen, S.; Hill, C. A.; Thomas, C. L. P., "Ion mobility spectrometry: A review. Part 1. Structural analysis by mobility measurement", *Analyst* **2004**, *129*, 984–994.
4. Weis, P., "Structure determination of gaseous metal and semi-metal cluster ions by ion mobility spectrometry", *Int J Mass Spectrom* **2005**, *245*, 1–13.
5. Bohrer, B. C.; Mererbloom, S. I.; Koeniger, S. L.; Hilderbrand, A. E.; Clemmer, D. E., "Biomolecule analysis by ion mobility spectrometry", *Annu Rev Anal Chem* **2008**, *1*, 293–327.
6. Mason, E. A.; McDaniel, E. W., *Transport properties of ions in gases*; Wiley: New York, 1988.
7. Mesleh, M. F.; Hunter, J. M.; Shvartsburg, A. A.; Schatz, G. C.; Jarrold, M. F., "Structural information from ion mobility measurements: Effects of the long-range potential", *J Phys Chem* **1996**, *100*, 16082–16086.
8. Wyttenbach, T.; von Helden, G.; Batka, J. J.; Carlat, D.; Bowers, M. T., "Effect of the long-range potential on ion mobility measurements", *J Am Soc Mass Spectrom* **1997**, *8*, 275–282.
9. Wyttenbach, T.; Witt, M.; Bowers, M. T., "On the stability of amino acid zwitterions in the gas phase: The influence of derivatization, proton affinity, and alkali ion addition", *J Am Chem Soc* **2000**, *122*, 3458–3464.
10. Shvartsburg, A. A.; Jarrold, M. F., "An exact hard-spheres scattering model for the mobilities of polyatomic ions", *Chem Phys Lett* **1996**, *261*, 86–91.
11. Gatland, I. R., "Analysis for ion drift tube experiments", *Case Stud Atom Phys* **1974**, *4*, 369.

12. Dwivedi, P.; Wu, C.; Matz, L. M.; Clowers, B. H.; Siems, W. F.; Hill, H. H., "Gas-phase chiral separations by ion mobility spectrometry", *Anal Chem* **2006**, *78*, 8200–8206.

13. Guharay, S. K.; Dwivedi, P.; Hill, H. H., "Ion mobility spectrometry: Ion source development and applications in physical and biological sciences", *IEEE T Plasma Sci* **2008**, *36*, 1458–1470.

14. Dass, C. *Fundamentals of contemporary mass spectrometry*; Wiley: Hoboken, 2007.

15. Hoaglund, C. S.; Valentine, S. J.; Clemmer, D. E., "An ion trap interface for ESI-ion mobility experiments", *Anal Chem* **1997**, *69*, 4156–4161.

16. Wyttenbach, T.; Kemper, P. R.; Bowers, M. T., "Design of a new electrospray ion mobility mass spectrometer", *Int J Mass Spectrom* **2001**, *212*, 13–23.

17. Shaffer, S. A.; Tang, K. Q.; Anderson, G. A.; Prior, D. C.; Udseth, H. R.; Smith, R. D., "A novel ion funnel for focusing ions at elevated pressure using electrospray ionization mass spectrometry", *Rapid Commun Mass Spectrom* **1997**, *11*, 1813–1817.

18. Tang, K.; Shvartsburg, A. A.; Lee, H. N.; Prior, D. C.; Buschbach, M. A.; Li, F. M.; Tolmachev, A. V.; Anderson, G. A.; Smith, R. D., "High-sensitivity ion mobility spectrometry/mass spectrometry using electrodynamic ion funnel interfaces", *Anal Chem* **2005**, *77*, 3330–3339.

19. Kemper, P. R.; Dupuis, N. F.; Bowers, M. T., "A new, higher resolution, ion mobility mass spectrometer", *Int J Mass Spectrom* **2009**, *287*, 46–57.

20. Dugourd, P.; Hudgins, R. R.; Clemmer, D. E.; Jarrold, M. F., "High-resolution ion mobility measurements", *Rev Sci Instrum* **1997**, *68*, 1122–1129.

21. Dakin, G., "Paschen curves for helium", *Electra* **1977**, *52*, 82–86.

22. Koppisetty, K.; Kirkici, H., "Breakdown characteristics of helium and nitrogen at kHz frequency range in partial vacuum for point-to-point electrode configuration", *IEEE T Dielect El In* **2008**, *15*, 749–755.

23. Baker, E. S.; Gidden, J.; Fee, D. P.; Kemper, P. R.; Anderson, S. E.; Bowers, M. T., "3-dimensional structural characterization of cationized polyhedral oligomeric silsesquioxanes (POSS) with styryl and phenylethyl capping agents", *Int J Mass Spectrom* **2003**, *227*, 205–216.

24. Buryakov, I. A.; Krylov, E. V.; Nazarov, E. G.; Rasulev, U. K., "A new method of separation of multi-atomic ions by mobility at atmospheric-pressure using a high-frequency amplitude-asymmetric strong electric-field", *Int J Mass Spectrom Ion Proc* **1993**, *128*, 143–148.

25. Purves, R. W.; Guevremont, R.; Day, S.; Pipich, C. W.; Matyjaszczyk, M. S., "Mass spectrometric characterization of a high-field asymmetric waveform ion mobility spectrometer", *Rev Sci Instrum* **1998**, *69*, 4094–4105.

26. Kolakowski, B. M.; Mester, Z., "Review of applications of high-field asymmetric waveform ion mobility spectrometry (FAIMS) and differential mobility spectrometry (DMS)", *Analyst* **2007**, *132*, 842–864.

27. Pringle, S. D.; Giles, K.; Wildgoose, J. L.; Williams, J. P.; Slade, S. E.; Thalassinos, K.; Bateman, R. H.; Bowers, M. T.; Scrivens, J. H., "An investigation of the mobility separation of some peptide and protein ions using a new hybrid quadrupole/travelling wave IMS/oa-ToF instrument", *Int J Mass Spectrom* **2007**, *261*, 1–12.

28. Shvartsburg, A. A.; Smith, R. D., "Fundamentals of traveling wave ion mobility spectrometry", *Anal Chem* **2008**, *80*, 9689–9699.

29. Kurulugama, R. T.; Nachtigall, F. M.; Lee, S.; Valentine, S. J.; Clemmer, D. E., "Overtone mobility spectrometry: Part 1. Experimental observations", *J Am Soc Mass Spectrom* **2009**, *20*, 729–737.

30. Valentine, S. J.; Stokes, S. T.; Kurulugama, R. T.; Nachtigall, F. M.; Clemmer, D. E., "Overtone mobility spectrometry: Part 2. Theoretical considerations of resolving power", *J Am Soc Mass Spectrom* **2009**, *20*, 738–750.

31. Kemper, P. R.; Weis, P.; Bowers, M. T., "Cr⁺(H₂)ₙ clusters: Asymmetric bonding from a symmetric ion", *Int J Mass Spectrom Ion Proc* **1997**, *160*, 17–37.

32. Hoaglund, C. S.; Valentine, S. J.; Sporleder, C. R.; Reilly, J. P.; Clemmer, D. E., "Three-dimensional ion mobility TOFMS analysis of electrosprayed biomolecules", *Anal Chem* **1998**, *70*, 2236–2242.

33. Kaneko, Y.; Megill, L. R.; Hasted, J. B., "Study of inelastic collisions by drifting ions", *J Chem Phys* **1966**, *45*, 3741.

34. Soto, C., "Unfolding the role of protein misfolding in neurodegenerative diseases", *Nat Rev Neurosci* **2003**, *4*, 49–60.

35. Stefani, M.; Dobson, C. M., "Protein aggregation and aggregate toxicity: New insights into protein folding, misfolding diseases and biological evolution", *J Mol Med* **2003**, *81*, 678–699.

36. Bernstein, S. L.; Wyttenbach, T.; Baumketner, A.; Shea, J. E.; Bitan, G.; Teplow, D. B.; Bowers, M. T., "Amyloid β-protein: Monomer structure and early aggregation states of Aβ42 and its Pro¹⁹ alloform", *J Am Chem Soc* **2005**, *127*, 2075–2084.

37. Baumketner, A.; Bernstein, S. L.; Wyttenbach, T.; Bitan, G.; Teplow, D. B.; Bowers, M. T.; Shea, J. E., "Amyloid β-protein monomer structure: A computational and experimental study", *Protein Sci* **2006**, 15, 420–428.

38. Teplow, D. B.; Lazo, N. D.; Bitan, G.; Bernstein, S.; Wyttenbach, T.; Bowers, M. T.; Baumketner, A.; Shea, J. E.; Urbanc, B.; Cruz, L.; Borreguero, J.; Stanley, H. E., "Elucidating amyloid beta-protein folding and assembly: A multidisciplinary approach", *Acc Chem Res* **2006**, *39*, 635–645.

39. Wyttenbach, T.; Bowers, M. T., "Intermolecular interactions in biomolecular systems examined by mass spectrometry", *Annu Rev Phys Chem* **2007**, *58*, 511–533.

40. Grabenauer, M.; Bernstein, S. L.; Lee, J. C.; Wyttenbach, T.; Dupuis, N. F.; Gray, H. B.; Winkler, J. R.; Bowers, M. T., "Spermine binding to Parkinson's protein alpha-synuclein and its disease-related A30P and A53T mutants", *J Phys Chem B* **2008**, *112*, 11147–11154.

41. Krone, M. G.; Baumketner, A.; Bernstein, S. L.; Wyttenbach, T.; Lazo, N. D.; Teplow, D. B.; Bowers, M. T.; Shea, J. E., "Effects of familial Alzheimer's disease mutations on the folding nucleation of the amyloid beta-protein", *J Mol Biol* **2008**, *381*, 221–228.

42. Bernstein, S. L.; Dupuis, N. F.; Lazo, N. D.; Wyttenbach, T.; Condron, M. M.; Bitan, G.; Teplow, D. B.; Shea, J.-E.; Ruotolo, B. T.; Robinson, C. V.; Bowers, M. T., "Amyloid-β protein oligomerization and the importance of tetramers and dodecamers in the aetiology of Alzheimer's disease", *Nat Chem* **2009**, 1, 326–331.

43. Murray, M. M.; Bernstein, S. L.; Nyugen, V.; Condron, M. M.; Teplow, D. B.; Bowers, M. T., "Amyloid beta protein: Aβ40 inhibits Aβ42 oligomerization", *J Am Chem Soc* **2009**, *131*, 6316–6317.

44. Wyttenbach, T.; Bushnell, J. E.; Bowers, M. T., "Salt bridge structures in the absence of solvent? The case for the oligoglycines", *J Am Chem Soc* **1998**, *120*, 5098–5103.

45. Babu, Y. S.; Bugg, C. E.; Cook, W. J., "Structure of calmodulin refined at 2.2 Å resolution", *J Mol Biol* **1988**, *204*, 191–204.

46. Fallon, J. L.; Quiocho, F. A., "A closed compact structure of native Ca²⁺-calmodulin", *Structure* **2003**, *11*, 1303–1307.

47. Wyttenbach, T.; Grabenauer, M.; Thalassinos, K.; Scrivens, J. H.; Bowers, M. T., "The effect of calcium ions and peptide ligands on the relative stabilities of the calmodulin dumbbell and compact structures", *J Phys Chem B* **2010**, *114*, 437–447.

48. Sawaya, M. R.; Sambashivan, S.; Nelson, R.; Ivanova, M. I.; Sievers, S. A.; Apostol, M. I.; Thompson, M. J.; Balbirnie, M.; Wiltzius, J. J. W.; McFarlane, H. T.; Madsen, A. O.; Riekel, C.; Eisenberg, D., "Atomic structures of amyloid cross-beta spines reveal varied steric zippers", *Nature* **2007**, *447*, 453–457.

49. Bleiholder, C.; Dupuis, N. F.; Wyttenbach, T.; Bowers M. T., "Conformational conversion from random assembly to beta-sheet in amyloid fibril formation", *Nat Chem* (submitted).

50. Gidden, J.; Bowers, M. T., "Gas-phase conformational and energetic properties of deprotonated dinucleotides", *Eur Phys J D* **2002**, *20*, 409–419.

51. Robinson, P. J.; Holbrook, K. A. *Unimolecular reactions*; Wiley: New York, 1972.

52. Gidden, J.; Wyttenbach, T.; Batka, J. J.; Weis, P.; Jackson, A. T.; Scrivens, J. H.; Bowers, M. T., "Poly (Ethylene terephthalate) oligomers cationized by alkali ions: Structures, energetics, and their effect on mass spectra and the matrix-assisted laser desorption/ionization process", *J Am Soc Mass Spectrom* **1999**, 10, 883–895.

53. Jarrold, M. F., "Helices and sheets in vacuo", *Phys Chem Chem Phys* **2007**, *9*, 1659–1671.

54. Wyttenbach, T.; Bowers, M. T., "Hydration of biomolecules", *Chem Phys Lett* **2009**, 480, 1–16.

55. Liu, D. F.; Wyttenbach, T.; Bowers, M. T., "Hydration of mononucleotides", *J Am Chem Soc* **2006**, 128, 15155–15163.

56. von Helden, G.; Wyttenbach, T.; Bowers, M. T., 1995. "Inclusion of a MALDI ion-source in the ion chromatography technique–conformational information on polymer and biomolecular ions", *Int J Mass Spectrom Ion Proc* **1995**, 146, 349–364.

2 Electronic State Chromatography
Ion Mobility of Atomic Cations and Their Electronic States

Peter B. Armentrout

CONTENTS

2.1 INTRODUCTION

The separation of molecular conformations using differences in their ion mobility is generally perceived to require substantial differences in their size and/or shape and thus their cross section for interaction with the buffer gas. Perhaps not as obviously, appreciable changes in cross section can also be induced by variations in the electronic state of an atomic ion, which otherwise would seem to have very similar sizes and shapes. Indeed the development of ion mobility mass spectrometry as we know it today has its roots in this particular application. Exciting prospects for studying electronic state–specific chemistry, especially that of heavier metal elements, is made possible by this interesting observation. In this chapter, the theory behind such separations and the historical development of this area along with recent advances are reviewed. Table 2.1 lists the mobilities of the various atomic cations discussed in the remainder of this chapter.

2.2 THEORETICAL BACKGROUND

The mobility of any ion, K, is given by Equation (2.1),

$$K = v_d/E = L/t_d\, E \tag{2.1}$$

where v_d is the ion drift velocity, E is the drift field (= V/L, where V is the drift voltage and L is the length of the drift cell), and t_d is the drift time (arrival time). These mobilities can be converted to reduced mobilities, K_0, using Equation (2.2) to remove the first-order effects of the density of the buffer gas used,

$$K_0 = K\,(P/760\text{ Torr})\,(273.15\text{ K}/\,T\,) \tag{2.2}$$

where P is the pressure of the buffer gas in Torr and T is its temperature in Kelvin. Finally, the reduced mobility is inversely related to the cross section for interaction between the ion and the buffer gas, Ω, by Equation (2.3),

$$K_0 = 3(2\pi/\mu\, k_B\, T)^{1/2}\, z\, e/16\, \Omega\, N \tag{2.3}$$

where μ is the reduced mass of the ion interacting with the buffer gas, k_B is Boltzmann's constant, $z\, e$ is the charge on the ion, and N is the number density of the buffer gas. Thus, ions that are physically large have smaller mobilities than more compact ions and therefore lower drift velocities and longer arrival times in a pulsed experiment.

2.2.1 ELECTRONIC STATES

In understanding how these equations can be related to the mobilities of atomic ions and their various electronic states, it is critical to focus on the cross section, which can be qualitatively understood by examining the interaction potential between atomic ions and a buffer gas like He. Figure 2.1 shows interaction potentials for two

TABLE 2.1

Reduced Mobilities of Atomic Cations in Different Electronic States[a]

Ion	Valence Configuration	Probable State[b]	K_0, cm²/(V s)
Kr⁺	$4s^24p^5$	$^2P_{3/2}$	0.848 ± 0.008^c
	$4s^24p^5$	$^2P_{1/2}$	0.876 ± 0.008^c
Xe⁺	$5s^25p^5$	$^2P_{3/2}$	0.531 ± 0.004^c
	$5s^25p^5$	$^2P_{1/2}$	0.562 ± 0.004^c
O⁺	$2s^22p^3$	^4S	24.3^d
	$2s^22p^3$	^2P	22.3^d
C⁺	$2s^22p^1$	^2P	25.5 ± 1.3^e
	$2s^12p^2$	^4P	20.5 ± 1^e
Ti⁺	$4s^13d^2$	^4F	$25.6\ (28.5),^f\ 25.5^g$
	$4s^13d^2$	^2F	$22.7\ (23.2)^f$
	$4s^13d^2$	$^2D,^4P$	$19.7\ (22.7)^f$
V⁺	$3d^4$	^5D	$16.7\ (15.0),^f\ 16 \pm 1^h$
	$4s^13d^3$	$^5F,^3F$	$22.0\ (22.2)^f$
	$4s^13d^3$		$25.1\ (23.9),^f\ 21.4 \pm 0.2^h$
Cr⁺	$3d^5$	^6S	$17.5\ (16.4),^f\ 18.0 \pm 0.3,^h\ 18.9 \pm 1.9^i$
	$4s^13d^4$	^6D	$21.4\ (23.0),^f\ 24.6 \pm 2.5^i$
	$4s^13d^4$	^4D	$20.4,^f\ {\sim}23^i$
Mn⁺	$4s^13d^5$	$^7S,^5S$	$23.1\ (22.4)^f$
	$4s^13d^6$	^5D	$16.2\ (15.2)^f$
Fe⁺	$4s^13d^6$	$^6D,^4D$	$23.7\ (22.9),^f\ 23 \pm 1^h$
	$3d^7$	$^4F,^4P,^2G$	$16.9\ (15.7)^f$
Co⁺	$3d^8$	^3F	$15.6\ (15.3),^f\ 15 \pm 2^h$
	$4s^13d^7$	$^5F,^3F$	$22.9\ (22.3),^f\ 22 \pm 2^h$
Ni⁺	$3d^9$	^2D	$16.3\ (14.8),^f\ 16.2 \pm 0.2^h$
	$4s^13d^8$	$^4F,^2F$	$24.2\ (21.9),^f\ 24 \pm 2^h$
Cu⁺	$3d^{10}$	^1S	$15.7,^f\ 15.8 \pm 0.6^h$
	$4s^13d^9$	^3D	22 ± 1^h
Zn⁺	$4s^13d^{10}$	^2S	$23.4,^f\ 24 \pm 2^h$
Mo⁺	$4d^5$	^6S	18.5 ± 1.9^i
	$5s^14d^4$	$^6D,^4D$	23.4 ± 2.3^i
Pd⁺	$4d^9$	^2D	18.2 ± 0.6^h
	$5s^14d^8$	$^4F,^4P,^2F$	23.2 ± 0.4^h
Ag⁺	$4d^{10}$	^1S	18 ± 1^h
Hf⁺	$6s^25d^1$	^2D	18 ± 1^h
Ta⁺	$6s^15d^3$	^5F	21 ± 1^h
W⁺	$6s^15d^4$	^6D	$22.7 \pm 2.3,^i\ 19.7 \pm 0.9^h$
	$5d^5$	^6S	20.5 ± 2.1^i
Re⁺	$6s^15d^5$	^7S	19.4 ± 0.4^h
Ir⁺	$6s^15d^7$	^5F	20.2 ± 0.5^h
Pt⁺	$5d^9$	^2D	20.6 ± 0.9^h
	$6s^15d^8$	$^4F,^4P,^2F$	23 ± 2^h
Au⁺	$5d^{10}$	^1S	19.5 ± 0.5^h
	$6s^15d^9$	^3D	22 ± 2^h

(continued)

TABLE 2.1 (continued)
Reduced Mobilities of Atomic Cations in Different Electronic States[a]

Ion	Valence Configuration	Probable State[b]	K_0, cm²/(V s)
Hg⁺	$6s^1 5d^{10}$	²S	19.1 ± 1,[h] 19.6[j]

[a] Values at room temperatures unless otherwise noted.
[b] Values in bold denote ground states.
[c] Helm, H. *J. Phys. B.* 1976, *9*, 2931–2943.
[d] Rowe, B. R.; Fahey, D. W.; Fehsenfeld, F. C.; Albritton, D. L. *J. Chem. Phys.* 1980, *73*, 194–205.
[e] Twiddy, N. D.; Mohebati, A.; Tichy, M. *Int. J. Mass Spectrom. Ion Processes* 1986, *74*, 251–263. Values are taken as the maximum versus E/N.
[f] Kemper, P. R.; Bowers, M. T. *J. Phys. Chem.* 1991, *95*, 5134–5146. Uncertainties are 7%. Values in parentheses correspond to temperatures of 140–180 K.
[g] Johnsen, R.; Castell, F. R.; Biondi, M. A. *J. Chem. Phys.* 1974, *61*, 5404–5407.
[h] Taylor, W. S.; Spicer, E. M.; Barnas, D. F. *J. Phys. Chem. A* 1999, *103*, 643–650. Temperature = 180–208 K for all third-row metal cations except Hg+.
[i] Iceman, C.; Rue, C.; Moision, R. M.; Chatterjee, B. K.; Armentrout, P. B. *J. Am. Soc. Mass Spectrom.* 2007, *18*, 1196–1205. Temperature = 120 K.
[j] Lindinger, W.; Albritton, D. L. *J. Chem. Phys.* 1975, *62*, 3517.

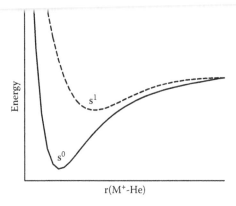

FIGURE 2.1 Qualitative potential energy curves for the interaction of He with metal cations having s^0 and s^1 electron configurations.

ions, one of which has an empty s orbital (s^0) and one in which this orbital is singly occupied (s^1). (The curves shown are quantitative interaction potentials for Li⁺-He ($^1\Sigma^+$) and Be⁺-He ($^2\Sigma^+$) calculated at the B3LYP/6-31G* level. Bond energies (D_e) and equilibrium bond distances (r_e) are 0.098 eV and 2.017 Å for Li⁺-He, and 0.035 eV and 2.648 Å for Be⁺-He.) For both ions, there is initially a long-range attractive interaction with He (or any other neutral), which becomes repulsive as the nuclei approach too closely. Clearly, the ion in the s^0 configuration becomes repulsive at much shorter internuclear distances than that having an s^1 configuration. A quantum mechanical view of these interactions points out that in M⁺(s^0)He, a pair of electrons occupies the σ bonding orbital, whereas ground state M⁺(s^1)He has a $\sigma^2 \sigma*^1$ configuration and therefore a much weaker bond and much longer bond length.

Naïvely, it might be thought that the "larger" s^1 ion might have a smaller mobility than the "smaller" s^0 ion; however, the attractive part of the interaction potential governs the cross section much more than the repulsive part of the potential. Namely, because the s^0 ion interacts more strongly with He, its interaction cross section is considerably larger than that of an s^1 ion. In contrast, because of the larger repulsive radius for the $M^+(s^1)$-He interaction, the particles never get close enough to interact strongly, and therefore the ion moves more rapidly through the buffer gas.

2.2.2 EFFECTS OF TEMPERATURE

Figure 2.2 directly illustrates the effects of temperature on the mobility of Mo^+ cations in different electronic states (see below).[1] It can be seen that the drift times increase as the drift cell temperatures decrease. Average drift times of the more intense peak increase from 165 μs at 175 K, to 190 μs at 135 K, to 215 μs at 115 K. This trend roughly follows the predicted $T^{-1/2}$ dependence for t_d that can be derived from Equations (2.1)–(2.3). As the temperature is further reduced below 115 K, the drift times change very little. As pointed out by Kemper and Bowers,[2] this plateau

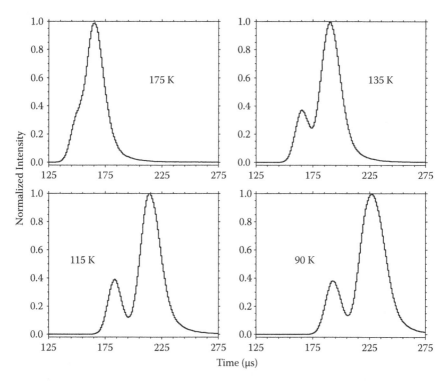

FIGURE 2.2 The temperature dependence of the arrival times and resolution of Mo^+ ion configurations in He (conditions: Ee = 40 V, T as indicated, P = 1.4–1.6 Torr, drift field = 6 V cm^{-1}). (Reprinted from Iceman, C.; Rue, C.; Moision, R. M.; Chatterjee, B. K.; Armentrout, P. B., *J. Am. Soc. Mass Spectrom.*, 18, 1196–1205, 2007. With permission from Elsevier.)

in the drift time occurs because ion mobilities at very low temperatures converge to a common value that depends only on the long-range ion-induced dipole potential, the Langevin mobility given by Equation (2.4),[3]

$$K_L \text{ (cm}^2/\text{V s)} = 13.876 \, (\alpha\mu)^{-1/2} \tag{2.4}$$

where α is the polarizability of the buffer gas in Å^3 and μ, the reduced mass, is in daltons.

In addition, temperature has a direct effect on the resolution of the ion mobility peaks. Kemper and Bowers derived that the resolution is given by Equation (2.5).[2]

$$\Delta t_d/t_d = (8k_B T/z \, e \, V)^{1/2} \tag{2.5}$$

By lowering the temperature of the buffer gas, the diffusion of the ions is reduced and better separation is achieved as illustrated in Figure 2.2. At temperatures ≥ 175 K, the ion mobility peak is clearly distorted by two populations corresponding to different electronic configurations (as verified below), but these are not completely resolved. As the temperature is lowered, the resolution improves, thereby allowing distinct peaks for each configuration to be observed. Below about 115 K, the resolution plateaus or decreases slightly at lower temperatures for the same reasons noted above. In the low temperature limit, separation can no longer be achieved because all ions begin to converge to K_L.

Another interesting effect of temperature is that higher ion intensities are achieved at lower temperatures. This is not apparent in Figure 2.2 because the peaks are normalized but the intensities increase by factors of two to five in going from room temperature to 115 K.[1] This is because lowering the temperature reduces the radial diffusion of the ions as they traverse the cell, thereby allowing a greater number of ions to reach the exit orifice.

2.3 FIRST OBSERVATIONS OF ELECTRONIC STATES

2.3.1 ATOMIC RARE GAS CATIONS

In 1976, Helm observed that the mobility of the atomic cations of the rare gases (Rg), krypton and xenon, in their parent gases exhibited splitting.[4] Analysis of the data demonstrated that the mobilities of the $^2P_{3/2}$ ground states of these ions were lower than those of the $^2P_{1/2}$ states, by factors of 3.3% and 5.8%, respectively, at room temperature (Table 2.1). Although not large, the differences in the mobilities of these states can be understood using the same concepts that lead to the potential energy surfaces shown in Figure 2.1. In these systems, both states correspond to a p^5 valence electron configuration, explaining why the difference in mobilities is small. Indeed, understanding the difference in the interaction potentials of the atomic cation in these two states with a neutral rare gas atom requires consideration of spin-orbit coupling effects. From a qualitative point of view, the interaction of $Rg^+(^2P, p^5)$ with Rg will be most attractive when the p orbital pointed at the neutral atom is singly occupied, whereas less attractive potentials will be formed when this orbital is doubly occupied. In essence, this p_z orbital acts in the same fashion as the s orbital in the discussion above. It can be shown that when this orbital is singly occupied, the surface adiabatically correlates with the $^2P_{3/2}$ state exclusively (for example, see the discussion in ref.[5]). More weakly bound states of the rare gas dimer cations in which this orbital is doubly occupied evolve from both the $^2P_{3/2}$ and $^2P_{1/2}$ states. Because the $^2P_{3/2}$ state interacts more strongly with another rare gas atom than the $^2P_{1/2}$ state, its cross section is larger and thus its mobility is smaller.

2.3.2 ATOMIC OXYGEN CATIONS

In 1980, two groups independently observed that the mobility spectrum of atomic oxygen ions in He contained two components. Johnsen and Biondi[6] examined the mobility of atomic oxygen ions generated by charge transfer with He^+, a process previously suggested to yield a substantial fraction of metastable O^+ in ^2D and ^2P states.[7-9] They found that the metastable state had a lower mobility by 8.5% compared to ground state $O^+(^4S)$. Using somewhat more sophisticated pulsing technology, Rowe et al. made comparable observations again generating O^+ by charge transfer with He^+.[10] Although not completely resolved, they were able to measure the mobilities of the states independently, finding that metastable O^+ has a mobility that is 7%–11% smaller than $O^+(^4S)$ (Table 2.1). A monitor ion technique (in this case, charger transfer with CO, which is exothermic for the ^2D and ^2P states but not the ^4S state) was also used to confirm the state separation and identity. In both studies,[6,10] the authors go on to examine the state-specific rate coefficients for reactions with N_2 and O_2. Both papers discuss evidence that the metastable ion present is largely in the ^2D state, but rely primarily on previous work by Gentry,[11,12] where the ions are not subject to repeated collisions as they are in the mobility cell. The identity of the metastable state observed in the mobility studies has been reassigned as the ^2P state by Simpson et al. on the basis of their calculated interaction potentials and mobilities.[13] They point out that even though charge transfer from He^+ generates both the ^2D and ^2P states, the ^2D state is quenched to the ^4S state under the experimental conditions used for mobility measurements (but not for Gentry's experiments). In contrast, the ^2P state is not quenched easily.

In the atomic oxygen ion system, the ^4S ground state and ^2D and ^2P excited states all have $2p^3$ electron configurations, such that the mobility differences are again small. However, the quartet spin of the ^4S state means that the p_x, p_y, and p_z orbitals must all be occupied, whereas the doublet spin states allow one of these orbitals to be empty (depending on the spin-orbit state). During a mobility experiment, as O^+ interacts with the He buffer gas (along the z-axis), the occupation of the p_z orbital takes on the same role as the s orbital in Figure 2.1. When p_z is occupied (as it must be in the ^4S state), the behavior is similar to the s^1 state and the mobility is high, whereas the $O^+(^2P)$ state can have a lower mobility (like the s^0 state) by keeping the p_z orbital empty. Explicit calculations of the interaction potentials of these states with He are able to reproduce the experimental results nicely.[13]

2.3.3 ATOMIC CARBON CATIONS

In 1986, Twiddy et al. observed similar phenomena for the atomic carbon cation, formed by electron ionization of CO.[14] Here the mobilities differed by 20% (Table 2.1) such that two clear peaks in the arrival time distribution could be resolved, with the more intense, higher mobility peak assigned to the $C^+(^2P)$ ground state and the weaker, lower mobility peak assigned to metastable $C^+(^4P)$. In this case, the conceptual aspects of how the electron configurations track with the interaction potentials is somewhat subtler because the 2s and $2p_z$ orbitals mix, but still follow the ideas outlined above. For the ^2P ground state, which has a $2s^2 2p^1$ configuration, interaction with He leads to states of $^2\Sigma^+$ (when the p_z orbital is occupied) and $^2\Pi$ (when the p_x or p_y orbital is occupied). The $C^+(^4P, 2s^1 2p^2)$ state interacts with He to form $^4\Sigma^-$ (when the carbon has a $2s^1 2p_x^1 2p_y^1$ occupation) and $^4\Pi$ states (from $2s^1 2p_z^1 2p_{x,y}^1$ occupations). One expects that occupation of the p_z orbital will lead to weakly bound states ($^2\Sigma^+$ and $^4\Pi$). Indeed, quantum chemical calculations of Grice et al.[15] found that these states are very weakly bound, $D_e = 0.018$ and 0.022 eV, respectively, with relatively long bond lengths (r_e) of 2.978 and 2.805 Å, respectively. Less obvious is the result that the $^2\Pi$ state is bound only slightly more strongly, $D_e = 0.050$ eV, with a somewhat shorter bond, $r_e = 2.329$ Å. This is because the doubly occupied 2s orbital on carbon mixes with the empty $2p_z$ orbital to form a hybrid orbital pointing away from the He, but still of antibonding character in the HeC$^+$ molecule. However, once this orbital is singly occupied, as in the $^4\Sigma^-$ state, the bond energy increases dramatically to 1.271 eV (relative to the ^4P asymptote) and the bond length decreases to 1.158 Å. As a consequence of these interactions, Grice et al. calculate that the mobilities of the more repulsive states, $^2\Sigma^+$, $^2\Pi$, and $^4\Pi$, are fairly similar (near 25 cm^2/Vs), whereas the strongly bound $^4\Sigma^-$ state has a much lower mobility (near 15 cm^2/Vs). By statistically weighting the contributions of the Σ and Π states (in a ratio of 1:2), Grice et al. are able to accurately reproduce the state-specific mobility data of Twiddy et al. Thus $C^+(^4P)$ has a lower mobility than the ^2P state because the mobility of the strongly bound $^4\Sigma^-$ state differs so dramatically from the other states.

2.4 TRANSITION METAL CATIONS: FIRST ROW

In 1990, Kemper and Bowers observed two well-resolved peaks in the arrival time distribution of atomic cobalt cations (Figure 2.3).[16] The relative intensities of the two peaks changed with the precursor used, $Co(CO)_3(NO)$ and $Co(C_5H_5)(CO)_2$ (labeled I and II in the figure), and with the electron energy (Ee) used to ionize the precursor. At low electron energies (near 15 eV), the lower mobility peak dominates the spectrum, clearly identifying this peak as corresponding to the $^3F(3d^8)$ ground state of Co$^+$. As the Ee was raised, a second, higher mobility peak increased in intensity, with the intensity of the lower mobility peak decreasing to approximately 35% and 85% of the total, respectively, for the two precursors. Strikingly, in contrast to the results for Rg$^+$, O$^+$, and C$^+$, which exhibited different mobilities between the electronic states of about 5%, 10%, and 20%, respectively, the mobility difference for Co$^+$ was found to be almost 50%. Kemper and Bowers tentatively identified the two peaks as corresponding to different electron configurations of the cobalt cation.

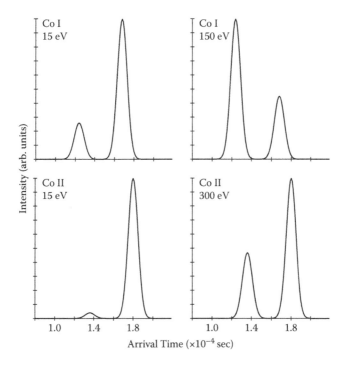

FIGURE 2.3 The dependence of the arrival time distributions for Co^+ in He on the electron energy used to ionize $Co(CO)_3(NO)$ (I) and $CoC_5H_5(CO)_2$ (II) (conditions: T = 160 K, P = 1.0 Torr). (Reprinted with permission from Kemper, P. R.; Bowers, M. T., *J. Am. Chem. Soc.* 1990, *112*, 3231. Copyright 1990, American Chemical Society. 21.)

In a subsequent comprehensive study, Kemper and Bowers went on to study most of the first-row transition metal ions, Ti^+–Zn^+.[2] Multiple peaks were observed in the mobility spectra for all metal cations except those for Cu^+ and Zn^+, where excited states lie very high in energy and thus are not easily generated. The reduced mobilities measured are summarized in Figure 2.4 and listed in Table 2.1. Assignment of peaks to ground states was achieved by varying the electron energy used to generate the atomic cations from their organometallic precursor, as illustrated in Figure 2.3. Additional state assignments were obtained by correlating this information with state-specific results in the literature, primarily from reaction studies of Elkind and Armentrout on the reactions of these atomic metal cations with dihydrogen.[17–23] Overall, the observations made in this work make a compelling case for the assignment of the different peaks observed to different electronic configurations of the metal cations.

2.4.1 Periodic Trends: Late Metals

The assignments made by Kemper and Bowers can be understood by starting with their measurements for Cu^+ and Zn^+, where formation of the ground states is overwhelmingly likely because of the high excitation energies of all excited states. This

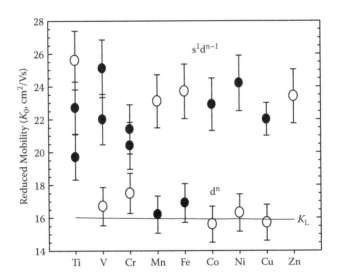

FIGURE 2.4   Reduced mobilities in cm²/(V s) of the first-row transition metal cations. Ground states are indicated by open symbols. The Langevin mobilities, K_L, are shown by the line. (Data are taken from Kemper and Bowers, *J. Phys. Chem.* 1991, 95, 5134–5146, except for the excited state of Cu⁺, which is taken from Taylor et al. *J. Phys. Chem. A* 1999, 103, 643–650.)

is because the ground states have very stable electron configurations of $3d^{10}$ (^1S) and $4s^13d^{10}$ (^2S), respectively. As shown in Figure 2.4, the mobilities of these two cations are quite distinct, 15.7 ± 1.1 and 23.4 ± 1.6 cm²/(V s), respectively. Here, the d^n configuration of Cu⁺ is the analogue of the s^0 potential shown in Figure 2.1, whereas the s^1d^{n-1} configuration of Zn⁺ is like the s^1 potential. This analogy holds because the principal quantum number for the s orbitals is one higher than for the d orbitals. Thus the charged nucleus differentially contracts the d orbitals such that the s orbital is the largest valence orbital. Therefore, Cu⁺ (^1S, $3d^{10}$) has a small mobility because of the strong attractive nature of its interaction with He, whereas Zn⁺ (^2S, $4s^13d^{10}$) has a high mobility characteristic of a weak interaction.

Most low-lying electronic configurations of the first-, second-, and third-row transition metal cations have either d^n or s^1d^{n-1} configurations (although there are some exceptional cases in the third row in which the s orbital contains two electrons, s^2d^{n-2}, e.g., the ground states of Lu⁺ and Hf⁺). Thus the other first-row transition metal ions to the right of the periodic table, Mn⁺–Ni⁺, exhibit two peaks with mobilities that correspond approximately to the values observed for Cu⁺ and Zn⁺ (Figure 2.4). Furthermore, for Mn⁺ and Fe⁺, the peaks that correspond to the ground states (on the basis of the Ee variations) have mobilities that characterize them as s^1d^{n-1} configurations, consistent with the known ^7S ($4s^13d^5$) and ^6D ($4s^13d^6$) ground states, respectively. Likewise, peaks corresponding to the ground states of Co⁺ and Ni⁺ have mobilities that characterize them as d^n configurations, consistent with the known ^3F ($3d^8$) and ^2D ($3d^9$) ground states, respectively. Interestingly, all the lower mobility peaks have values that correspond closely with the Langevin value defined above.[3] Given α (He) = 0.206 Å³,[24] K_L is approximately 15.9 ± 0.05 cm²/(V s) for the first-row transition metal cations interacting with He. This is consistent with the strong attractive interaction associated with these configurations.

2.4.2 PERIODIC TRENDS: EARLY METALS

Before the work of Kemper and Bowers,[2,16] the only transition metal whose mobility had previously been measured was Ti⁺.[25] Kemper and Bowers reproduced this value and were able to assign it to the electronic ground state by Ee variation studies. In addition, the three early first-row transition metal cations, Ti⁺–Cr⁺, exhibited three peaks in their ion mobility data at higher Ees. For V⁺ and Cr⁺, Kemper and Bowers assign the low mobility peak to the respective d^n configuration ground states, V⁺(^5D, $3d^4$) and Cr⁺(^6S, $3d^5$). In both cases, the mobilities are quite consistent with those of the d^n configurations of the late transition metal cations (Figure 2.4). The remaining peaks are all assigned to s^1d^{n-1} configurations on the basis of a comparison of the observed electron energy dependences with literature information regarding electronic state distributions. Again, correlation of these higher mobilities with the analogous assignments for the s^1d^{n-1} configurations of the late metals is reasonable, but there are clearly broader variations in the mobilities of these excited states compared with those of the late metals. Kemper and Bowers suggest that these variations may be attributed to changes in the size of the occupied 4s orbital, which could influence both the repulsive (e.g., nuclear-nuclear potential) and attractive (e.g., 1/r^6 induced dipole-induced dipole potential) parts of the M⁺-He interaction potential, as well as the dominant ion-induced dipole potential (varying as 1/r^4). The contribution of these potentials depends on the ion's polarizability and size, which will vary with the ion electronic state and electronic configuration. In particular, because of the increasing nuclear charge, the size of the orbitals changes appreciably in moving across the periodic table, e.g., the s orbital has a calculated radius of 1.89 Å for Ti⁺ and only 1.49 Å for Zn⁺.[26] This could explain why the highest mobility peaks observed drop from about 25 cm²/(V s) for Ti⁺ and V⁺ to about 23 cm²/(V s) for the later metal cations.

We also suggest here that some of these variations are dependent on what other states (having different configurations) may mix with the states under consideration as the ions interact with He. For instance, the two peaks assigned to the $^6D(s^1d^4)$ and $^4D(s^1d^4)$ excited states of Cr⁺ have mobilities of 21.4 and 20.4 cm²/(V s), respectively (Table 2.1). The higher mobility peaks have mobilities slightly lower than those of the late metal s^1d^{n-1} states, which could indicate that the ^6D state mixes in some $^6S(d^5)$ ground state character as it interacts with He. Note that for heavier metals, this is not possible because there are no empty orbitals that permit such mixing while retaining the same spin. For the quartet state, there are several quartet states having d^5 configurations that lie only 0.12–0.68 eV higher in energy.[27] Because they are close in energy, these states can mix more thoroughly with the $^4D(s^1d^4)$ state to lower its mobility in He even further.

2.4.3 CLUSTERING

One particularly interesting observation made by Kemper and Bowers was the formation of M^+(He) complexes.[2] Notably, these complexes were only observed in conjunction with specific electronic species, namely $V^+(^5D, 3d^4)$, $Cr^+(^6S, 3d^5)$, $Co^+(^3F, 3d^8)$, and $Ni^+(^2D, 3d^9)$. Note that these species are four of the five elements (along with Cu^+) for which the d^n configuration is the ground state. Thus, clustering is observed only for the strongly bound d^n states, which is an experimental verification of the qualitative potential energy curves analogous to those shown in Figure 2.1. Clustering with $Cu^+(^1S, 3d^{10})$ was not observed because of low signal intensities of the atomic ion. This is presumably true for excited states of d^n configuration as well.

By measuring the relative intensities of the M^+ and M^+(He) ions emitted from the drift cell, Kemper and Bowers could extract thermodynamic information for the interaction between the metal cations and helium. When temperature-dependent data were available, both the enthalpies and entropies of binding could be extracted and are listed in Table 2.2. The binding enthalpies are stronger for the later metals because the increased nuclear charge decreases the ionic radius, such that the He binds closer to the metal center, enhancing this largely electrostatic bond. The binding enthalpy of V^+ $(^5D, 3d^4)$ is stronger than that of Cr^+ $(^6S, 3d^5)$ because the high spin state and half-filled shell of $Cr^+(^6S, 3d^5)$ requires that the d orbital pointed at the He is occupied, whereas the resulting Pauli repulsion can be reduced by leaving this orbital empty, which is possible for V^+ $(3d^4)$. The entropies found in these cases are consistent with a simple statistical mechanical estimate for a diatomic molecule.[2]

2.4.4 QUENCHING

Kemper and Bowers also observed that some excited states could be quenched to the ground state by collisions with the He buffer gas.[2] Evidence for this phenomenon was a filling in of observed intensity at drift times in between those characteristic of the d^n and s^1d^{n-1} configurations. However, only two ions exhibited such evidence, Mn^+ and Fe^+. In both cases, quenching is inefficient with rates of $5 \pm 3 \times 10^{-14}$ and $4 \pm 2 \times 10^{-14}$ cm^3/s, respectively (approximately 0.009% and 0.007%, respectively, of

---

TABLE 2.2
Enthalpies and Entropies of Binding between $M^+(3d^n)$ Ions and He[a]

M^+(State, Configuration)	ΔH_0 (kJ/mol)	ΔS (J/mol K)
$V^+(^5D, 3d^4)$	10.9[b]	
$Cr^+(^6S, 3d^5)$	5.6 ± 0.1	−55 ± 6
$Co^+(^3F, 3d^8)$	15.4 ± 0.8	−73 ± 2
$Ni^+(^2D, 3d^9)$	13.0 ± 0.4	−59 ± 2

[a] Kemper, P. R.; Bowers, M. T. *J. Phys. Chem.* 1991, 95, 5134–5146.

[b] Because only one temperature was available, the enthalpy was calculated assuming a binding entropy of −61 J/(mol K).

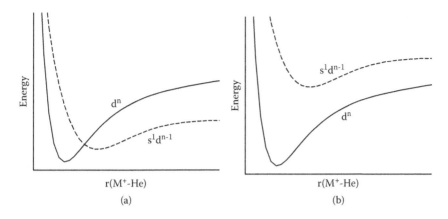

FIGURE 2.5 Qualitative potential energy curves for the interaction of He with metal cations having d^n and s^1d^{n-1} electron configurations. In part (a), the ground state of M^+ has the s^1d^{n-1} electron configuration, whereas part (b) shows the situation for a d^n ground state configuration.

the collision rate). Note that these two ions are the only ones with s^1d^{n-1} ground states and well-resolved excited state mobilities (Figure 2.4). This allows a straightforward explanation for why these two species alone exhibit excited state deactivation, as originally outlined by Loh et al.[28] This returns to the potential energy curves shown in Figure 2.1, but now organizes the asymptotes to properly reflect the excitation energies. In Figure 2.5a, the situation appropriate for Mn^+ or Fe^+ is shown, namely the weakly bound s^1d^{n-1} configuration is the ground state with the more strongly bound d^n state lying higher in energy. As the metal ion approaches He, note that the excited state potential energy surface crosses that of the ground state surface, which allows for mixing of the states and a pathway for quenching. Figure 2.5b exhibits the alternative situation where the strongly bound d^n configuration is the ground state along with an s^1d^{n-1} excited state, as appropriate for V^+, Cr^+, Co^+, Ni^+, and Cu^+. Now no obvious pathway for coupling of the two surfaces is found, explaining why quenching is not observed in these systems. Because Ti^+ also has an s^1d^{n-1} ground state, it could also exhibit quenching (as evidenced by intensity between well-resolved peaks), but here the peaks are more tightly spaced (Figure 2.4) such that the observation of deactivation is hindered.

2.4.5 ION SOURCE VARIATIONS

As noted above, the work of Kemper and Bowers utilized electron ionization of volatile organometallic compounds to generate the atomic metal cations.[2,16] This has the advantage of being able to control the energy available to the ions, thereby allowing independent identification of the ground electronic state. A disadvantage of this source is the availability of appropriately volatile organometallic precursors. Using a different source, a glow discharge, Taylor et al.[29] reproduced many of the results of Kemper and Bowers for most of the first-row transition metal cations, V^+–Zn^+. Taylor et al. observed excited states for all the metals except for Zn^+, in agreement with Kemper and Bowers. However, for Cr^+ and Fe^+, the excited mobility peaks were not readily resolved from the dominant ground state peak, and for V^+, only one excited state was resolved. Some of these differences are because the glow discharge source operates at much higher pressures than the electron ionization source, conditions that can lead to more efficient deactivation of many excited states in the source itself. Indeed, Taylor et al. show that the extent of excited configurations observed was highly dependent on the distance between the source cathode and the sampling orifice as well as the pressure and identity (Ar or Ne) of the discharge gas. Further, because these rare gases have higher polarizabilities than He, their interactions with M^+ are stronger, which can alter the quenching probabilities appreciably.

In contrast to most of the first-row metal cations, Kemper and Bowers observed no excited states for Cu^+, whereas Taylor et al. were able to resolve both the $^1S(3d^{10})$ ground state as well as an excited state. The excited state had a mobility consistent with a $4s^13d^9$ configuration (Figure 2.4 and Table 2.1) and was assigned to the ^3D metastable state. Here, the excited state lies fairly high in energy such that generation by electron ionization is suppressed. Also the large excitation energy means that coupling between states by collisions (leading to quenching) may be less efficient in the glow discharge source.

2.4.6 STATE-SPECIFIC REACTIVITY STUDIES

In subsequent work, Bowers and coworkers used this separation technique to examine state-specific reactions of Co^+ with C_3H_8, CH_3I, and rare gases,[30–33] as well as Fe^+ with C_3H_8,[31] and Ni^+ and Cr^+ with rare gases,[33] all at thermal energies. Likewise Taylor et al. adopted this technique to study the state-specific reactions of Cu^+ with CH_3Cl, CH_3Br, CH_2ClF, $CHClF_2$, and $CClF_3$,[34,35] again at thermal energies. One potential difficulty with these studies is that the reactions and separation are both conducted simultaneously in the drift cell. Thus, quenching processes by the reactant molecule and other secondary effects may influence the state-specific reactivity observed. Furthermore, reactions are limited to thermal energy conditions as they are in equilibrium with the temperature of the buffer gas.

2.5   TRANSITION METAL CATIONS: SECOND AND THIRD ROW

2.5.1   GLOW DISCHARGE SOURCE

One major limitation associated with the use of electron ionization of volatile organometallics is that many fewer such compounds exist for heavier metals. However, the glow discharge source utilized by Taylor et al. is easily extended to heavier metals, as the cathode can be formed of most metallic substances. Thus, Taylor et al. successfully extended their atomic cation mobility results to second-row (Pd^+ and Ag^+) and third-row (Hf^+, Ta^+, W^+, Re^+, Ir^+, Pt^+, Au^+, and Hg^+) transition metal ions.[29] Unlike their results for first-row transition metal cations, most of these systems yielded only a single mobility peak that could be attributed to the ground state of the cation, presumably a result of deactivation of most excited states in the glow discharge source.

However, Pd^+, Pt^+, and Au^+ all exhibited multiple peaks associated with different electronic states, but in no case were these well resolved, even though low temperatures were used to enhance the resolution. As noted by Taylor et al., this is partly because the 6s and 5d orbitals differ in size by less (~30%) than the 4s and 3d orbitals.[36,37]

2.5.2 ELECTRON IONIZATION SOURCE

Notable exceptions to the lack of volatile organometallics containing heavier metals are the group 6 hexacarbonyls, $Cr(CO)_6$, $Mo(CO)_6$, and $W(CO)_6$. Iceman et al.[1] utilized these species to generate the atomic cations of all three metals using electron ionization. Their results for Cr^+ reproduce those of Kemper and Bowers, including all three electronic states observed, although the two excited states are not well resolved. Extensions to the heavier metals yielded results that parallel those of the first row. As shown in Figure 2.6a, when Mo^+ is formed by electron ionization of $Mo(CO)_6$ at an electron energy (Ee) of 20 V, which is just above the appearance energy for production of the bare metal ions,[38,39] only one peak in the ion mobility spectrum is observed. Therefore, this peak can be assigned to the ^6S ($4d^5$) ground state.[27] As the Ee is further increased, another peak at earlier drift times increases in intensity, identifying this peak as being associated with an excited state configuration of Mo^+, specifically $5s^1 4d^4$, for which a ^6D state is the lowest-lying state.[40]

For the third-row group 6 metal cation, W^+, similar studies of varying the Ee were conducted and are shown in Figure 2.6b. With an Ee just above the appearance energy of the atomic cation, 18 V, only a single peak is observed. This identifies the peak as corresponding to the ^6D ($6s^1 5d^4$) ground state of W^+. Increases in Ee yield a new peak at *longer* drift times, which becomes more intense than the ground state peak at the highest Ee values. The relative mobility of this peak identifies it as being associated with an electronic state having a $5d^5$ electronic configuration, including the ^6S first excited state. This change in the relative mobilities of the ground state versus excited state peaks is a nice confirmation that the mobilities of the d^n and $s^1 d^{n-1}$ configurations are properly identified. For the first-, second-, and third-row group 6 congeners, states having $s^1 d^4$ configurations appear at shorter times than those having d^5 configurations, although now the ground state for W^+ has switched configurations compared to Cr^+ and Mo^+.

2.5.3 PERIODIC TRENDS

The mobilities measured for second- and third-row transition metal cations are shown in Figure 2.7 and listed in Table 2.1.[1,29] For the second-row metals, Mo^+, Pd^+, and Ag^+, the mobilities fall into the same two main categories found for the first-row metals, a low mobility associated with a d^n configuration and a high mobility for the $s^1 d^{n-1}$ configurations. Note that the relative mobilities of the two configurations differ by 27% for the second-row metal cations (Table 2.1), whereas a difference of about 50% is observed for the first row. This is mainly because the mobilities of the d^n configurations are now larger and no longer clustered around K_L as they were for the first-row metals. This is presumably because the larger size of the second-row

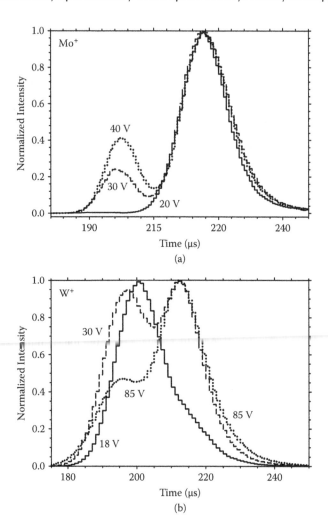

FIGURE 2.6 The dependence of the arrival time distributions for Mo⁺ (part a) and W⁺ (part b) on the electron energy used to ionize $M(CO)_6$ (conditions: part a, Ee as shown, T = 95 K, P = 1.4 Torr, drift field = 6 V cm⁻¹; part b, Ee as shown, T = 120 K, P = 1.4 Torr, drift field = 7 V cm⁻¹). (Adapted from Iceman, C.; Rue, C.; Moision, R. M.; Chatterjee, B. K.; Armentrout, P. B., *J. Am. Soc. Mass Spectrom.*, 18, 1196–1205, 2007.)

metals no longer allows them to mimic the point charge induced-dipole potential in their interactions with He.

The trends in the third-row metal cation mobilities are considerably more complex. Focusing first on the three metals for which excited state configurations are observed, W⁺, Pt⁺, and Au⁺, one again finds that the d^n configurations (ground state for Pt⁺ and Au⁺, and excited state for W⁺) have lower mobilities than the s^1d^{n-1} configurations. The mobilities of the latter configurations are comparable to those of the first- and second-row transition metal cations; however, the mobilities of the d^n

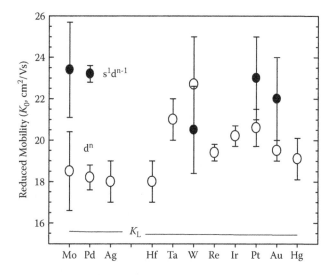

FIGURE 2.7 Reduced mobilities in $cm^2/(V\ s)$ of second- and third-row transition metal cations. Assigned ground states are indicated by open symbols. The Langevin mobilities, K_L, are shown by the line. (Data are taken from Taylor et al., *J. Phys. Chem. A* 1999, 103, 643–650, except those for Mo^+ and W^+, which are from Iceman et al. *J. Am. Soc. Mass Spectrom.* 2007, 18, 1196–1205.)

configurations have again increased such that the relative mobilities differ by only about 12% (Table 2.1). Again this suggests that the increase in the size of the third row metal cations in d^n configurations does not allow as strong an interaction with He as for the first- and second-row metal cations.

The mobilities of Ta^+, Re^+, Ir^+, and Hg^+ are all around 19–21 $cm^2/(V\ s)$, similar to the values for the d^n configurations of W^+, Pt^+, and Au^+. However, Taylor et al. argue that most of the excited states should be quenched to either the ground state or excited states having similar mobilities as the ground state. Thus, in all cases, the mobilities of these ions were assigned to s^1d^{n-1} configurations. It clearly would be valuable to have an independent means of assessing the configurations associated with these cases, especially as the mobility measured by Taylor et al. for W^+, 19.7 ± 0.9 $cm^2/(V\ s)$,[29] is closer to the value Iceman et al. assign to the $^6S(5d^5)$ excited state, 20.5 ± 2.1 $cm^2/(V\ s)$, than that of the $^6D(6s^15d^4)$ ground state, 22.7 ± 2.3 $cm^2/(V\ s)$.[1] Finally, as shown in Figure 2.7, the third-row metal cation having the lowest mobility was found to be Hf^+,[29] which has a $6s^25d^1$ ground state electronic configuration. This result is counterintuitive as the filled 6s shell might be expected to lead to an even less attractive potential than a $6s^15d^{n-1}$ configuration.

2.5.4 State-Specific Reactivity Studies

To date the only state-specific reactivity study for the second- and third-row transition metal cations that utilizes ion mobility state separation is the reaction of $Au^+(^1S, ^3D)$ with CH_3Br as conducted by Taylor et al.[34] At room temperature, they observed association products and formation of $AuCH_2^+$ for the $^1S(5d^{10})$ state, whereas the

$^3D(6s^1 5d^9)$ state of Au^+ abstracted either Br or CH_3. Note that the association reaction is consistent with having an unoccupied 6s orbital in the ^1S state, and the latter reactivity is consistent with the radical character of the $6s^1$ electron configuration.

In unpublished work, Iceman[41] has also demonstrated that ion mobility separation can be used as a source for state-specific studies in a guided ion beam tandem mass spectrometer, which allows the reactions to be studied as a function of kinetic energy. By gating the ions following the ion mobility source, an ion beam containing a specific electronic configuration can be generated. Figure 2.8 shows such results for the case of Mo^+ ions reacting with methane. The lines indicate data taken from the literature for reactions of ground state $Mo^+(^6S, 4d^5)$,[42] whereas the symbols show data for mobility selected $Mo^+(4d^5)$ ions generated by electron ionization of $Mo(CO)_6$ at Ee = 50 V. The high-energy features in the latter cross sections for both the MoH^+ and $MoCH_2^+$ product channels match those of the ground state reactivity. However, the mobility selected $Mo^+(4d^5)$ data also exhibit low-energy features in both cross sections, which disappear when an Ee = 30 V is used to generate the ions. The difference in the onsets of the two endothermic features in the MoH^+ cross section indicate an excitation energy of approximately 2 eV, which matches the 1.88 eV excitation energy of the ^4G state, the lowest excited state having a $4d^5$ configuration.

Such kinetic-energy-dependent studies have the advantage of providing additional energetic information that can be used to identify the states present, as performed in previous state-specific experiments in our laboratory for first-row transition

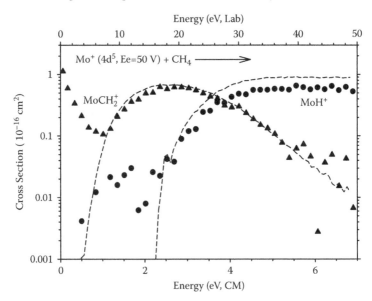

FIGURE 2.8   Cross sections for reactions of Mo^+ with CH_4 forming $MoCH_2^+$ (triangles) and MoH^+ (circles) as a function of kinetic energy in the center-of-mass (lower axis) and laboratory (upper axis) frames. Data for ground state $Mo^+(^6S, 4d^5)$ is shown by the lines and is taken from ref. 42. Symbols show data for ion mobility selected $4d^5$ configurations of Mo^+ formed by 50-V electron ionization of $Mo(CO)_6$. (Adapted from Iceman, C. PhD thesis, University of Utah, 2008.)

metal cations.[17-23,28,43-55] This method of state identification could permit ambiguities in state assignments of ion mobility peaks having identical configurations (see Table 2.1) to be resolved. Complementarily, the ion mobility characterization provides quantification of the amounts of individual configurations, which would allow more accurate absolute cross sections for ground and excited states to be extracted from such studies.

ACKNOWLEDGMENTS

PBA thanks Drs. Chad Rue and Chris Iceman for their substantive contributions to our ion mobility studies, which have been funded by the National Science Foundation.

REFERENCES

1. Iceman, C.; Rue, C.; Moision, R. M.; Chatterjee, B. K.; Armentrout, P. B., "Ion mobility studies of electronically excited states of atomic transition metal cations: Development of an ion mobility source for guided ion beam experiments", *J. Am. Soc. Mass Spectrom.* 2007, *18*, 1196–1205.
2. Kemper, P. R.; Bowers, M. T., "Electronic-state chromatography: Application to first-row transition-metal ions", *J. Phys. Chem.* 1991, *95*, 5134–5146.
3. McDaniel, E. W.; Mason, E. A. *The mobility and diffusion of ions in gases*; Wiley: New York, 1973.
4. Helm, H. "The mobilities of atomic krypton and xenon ions in the $^2P_{1/2}$ and $^2P_{3/2}$ state in their parent gas", *J. Phys. B* 1976, *9*, 2931–2943.
5. Ervin, K. M.; Armentrout, P. B. "Spin-orbit State-selected Reactions of $Kr^+(^2P_{3/2}$ and $^2P_{1/2})$ with H_2, D_2, and HD from Thermal Energies to 20 eV c.m." *J. Chem. Phys.* 1986, *85*, 6380–6395.
6. Johnsen, R.; Biondi, M. A., "Charge transfer coefficients for the $O^+(^2D) + N_2$ and $O^+(^2D) + O_2$ excited ion reactions at thermal energy", *J. Chem. Phys.* 1980, *73*, 190–193.
7. Stebbings, R. F.; Turner, B. R.; Rutherford, J. A. "Low-Energy Collisions between Some Atmospheric Ions and Neutral Particles", *J. Geophys. Res.* 1966, *71*, 771–784.
8. Mauclaire, G.; Dera, D.; Fenistein, S.; Marx, R.; Johnsen, R. "Thermal energy charge transfer from He^+ to O_2: Kinetic energy, nature, and reactivity of the O^+ product ions", *J. Chem. Phys.* 1979, *70*, 4023–4026.
9. Johnsen, R.; Macdonald, J. A.; Biondi, M. A. "Product ion distributions for the charge transfer reactions He^++O_2 and He^++N_2 at thermal energy", *J. Chem. Phys.* 1977, *66*, 4718–4719.
10. Rowe, B. R.; Fahey, D. W.; Fehsenfeld, F. C.; Albritton, D. L., "Rate constants for the reactions of metastable O^{+*} ions with N_2 and O_2 at collision energies 0.04 to 0.2 eV and the mobilities of these ions at 300 K", *J. Chem. Phys.* 1980, *73*, 194–205.
11. Gentry, W. R. "Molecular beam techniques: Applications to the study of ion-molecule collisions", In *Gas Phase Ion Chemistry* Bowers, M. T., Ed.; Academic: New York, 1979; Vol. 2, p 221.
12. Gentry, W. R. In *Kinetics of Ion-Molecule Reactions*; Ausloos, P., Ed.; Plenum: New York, 1979; p 148.
13. Simpson, R. W.; Maclagan, R. G. A. R.; Harland, P. W. "Interaction potentials and mobility calculations for the HeO^+ system", *J. Chem. Phys.* 1987, *87*, 5419–5424.
14. Twiddy, N. D.; Mohebati, A.; Tichy, M. *Int. J. Mass Spectrom. Ion Processes* 1986, *74*, 251–263.

15. Grice, S. T.; Harland, P. W.; Maclagan, R.; Simpson, R. W. "Ab initio calculation of the mobility of $C^+(^2P)$ and $C^+(^4P)$ in helium", *Int. J. Mass Spectrom. Ion Processes* 1989, *87*, 181–186.
16. Kemper, P. R.; Bowers, M. T. "State-selected mobilities of atomic cobalt ions", *J. Am. Chem. Soc.* 1990, *112*, 3231–3232.
17. Elkind, J. L.; Armentrout, P. B., "Effect of kinetic and electronic energy on the reaction of V^+ with H_2, HD and D_2", *J. Phys. Chem.* 1985, *89*, 5626–5636.
18. Elkind, J. L.; Armentrout, P. B., "Effect of kinetic and electronic energy on the reactions of Fe^+ with H_2, HD and D_2: State-specific cross sections for $Fe^+(^6D)$ and $Fe^+(^4F)$", *J. Phys. Chem.* 1986, *90*, 5736–5745.
19. Elkind, J. L.; Armentrout, P. B. "Effect of Kinetic and Electronic Energy on the Reactions of Mn^+ with H_2, HD and D_2", *J. Chem. Phys.* 1986, *84*, 4862–4871.
20. Elkind, J. L.; Armentrout, P. B. "Effect of Kinetic and Electronic Energy on the Reactions of Co^+, Ni^+ and Cu^+ with H_2, HD and D_2", *J. Phys. Chem.* 1986, *90*, 6576–6586.
21. Elkind, J. L.; Armentrout, P. B., "State-specific reactions of atomic transition metal ions with H_2, HD and D_2: Effects of d orbitals on chemistry", *J. Phys. Chem.* 1987, *91*, 2037–2045.
22. Elkind, J. L.; Armentrout, P. B. "Effect of Kinetic and Electronic Energy on the Reactions of Cr^+ with H_2, HD and D_2", *J. Chem. Phys.* 1987, *86*, 1868–1877.
23. Elkind, J. L.; Armentrout, P. B., *Int. J. Mass Spectrom. Ion Processes* 1988, *83*, 259.
24. Rothe, E. W.; Bernstein, R. B. "Total Collision Cross Sections for the Interaction of Atomic Beams of Alkali Metals with Gases", *J. Chem. Phys.* 1959, *31*, 1619–1627.
25. Johnsen, R.; Castell, F. R.; Biondi, M. A. "Rate coefficients for oxidation of Tl⁺ and Tl⁺ by O_2 and NO at low energies", *J. Chem. Phys.* 1974, *61*, 5404–5407.
26. Schilling, J. B.; Goddard, W. A.; Beauchamp, J. L. "Theoretical studies of transition-metal hydrides. 2. CaH^+ through ZnH^+", *J. Phys. Chem.* 1987, *91*, 5616–5623.
27. Sugar, J.; Corliss, C., "Atomic energy levels of the iron-period elements: Potassium through nickel", *J. Phys. Chem. Ref. Data* 1985, *14*, (Suppl. 2) 1–664.
28. Loh, S. K.; Fisher, E. R.; Lian, L.; Schultz, R. H.; Armentrout, P. B., "State-specific reactions of $Fe^+(^6D, ^4F)$ with O_2 and cyclo-C_2H_4O: $D^o\,_0(Fe^+\text{-}O)$ and effects of collisional relaxation", *J. Phys. Chem.* 1989, *93*, 3159–3167.
29. Taylor, W. S.; Spicer, E. M.; Barnas, D. F., "Metastable metal ion production in sputtering dc glow discharge plasmas: Characterization by electronic state chromatography", *J. Phys. Chem. A* 1999, *103*, 643–650.
30. van Koppen, P. A. M.; Kemper, P. R.; Bowers, M. T., "Reactions of state-selected Co^+ with C_3H_8", *J. Am. Chem. Soc.* 1992, *114*, 1083–1084.
31. van Koppen, P. A. M.; Kemper, P. R.; Bowers, M. T., "Electronic state-selected reactivity of transition metal ions: Co^+ and Fe^+ with propane", *J. Am. Chem. Soc.* 1992, *114*, 10941–10950.
32. van Koppen, P. A. M.; Kemper, P. R.; Bowers, M. T., "Fundamental studies of the energetics and dynamics of state-selected Co^+ reacting with CH_3I. The Co^+-CH_3 and Co^+-I bond energies", *J. Am. Chem. Soc.* 1993, *115*, 5616–5623.
33. Kemper, P. R.; Hsu, M.-T.; Bowers, M. T., "Transition-metal ion—Rare gas clusters: Bond strengths and molecular parameters for $Co^+(He/Ne)_n$, $Ni^+(He/Ne)_n$, and $Cr^+(He/Ne/Ar)$", *J. Phys. Chem.* 1991, *95*, 10600–10609.
34. Taylor, W. S.; May, J. C.; Lasater, A. S., "Reactions of $Cu^+(^1S,^3D)$ and $Au^+(^1S,^3D)$ with CH_3Br", *J. Phys. Chem. A* 2003, *107*, 2209–2215.
35. Taylor, W. S.; Matthews, C. C.; Parkhill, K. S., "Reactions of $Cu^+(^1S, ^3D)$ with CH_3Cl, CH_2ClF, $CHClF_2$, and $CClF_3$", *J. Phys. Chem. A* 2005, *109*, 356–365.
36. Barnes, L. A.; Rosi, M.; Bauschlicher, C. W. "Theoretical studies of the first- and second-row transition-metal mono- and dicarbonyl positive ions", *J. Chem. Phys.* 1990, *93*, 609–624.

37. Irikura, K. K.; Beauchamp, J. L. "Electronic structure considerations for methane activation by third-row transition-metal ions", *J. Phys. Chem.* 1991, *95*, 8344–8351.

38. Das, P. R.; Nishimura, T.; Meisels, G. G., "Fragmentation of energy-selected hexacarbonylchromium ion", *J. Phys. Chem.* 1985, *89*, 2808–2812.

39. Michels, G. D.; Flesch, G. D.; Svec, H. J., "Comparative mass spectrometry of the group 6B hexacarbonyls and pentacarbonyl thiocarbonyls", *Inorg. Chem.* 1980, *19*, 479–485.

40. Moore, C. E. *Atomic Energy Levels, NSRDS-NBS 35* Washington, DC, 1971; Vol. III.

41. Iceman, C. R. "Ionic reactions using a guided ion beam tandem mass spectrometer: Collision induced dissociation and electronic state chromatography", PhD thesis, Chemistry Department, University of Utah, 2008.

42. Armentrout, P. B. "Activation of CH4 by gas-phase Mo$^+$ and the thermochemistry of Mo-ligand complexes", *J. Phys. Chem. A* 2006, 110, 8327–8338.

43. Aristov, N.; Armentrout, P. B., "Reaction mechanisms and thermochemistry of V$^+$ + C$_2$H$_{2p}$ (p = 1,2,3)", *J. Am. Chem. Soc.* 1986, *108*, 1806–1819.

44. Schultz, R. H.; Elkind, J. L.; Armentrout, P. B., "Electronic effects in C-H and C-C bond activation: State-specific reactions of Fe$^+$(^{6}D, ^{4}F) with methane, ethane and propane", *J. Am. Chem. Soc.* 1988, *110*, 411–423.

45. Armentrout, P. B., "Chemistry of excited electronic states", *Science* 1991, *251*, 175–179.

46. Fisher, E. R.; Armentrout, P. B., "Electronic effects in C-H and C-C bond activation: Reactions of excited state Cr$^+$ with propane, butane, methylpropane, and dimethylpropane", *J. Am. Chem. Soc.* 1992, *114*, 2049–2055.

47. Clemmer, D. E.; Chen, Y.-M.; Khan, F. A.; Armentrout, P. B., "State-specific reactions of Fe$^+$(a^{6}D, a^{4}F) with D$_2$O and reactions of FeO$^+$ with D$_2$", *J. Phys. Chem.* 1994, *98*, 6522–6529.

48. Clemmer, D. E.; Chen, Y.-M.; Aristov, N.; Armentrout, P. B., "Kinetic and electronic energy dependence of the reaction of V$^+$ with D$_2$O", *J. Phys. Chem.* 1994, *98*, 7538–7544.

49. Chen, Y.-M.; Clemmer, D. E.; Armentrout, P. B., "Kinetic and electronic energy dependence of the reactions of Sc$^+$ and Ti$^+$ with D$_2$O", *J. Phys. Chem.* 1994, *98*, 11490–11498.

50. Kickel, B. L.; Armentrout, P. B., "Guided ion beam studies of the reactions of Ti$^+$, V$^+$, and Cr$^+$ with silane. Electronic state effects, comparison to reactions with methane, and M$^+$-SiH$_x$ (x = 0 – 3) bond energies", *J. Am. Chem. Soc.* 1994, *116*, 10742–10750.

51. Kickel, B. L.; Armentrout, P. B., "Reactions of Fe$^+$, Co$^+$ and Ni$^+$ with silane. Electronic state effects and M$^+$-SiH$_x$ (x = 0 – 3) bond energies", *J. Am. Chem. Soc.* 1995, *117*, 764–773.

52. Kickel, B. L.; Armentrout, P. B., "Guided ion beam studies of the reactions of group 3 metal ions (Sc$^+$, Y$^+$, La$^+$, and Lu$^+$) with silane. Electronic state effects, comparison to reactions with methane, and M$^+$-SiH$_x$ (x = 0 – 3) bond energies", *J. Am. Chem. Soc.* 1995, *117*, 4057–4070.

53. Rue, C.; Armentrout, P. B.; Kretzschmar, I.; Schroder, D.; Harvey, J. N.; Schwarz, H., "Kinetic-energy dependence of competitive spin-allowed and spin-forbidden reactions: V$^+$ + CS$_2$", *J. Chem. Phys.* 1999, *110*, 7858–7870.

54. Rodgers, M. T.; Walker, B.; Armentrout, P. B., "Reactions of Cu$^+$(^{1}S and ^{3}D) with O$_2$, CO, CO$_2$, NO, N$_2$O, and NO$_2$ studied by guided ion beam mass spectrometry ", *Int. J. Mass Spectrom.* 1999, *182/183*, 99–120.

55. Rue, C.; Armentrout, P. B.; Kretzschmar, I.; Schroder, D.; Schwarz, H., "Guided ion beam studies of the state-specific reactions of Cr$^+$ and Mn$^+$ with CS$_2$ and COS", *Int. J. Mass Spectrom.* 2001, *210/211*, 283–301.

3 Measuring Ion Mobility in a Gas Jet Formed by Adiabatic Expansion

Gökhan Baykut, Oliver von Halem,
Jochen Franzen, and Oliver Raether

CONTENTS

3.1 INTRODUCTION

Ion mobility spectrometry (IMS) using drift tube technology has a history of almost half a century. Initially, drift tube studies were used in chemical physics research for acquiring information about ion-molecule reactions.[1-3] Ion mobility spectrometry coupled to mass spectrometry (IMS-MS) has also been known for a very long time. IMS systems with an orthogonal acceleration time-of-flight spectrometer as ion detector were already reported in the 1970s.[1] In recent years mass spectrometric analysis of mobility-separated ions became very popular not only due to the extra dimension of information that helps in the analytical differentiation of compounds, but also due to the useful option of separating components of a mixture. Ionic cross section information directly obtained from ion mobility experiments is very important feedback for structural studies and conformational analysis in particular for biopolymers.[4-13]

Conventional gas phase ion mobility measurements are performed in a drift tube filled with a stationary buffer gas, and an electric field is used to accelerate the ions.

The frictional force due to the collisions of ions with the buffer gas molecules (or atoms) acts against their acceleration due to the electric field. Thus an equilibrium state is quickly reached, and the ions start moving with a constant velocity υ, which is proportional to the applied electric field E. The proportionality constant K is the gas phase mobility of an ion:[1]

$$\upsilon = K E \qquad (3.1)$$

The mobility is a function of the ion's collision cross section in the used buffer gas, its mass, and its electrical charge. Furthermore, the mobility depends on the temperature and on the voltage applied in the drift tube, as well as on the mass of the buffer gas particles.

In the IMS-MS instruments, the mobility-separated ions are introduced into a mass spectrometric analyzer. Currently, quadrupole mass spectrometers and time-of-flight mass spectrometers with orthogonal-acceleration (OTOF-MS) are the most frequently used mass analyzers in the IMS-MS technology. Mobility-based separation of a pulsed group of ions in a drift tube leads to a sequential arrival of the ions, and this "arrival time distribution" represents the ion mobility spectrum. Using long drift tubes (≥ 2 m) ion mobility resolving powers $R_{mob} = K/\Delta K$ of 100 or higher have recently been achieved,[14–16] with ΔK being the full width at half the height of the ion mobility peak.

Although most of the conventional ion mobility measurements are performed in a stationary buffer gas, the experiments can also be performed in a flowing gas. Mobility experiments of ions flowing against a buffer gas in counter flow[17] and experiments in a buffer gas flowing perpendicular to the ion's flight path are known.[18]

In a classical drift tube, the frictional force due to the collisions of the moving ions with the stationary buffer gas molecules (or atoms) acts against the electrical force that drives the ions. The frictional force on the ions depends on their mobility, i.e., their collision cross sections, on their electrical charges, and on their masses. Highly charged ions are driven more easily by an electric field than singly charged ones, and ions with larger cross sections can be retarded more easily by collisions with the stationary buffer gas in the drift tube. If there is no electric field that drives the ions, and if the buffer gas is not stationary but moving, the ions will only be dragged into the flow direction of the buffer gas. In order to discriminate mobilities, an electric field directed against the flight direction of the ions can be introduced. This can be implemented by applying a positive voltage (to an electrode) in order to retard positive ions driven by the gas stream. The generated potential shows an increasing, steplike curve, a potential barrier. The slope of the potential barrier represents the corresponding electric field barrier, the height of which is defined by the slope of the steepest point of the potential barrier. The barrier voltage is defined as the voltage of the potential barrier, which generates a proportionally high electric field barrier. The electric field tries here to block the ions driven by the gas stream. In a given gas stream and at any given level of the barrier voltage, ions with mobilities above a certain threshold can be successfully blocked by the electric field (Figure 3.1). Ions with mobilities below this threshold are carried by the gas flow and penetrate the barrier. If the barrier is progressively increased the threshold gradually moves from high to

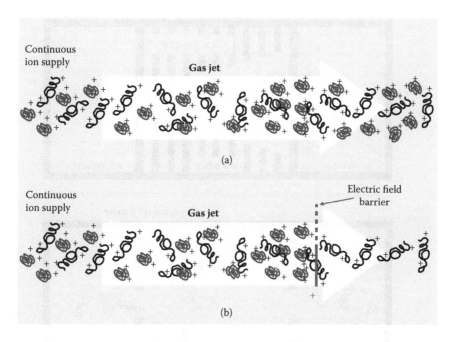

FIGURE 3.1 (a) Ions continuously produced by an electrospray ion source in a gas jet. A mobility separation cannot be observed here. However, if an electric field barrier can be introduced, these ions can be discriminated with regard to their mobility differences (b). At a certain level of the barrier the ions with large cross sections are still dragged by the gas stream and penetrate the barrier, while those with smaller cross sections are blocked by the electric barrier.

low mobility ranges. This way, ions within a corresponding "mobility window" can be observed during such a barrier voltage scan.

A quick comparison of classical ion mobility experiments with those in a gas jet can be summarized as follows: In the classical ion mobility tubes ions are driven by the electric field and retarded by collisions with a stationary gas. In the combination of a gas jet with a potential barrier, ions are driven by the flowing gas and retarded by the electric field. The electric force and the frictional force by collisions are in competition in both cases (Figure 3.2).

In order to measure ion mobilities in an existing mass spectrometer without introducing mechanical or electronic hardware changes, an adequate gas jet is required at a location where an electric field barrier can be introduced. In the quadrupole orthogonal acceleration time of flight (Q-OTOF) mass spectrometer we used for these experiments, the vacuum interface of the electrospray ion source consists of a dual ion funnel system (Figure 3.2). Ion funnels have been used in mass spectrometers since the late 1990s in order to increase the ion transmission efficiency between the atmospheric pressure ion sources and the mass spectrometric high vacuum.[19] They operate at relatively high pressures, e.g., 1–10 mbar, and usually consist of a radio frequency (RF) ion guide made of stacked ring electrodes with decreasing inner diameters. They often have a superimposed DC potential gradient along the ring

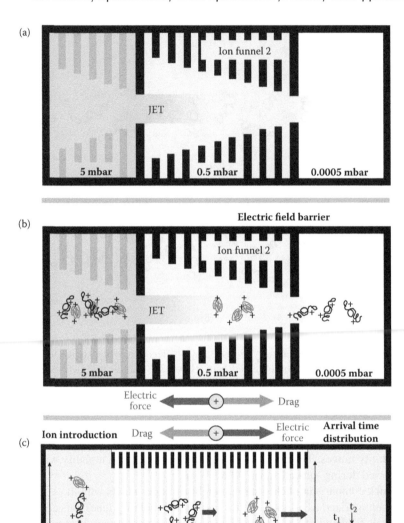

FIGURE 3.2   (a) Adiabatically expanding gas jets are formed in the ion funnel system due to the pressure differences between pumping stages. (b) If an electric field barrier is established in ion funnel 2, it is possible to discriminate mobilities of the ions. The drag force by collisions with the atoms or molecules of *the gas jet drives the ions* and the electric force of the barrier tries to retard them. (c) The classic method of performing ion mobility spectrometry in a drift tube, in which *the electric force drives the ions*, and the frictional force due to collisions with the stationary gas tries to retard them.

stack. In the instrument we used for this study, the pressure in the first ion funnel was about 5 mbar. In reference to the flight path of the ions, this ion funnel was located directly after the exit of the electrospray transfer capillary. The second ion funnel was placed right behind the first one in the next pumping stage with a pressure of about 0.5 mbar. The pressure difference of one order of magnitude leads to a gas flow from the first to the second ion funnel. The gas entering the second ion funnel adiabatically expands as it passes the conductance-limiting orifice. A cold gas jet carrying the ions is formed.

As the ion funnels are made of stacked ring electrodes, in both of the funnels efficient pumping is expected between the electrodes. Due to the adiabatic expansion through the orifice between the two pressure stages the expanded gas cools down, and this gas jet tends to conserve its integrity for a relatively long time (and distance) in the second ion funnel. It does not easily dissipate—despite the efficient pumping. This cold gas jet enables the measurement of ion mobility in the ion funnel with a sufficient resolution.

For measuring the ion mobility in the gas jet, an electric field barrier was introduced in the second ion funnel and the barrier voltage was programmed to be gradually increased or decreased. In combination with the constant gas flow, which drags the ions in the funnel, the electric barrier permits only ions below a certain mobility threshold to pass the funnel base orifice and enter the mass spectrometer. As the barrier voltage progressively increases, this threshold moves from high to low ion mobility ranges. When the threshold passes a value that corresponds to an ion's mobility, the ion's signal intensity decreases to zero. This ion no longer appears in the mass spectrum. If the mobility of an ion just corresponds to the current position of the mobility threshold, it shows a decreasing signal curve on the barrier voltage scale. The position and form of this decreasing signal curve are already related to the mobility of the corresponding ion. As it is more convenient to work with peaks, we prefer to take the first derivative (the slope) of this curve. The negative derivative of this decreasing intensity curve leads to a positive peak. The tip of the peak corresponds to the inflection point of the signal drop curve. Positions of these peaks on the barrier voltage scale are strongly related to the mobility of the corresponding ions. An internal calibration of the barrier voltage scale with known mobility values helps determine the unknown mobilities for compounds of interest.[20]

An ion funnel used as an adjustable low-mass filter was first reported in a paper by Page et al.[21] They described a way of eliminating small ions up to mass-to-charge ratio (m/z) 500 by applying an adjustable barrier voltage to the conductance-limiting plate of the electrodynamic ion funnel interfacing the electrospray source with the mass spectrometer. Unwanted background species usually originating from low mass ions were prohibited from entering the mass spectrometer. The authors explained the mechanism for this filtering action by referring to the ions' mobilities. The barrier voltage in the work of Page et al.[21] was used for filtering ions but not for spectrometric measurement of ion mobilities.

3.2   DETAILS OF THE ION MOBILITY EXPERIMENTS

Ion mobility experiments were performed in a prototype Q-OTOF mass spectrometer of the series "Maxis" (Bruker Daltonik GmbH, Bremen, Germany). Compounds were dissolved in a 1:1 methanol-water solution to which 1% formic acid was added. Using the syringe pump infusion method, samples were electrosprayed at a flow rate of 3 µL/min. Analyte concentrations in the sprayed peptide sample solutions were in the range of 1–10 µg/mL. In the sprayed mixtures of PEG 400 and prostaglandin E_1 the concentrations were between 0.1 and 1 µg/mL. The angled sprayer (Figure 3.3) was kept at ground potential. Positive ions generated at atmospheric pressure entered the mass spectrometer through a glass capillary with metallized ends. At the entrance, the spray shield voltage was –4000 V and the capillary entrance voltage was –4500 V. Nitrogen was used as a drying gas at a capillary entrance temperature of approximately 120°C. Beyond the glass capillary, ions entered a dual ion funnel system (ion funnels 1 and 2) made of tapered stacked ring electrodes. The pressure in ion funnel 1 was about 5 mbar, and in ion funnel 2 it was about 0.5 mbar. Ions exiting funnel 2 through its conductance-limiting orifice entered a multipole ion guide in a pressure region of 5×10^{-4} mbar. Due to these pressure differences the gas carrying the ions flows from funnel 1 into funnel 2 through the orifice (Figure 3.2a) and adiabatically expands, forming a cold gas jet.

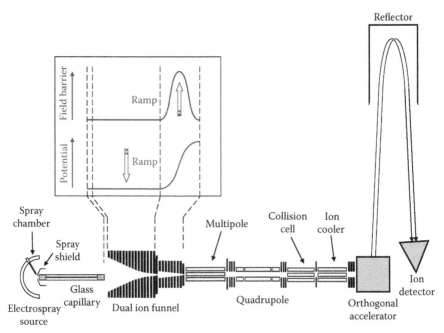

FIGURE 3.3   Schematic view of the Q-OTOF "Maxis" and description of the generation of the electric field barrier in the second ion funnel. To elevate the barrier, the base plate and the voltage of the base plate of funnel 2 were kept constant and the voltages of the upstream ion guiding system were progressively lowered.

Another gas jet appears due to the expanding gas through the base plate orifice of funnel 2 into the multipole ion guide. The ion mobility experiments reported here were performed exclusively in ion funnel 2. The peak-to-peak amplitude of the guiding RF field in funnel 2 was 400 V. The electric field barrier was formed in funnel 2 by using a repelling DC voltage (barrier voltage) in addition to the existing DC bias voltage of the ion funnel. The barrier voltage was increased by 0.1 volt every 6 seconds.

The electric field barrier can be introduced by different methods. The simplest method would be to apply the barrier voltage to the base plate of funnel 2 and increase it while the electrical conditions in the rest of the system are maintained. However, a more elegant way to apply the barrier voltage was to maintain the voltage of the conductance-limiting base plate of funnel 2 constant and progressively lower the DC bias voltages of the upstream ion guiding system between the exit of the electrospray capillary and the entrance of funnel 2 (capillary exit voltage, funnel 1 voltages, and funnel 2 entrance voltage) as shown in Figure 3.3. Since the stacked ring electrodes of funnel 2 are connected to each other via a resistor chain, a voltage gradient existed along the axis of the funnel and also increased with increasing barrier voltage. Thus ions entering funnel 2 flew against an uphill electric potential. An advantage of keeping the funnel 2 base plate at the same voltage during the ion mobility experiments was that the potential difference between the end of funnel 2 and the multipole ion guide was kept constant during a barrier voltage scan. This way, the changing barrier voltage did not continuously vary the penetrating field across the base plate of ion funnel 2, and the internal field conditions at the funnel-multipole transition zone were maintained.

In the experiments the electric field barrier was progressively increased in ion funnel 2. The barrier voltage, at which a particular ion can be prevented from entering the mass spectrometer, will in the following be referred to as the "blocking voltage" for this ion. The blocking voltage correlates to the mobility of the ion. Ion mobility information can be obtained by ramping up the barrier voltage, plotting the characteristic intensity drop of the ion signals at their blocking voltages, and taking the negative derivative (slope) of these curves to represent the change of the signal as a conventional peak. If the barrier voltage is scanned from high to low voltages, the signal intensity change is a rising curve, and its positive derivative can be used. The derivatives of noisy or fluctuating ion signal drop curves show very strong oscillations between positive and negative values. To eliminate or reduce this unwanted effect, the signal plots are artificially "smoothed" before taking their derivative using a method based on the Savitzky-Golay algorithm.[22]

The electrospray chamber was flooded with nitrogen as a drying gas that enters the spray chamber between the front end of the electrospray capillary and the spray shield (Figure 3.3). Due to the specially tailored gas flow around the capillary entrance the drying gas keeps the solvent vapors off the electrospray capillary. Only the ionized analyte, carried by nitrogen, is expected to enter the capillary. The sprayer itself is positioned not axial but at an angle to the capillary entrance in order to keep uncharged droplets from entering the vacuum system.

3.3 RESULTS OF THE EXPERIMENTS

3.3.1 MEASUREMENT OF ION MOBILITIES

In order to obtain quantitative information about mobilities of ions the barrier voltage scale has to be calibrated. A calibration can be performed by using ions with known mobilities or cross sections in the same buffer gas as was used in our experiments. As the electrospray ion source was continuously flushed with nitrogen, the gas entering the electrospray capillary and the ion funnels was practically nitrogen. Since the cross sections are buffer-gas-type dependent, reference data from classical ion mobility experiments in nitrogen buffer gas was used. Cross section data for ions generated from bradykinin, angiotensin I, fibrinopeptide A, and neurotensin in nitrogen obtained in classical drift tube experiments were kindly provided by Dr. Erin S. Baker[23] as listed in Table 3.1. Using the cross section values σ the ion mobilitiy K could be calculated with the equation[1]

$$K = \frac{3ze}{16N}\left(\frac{2\pi}{\mu k_B T}\right)^{1/2}\frac{1}{\sigma} \qquad (3.2)$$

in which z is the charge of the ion, N is the number density, k_B is the Boltzmann constant, T is the absolute temperature, and μ is the reduced mass $\mu = mM/(m + M)$. M is here the mass of the analyte ion and m is the mass of the collision gas molecule (nitrogen). Reduced mobilities K_0 were obtained using the temperature T and the pressure p for correction:

$$K_0 = K \times \frac{T_0}{T} \times \frac{p}{p_{atm}} \qquad (3.3)$$

TABLE 3.1

List of the Selected Reference Peptides with Known Cross Section Values in Nitrogen Used for Calibrating the Acquired Barrier Voltage Spectra

Reference Peptide Ion	m/z	Cross Section σ in Nitrogen (Å^2)	Reduced Mobility K_0 (cm^{2}V^{-1}s^{-1})
Angiotensin I + 3H$^+$	432.9	473[a]	1.297
Bradykinin + 2H$^+$	530.8	332, 340[a]	1.234, 1.205
Fibrinopeptide A + 2H$^+$	768.9	401, 413[a]	1.018, 0.988
Fibrinopeptide A + 3H$^+$	513.3	481[a]	1.273
GRGDS + H$^+$	491.2	185[b]	0.85
Neurotensin + 3H$^+$	558.3	494, 522[a]	1.239, 1.172

Note: Reduced mobility values (K_0) are calculated from cross sections using the experimental conditions.
[a] These cross section values were kindly provided by Erin S. Baker, Pacific Northwest National Laboratories.
[b] From Wu et al. *Anal. Chem.* 2000 72, 391–395.

---

where p_{atm} is the atmospheric pressure, 1013 mbar, and T_0 is 273.16 K. In addition to the ions formed from four reference peptides mentioned above, the penta-peptide Gly-Arg-Gly-Asp-Ser (GRGDS) was also used for calibration. The singly protonated GRGDS molecule offered a convenient reference point for calibration in the lower mobility range.[6] The list of these reference peptides, their cross sections in nitrogen as buffer gas, and the reduced mobilities calculated from their cross sections are included in Table 3.1.

As an example we would like to display here the measurement of the ion mobilities of angiotensin II, substance P, bombesin, and melittin. The cross section data of angiotensin I + $3H^+$, bradykinin + $2H^+$, fibrinopeptide A + $2H^+$, fibrinopeptide A + $3H^+$, neurotensin + $3H^+$ were available[23] from conventional ion mobility–mass spectrometry measurements in nitrogen as buffer gas at 5.33 mbar pressure and 298 K temperature. Hence, K_0 values were calculated using the conditions of the experiments in a classical drift tube. The cross section value for GRGDS + H^+ was also available from the reference Wu et al.[6] as obtained by a measurement in nitrogen at 523 K, and the corresponding K_0 was also listed. Negative derivative peaks of the ions angiotensin I + $3H^+$, bradykinin + $2H^+$, fibrinopeptide A + $2H^+$, fibrinopeptide A + $3H^+$, neurotensin + $3H^+$, and GRGDS + H^+ were used as internal calibration peaks (Table 3.1).

Figure 3.4a shows the decreasing signal intensity curves of the ions and Figure 3.4b shows the corresponding negative derivative plots versus the barrier voltage. Barrier voltage values at the inflection points of the intensity curves (= peak tips in derivative plots) are the blocking voltages for the corresponding compounds, which are related to the mobilities. Using the calibration curve (Figure 3.5) with the ions of the reference peptides, K_0 values could be determined for doubly protonated molecules of angiotensin II, substance P, bombesin, and for the quadruply protonated molecule of melittin in this particular experiment.

Three of the calibrant ions (bradykinin + $2H^+$, fibrinopeptide A + $2H^+$, and neurotensin + $3H^+$) had more than one cross section value (Table 3.1) obtained in the classical ion mobility drift cells.[23] Since these multiple cross sections were close to each other, we chose to use the average values of them in our calibrations. As in this example calibration, the K_0 values of the reference peptides showed a linear dependence on the blocking voltages in our experiments. Therefore, in this mobility range of compounds we used this straight calibration line (Figure 3.5) for determining the unknown K_0 values. We calculated the σ values for the tested compounds by considering the experimental conditions initially defined for the reference compounds (5.33 mbar and 298 K). K_0 values determined this way correspond to ion mobilities measured in nitrogen in classical drift tubes at 298 K. Ion mobilities and calculated collision cross sections from this experiment and other experiments are displayed in Table 3.2.

3.3.2   DOUBLY CHARGED IONS WITH VERY CLOSE MASSES BUT DIFFERENT SHAPES

As mentioned above, the potential barrier in an ion funnel has been used before for the elimination of low mass ions.[21] This is because lower mass ions of the same class of compounds also have smaller cross sections, i.e., higher mobilities. However, compounds not belonging to the same class cannot be easily compared.

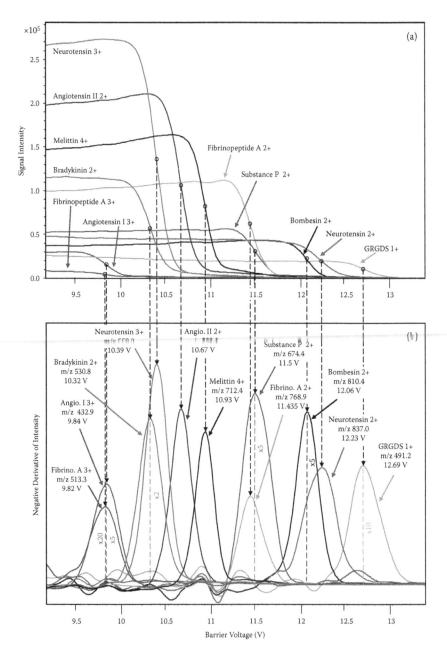

FIGURE 3.4   An example for signal intensity plots (a) and corresponding negative derivative plots (b). Ions were generated from substance P, angiotensin II, bombesin, and melittin, as well as from internal calibrant peptides angiotensin I, bradykinin, fibrinopeptide A, neurotensin, and GRGDS. The ions angiotensin I + 3H$^+$, bradykinin + 2H$^+$, fibrinopeptide A + 2H$^+$, fibrinopeptide A + 3H$^+$, neurotensin + 3H$^+$, and GRGDS + H$^+$ were used for calibration. In order to keep the diagram simple only the charge states of the compounds are indicated in the ion mobility spectrum, and not the protons.

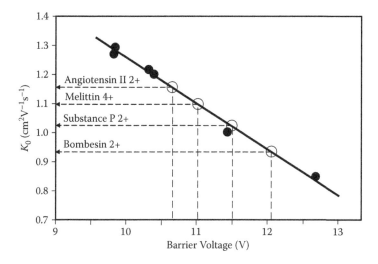

FIGURE 3.5 The calibration curve to determine mobilities of doubly protonated peptides bombesin, angiotensin II, and substance P, as well as quadruply protonated melittin. In the diagram full circles indicate the values obtained from internal calibrant peptides. These were angiotensin I + 3H+, bradykinin + 2H+, fibrinopeptide A + 2H+, fibrinopeptide A + 3H+, neurotensin + 3H+, and GRGDS + H+.

TABLE 3.2
List of Ion Mobilities Determined from Internally Calibrated Barrier Voltage Spectra

Measured Peptide Ion	m/z	Reduced Mobility K_0 $(cm^2V^{-1}s^{-1})$ Determined by Calibration	Calculated Cross Section σ in Nitrogen (Å^2)
Angiotensin I + 2H+	648.9	1.065	384
Angiotensin II + 2H+	523.8	1.155	355
[Arg^8]-Vasopressin + 2H+	542.7	1.259	326
Bombesin + 2H+	810.4	0.932	438
[Des-Glu^1]-LHRH + 2H+	536.3	1.152	356
Melittin + 4H+	712.4	1.097	741
Neurotensin + 2H+	837.0	0.912	447
Substance P + 2H+	674.4	1.022	400

Peptides and proteins have, for instance, more compact molecular structures than lipids and therefore have smaller cross sections than the lipids of the same mass. Also, peptides with an open chain structure and those with a cyclic structure may differ in their cross sections. The following example shows that the separation by the electric field barrier in the expanding gas jet is due to the ion mobilities. Doubly

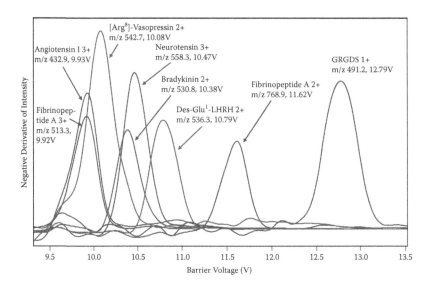

FIGURE 3.6   Negative derivative plots of [Arg^8]-vasopressin + 2H$^+$ and [des-Glu^1]-LHRH + 2H$^+$ (red peaks) showing that [Arg^8]-vasopressin + 2H$^+$ has a higher mobility (smaller cross section) than [des-Glu^1]-LHRH + 2H$^+$. The blue peaks are obtained by the reference compounds of known cross sections.

protonated ions of the peptides [des-Glu^1]-LHRH (average MW: 1071.2 Da) and [Arg^8]-vasopressin (average MW: 1083.4 Da) have m/z values of 536.3 and 542.7, respectively. Measurements showed that the [Arg^8]-vasopressin + 2H$^+$ ion could be blocked at a lower barrier voltage than [des-Glu^1]-LHRH + 2H$^+$: the ion with the slightly larger m/z value was clearly more mobile. Figure 3.6 shows the corresponding negative derivative plots together with the six reference peptide ions for internal calibration, as used before.

The peptide [des-Glu^1]-LHRH has an open chain structure, while [Arg^8]-vasopressin contains a ring structure due to two cysteine residues that form an intramolecular S-S bond. If the ionic structures do not undergo a complicated folding, a noticeable difference between collision cross sections of these two ions can be predicted. [Arg^8]-vasopressin was expected to have the more compact structure. As both of these peptides contain sufficient polar amino acid residues, both of the structures apparently open up due to Coulomb repulsion. Thus [Arg^8]-vasopressin + 2H$^+$ containing the internal ring in fact turns out to be more compact. The negative derivative plots show that [Arg^8]-vasopressin + 2H$^+$ has a higher mobility (smaller cross section) than [des-Glu^1]-LHRH + 2H$^+$. It could be blocked at about 0.7V lower barrier voltage than [des-Glu^1]-LHRH + 2H$^+$.

3.3.3   MOBILITY-DEPENDENT SELECTION

Polyethylene glycols (PEGs) are frequently used in addition to pharmaceutically active compounds in medical drugs. In an experiment we wanted to see how the polyethylene glycols PEG 400 (MW distribution around 400) and prostaglandin E$_1$

(average MW: 354.4) behave if the barrier voltage is increased. In the electrospray mass spectrum, sodiated and potassiated molecules of prostaglandin E$_1$ are produced. The sodiated one is the more abundant signal and appears at m/z 377.23. Similarly, PEGs also formed sodiated and potassiated molecules, and the peaks of the sodiated ones had significantly higher abundances. Results of the barrier voltage experiments showed that the PEG peaks around the sodiated prostaglandin peak disappeared at much lower voltages than the blocking voltage of the sodiated prostaglandin ion. In these particular experiments with the PEGs and prostaglandin the barrier voltage is applied only to the funnel 2 base plate and not to the entire funnel 2. Sodium and potassium ion-attached prostaglandin molecules were blocked at a slightly higher voltage than the sodiated PEG peak with m/z 569. Mobilities of the PEG ions of m/z 305, m/z 349, m/z 393, m/z 437, m/z 481, m/z 525, and m/z 569 turned out to be higher than the mobility of the sodiated prostaglandin E$_1$ at m/z 377. PEG molecules are known to fold around the alkaline metal ions analogous to crown ethers to produce quite compact and spherical structures.[24] Prostaglandin E$_1$ does not have as many polar groups as polyethylene glycols, it has long side chains, and with the ring it is less flexible to form folded structures. Thus its cross section turns out to be larger than the alkaline ion-attached PEGs of similar m/z. Figure 3.7a shows the negative derivative plot for sodiated PEG 400 ions and sodiated prostaglandin E$_1$ obtained in one of the experiments. The blocking voltage for the prostaglandin E$_1$ was even higher than that of the PEG peak with m/z 569. A two-dimensional (2D) display of m/z versus the barrier voltage for these ions is shown in Figure 3.7b. The sodiated and potassiated PEG homologue series describe here two separate diagonal patterns. The prostaglandin appears in the 2D display at an entirely different location clearly showing that it belongs to a different class of compounds. On top of the 2D display a mass spectrum recorded at the beginning of the barrier voltage scan is shown. This particular spectrum is taken at a low barrier voltage. Therefore, all PEG ions as well as the prostaglandin ions are represented in this spectrum.

3.4   DISCUSSION

Ion mobility can be measured in an ion funnel interface of an electrospray-orthogonal time-of-flight mass spectrometer by introducing a barrier voltage. As previously mentioned, the original hardware of the mass spectrometer has not been changed for performing these experiments. Solely, a new acquisition method for scanning the barrier voltage has been implemented. The electrospray ion source continuously produced ions as in standard experiments, while the barrier voltage was increased. An internal calibration using selected compounds of known ion mobilities or cross sections made it possible to determine ion mobilities of compounds under investigation.

The calibration plots can be used to convert the barrier voltage axis into a mobility axis, and from the widths of the negative derivative peaks at the half height, ion mobility resolving power ($R_{mob} = K_0/\Delta K_0$) can be determined. In the presented data some of the routinely obtained mobility resolving power (R_{mob}) values observed at various peaks were $R_{mob} = 30$ at angiotensin II 2+, $R_{mob} = 25$ at fibrinopeptide A 3+, $R_{mob} = 25$ at angiotensin I 3+, $R_{mob} = 21$ at bombesin, and $R_{mob} = 25$ at melittin 4+. The relatively high resolving powers for ion mobilities determined in such a compact

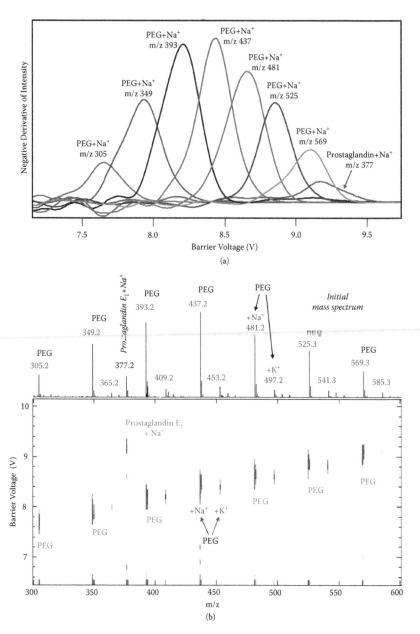

FIGURE 3.7 (a) The intensity derivative plots (ion mobility spectrum) for the mixture of PEG 400 with prostaglandin E₁. (b) A two-dimensional diagram of barrier voltages versus mass-to-charge ratios in a mixture of PEG 400 and prostaglandin E₁. Signal intensities in this 2D diagram are shown by colors. Red shows the higher intensity signals, green the lower intensity ones. The pattern of polyethylene glycols is completely separated from prostaglandin E₁. The main pattern of the sodium ion-attached PEGs is accompanied by a secondary and weaker pattern of potassium ion-attached PEG peaks. The sample spectrum on the top provides an idea of how the initial mass spectrum looked before elevating the barrier.

part of a mass spectrometer as an ion funnel can be explained by the adiabatically expanding gas jet into ion funnel 2. In order to achieve even better resolutions the molecular (ionic) speeds across the jet need to be equal, and the height of the electric barrier has to be ideally distributed across the gas jet. In the current setup there is no ideal distribution of the barrier height across the gas jet. It is expected that the ion mobility resolving powers in the gas jet can be increased by a factor of 2–3 after adequate improvements.

The areas of the ion mobility peaks in the barrier voltage experiments are related to the signal intensities of the corresponding ions since these are obtained by taking the (negative) derivative signal intensity drop curves. The peak height of a derivative peak is the slope of the ion intensity plot at the inflection point. Therefore, the peak heights are related to the steepness of the ion intensity plots, which in turn means that they are a measure for the resolving power of mobility. Nevertheless, we chose to determine the mobility resolving power from the FWHH (full width at half height) of the derivative peaks. To use the peak heights as a measure for the resolving power, one has to normalize the widths of the peaks (FWHH). In the case of ions having only one conformational isomer (only one cross section) it would produce a simple intensity drop curve (ideally in the shape of a complementary error function "*erfc*," and the derivative of it would be a Gauss function). However, in the case of multiple conformations, multiple steps will appear in the decreasing signal curve. If the mobility differences (cross sectional differences) between isomers are too small to be resolved by this technique, the complementary error functions overlap, and the derivative may show an unresolved number of peaks (a series of unresolved Gauss curves), i.e., a broad mobility peak.

The experiments with prostaglandin and PEGs show that ion-mobility-related filtering is not only useful for filtering the low mass ions,[21] but it can also more generally be used for filtering experiments if a less mobile analyte is contaminated with more mobile additives in the same mass range. It is possible to clean up more mobile components from the neighborhood of a less mobile analyte peak without a mass preselection, i.e., without MS/MS.

3.5   CONCLUSIONS AND PERSPECTIVES

The new operational mode of the ion funnel selected for mobility experiments allowed the determination of mobilities for ions of a series of compounds. Resolving powers obtained with this method for ion mobility were between 20 and 30. Two-dimensional displays of ion mobility peaks (derivatives of ion signals) on the barrier voltage scale versus m/z of the measured ions represent a convenient way of giving an overview of the substance classes, as shown with PEG 400 and prostaglandin E_1. The examples can be extended and substance classes with generally compact structures will be separately displayed from those that have larger and rather extended structures.

In the current method for performing experiments, a real physical separation of compounds is limited. At very high barrier voltages, ions with the lowest mobility are the only ones passing the funnel base plate and entering the mass spectrometer. These ions are physically separated from the rest. However, if the barrier voltage is dropped, ions with higher mobilities also start passing the funnel base. Thus instead of having a "moving window" that allows one mobility at a time to enter the mass spectrometer, with this method we have a "curtain" that opens up from the lower end of the mobility spectrum. When the barrier voltage is completely lowered, the curtain is fully open and all ions can enter the mass spectrometer.

At higher barrier voltages, ions with higher mobility are blocked. As the electrospray source continues to produce ions, the total number of ions can increase in the second ion funnel at increased barrier voltages. Coulombic repulsion in the dense ion cloud may lead to some losses and may therefore limit the ion number. However, space charge effects are expected to exist during the experiments.

It is possible to design other experiments, in which a predefined ion ensemble can be formed, kept in a trap using a high barrier voltage, and then gradually released by progressively lowering the barrier.[25] No new ions would be added to the ion ensemble during this process. These experiments would permit in the trapped ensemble a separation of ions with regard to their mobility: Initially, all ions with lowest mobility would exit the trap. Later, ions with higher mobility would follow them. The ion extraction would continue until the ions with the highest mobility are released. However, as the trapped ion ensemble would have only a limited number of ions, the sensitivity issue should be carefully examined in these types of experiments. These trap-and-pulse type experiments would require significant changes in the mass spectrometric equipment.

Since the electronic hardware for operating the mass spectrometer was not modified, the current state of the technique allows experiments with 0.1V increments for every six seconds. Thus these experiments are currently not compatible with faster preseparation methods, i.e., liquid chromatography. A modification would allow the speeding up of the barrier voltage scan by a factor of ten.

Ion mobility experiments have also been tried in the first ion funnel of the mass spectrometer. The resolving powers achieved here were not as good as in the second funnel. The gas jet blowing into the first ion funnel originates directly from the electrospray transfer capillary. However, the electrospray capillary is not coaxial with the axis of the ion funnel as can be seen in Figure 3.3. It is mounted off-axis in order to prevent the neutral gas stream from directly entering the second ion funnel and vacuum system. As a result, ions are guided through the first ion funnel into the second, but the neutral beam soon hits the stacked ring electrodes of the first funnel, becomes turbulent, breaks, and gets pumped away. In mass spectrometry, not letting an intact neutral beam enter the further ion optics is a good practice as it increases the sensitivity. Otherwise a molecular beam tends to continue its way through the orifices into the high vacuum and—although it gets weaker—adversely affects the mass spectrometric detection. Also, the possibility of uncharged droplets getting into the mass spectrometric vacuum is eliminated by using an offset capillary. Since an off-axis gas jet becomes severely turbulent by hitting the funnel electrodes, it is therefore not a good environment for ion mobility measurements. As mentioned above, an undisturbed gas jet is required for performing successful ion mobility separations. The gas jet in the second ion funnel results from the pressure difference between the first and second ion funnel and is perfectly coaxial with the second funnel. Nevertheless, imperfections of conduction-limiting orifices in

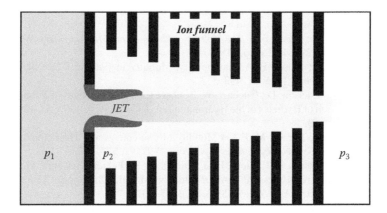

FIGURE 3.8 Schematic illustration of an ion funnel at the pressure p_2 into which a gas from the pumping stage at pressure p_1 enters through a Laval nozzle instead of a simple orifice. The entering gas expands and forms a cold supersonic jet that conserves its integrity for a long time and distance. This gas jet is promising for a high-resolution ion mobility separation of ions.

the ion funnel systems can completely spoil the expanding gas jet and may induce turbulences. One of the measures to prevent turbulent expansions is to use a specially formed conduction-limiting orifice, e.g., a Laval nozzle. The gas entering a Laval nozzle first passes through the compression end of the nozzle and expands at the decompression end. Laval nozzles offer the possibility of forming a controlled gas jet into the lower pressure side of a conduction limit as the gas undergoes a supersonic expansion. The supersonic gas jet would sustain its speed and integrity for a relatively long distance. If the simple entrance orifice of an ion funnel can be replaced with an adequate Laval nozzle (Figure 3.8), this system would become a better environment to measure ion mobility.

The ion mobility resolving powers currently obtained in the ion funnel experiments are surprisingly high. The fact that these were obtained in a very compact unit of an unmodified mass spectrometer opens new perspectives in the field of ion mobility–mass spectrometry. Improvements of this technique can lead to fairly compact instrumentation with ion mobility resolving powers comparable to those of contemporary high-resolution IMS-MS systems with classical drift tubes.

ACKNOWLEDGMENTS

We thank Dr. Erin Shammel Baker for the information on absolute cross sections of angiotensin I 3+, fibrinopeptide A 2+, fibrinopeptide A 3+, and neurotensin 3+ measured using drift tube ion mobility–mass spectrometry with nitrogen as buffer gas. Our appreciation also goes to Prof. Ryan Julian for initial discussions on ion mobility measurements using a barrier voltage in an OTOF mass spectrometer. We also thank Dr. Melvin Park, Dr. Ruediger Frey, Dr. Armin Holle, Dr. Ian Sanders, and Dr. Michael Schubert for fruitful discussions.

REFERENCES

1. McDaniel, E. W.; Mason E. A. *The Mobility and Diffusion of Ions in Gases.* Wiley, New York, 1973, pp. 29–84, also pp. 118–235.
2. Viehland, L. E.; Mason E. A. Gaseous Ion Mobility in Electric Fields of Arbitrary Strength. *Ann. Phys.* 1975, *91*, 499–533.
3. Young, C. E.; Edelson, D.; Falconer, W. E. Water Cluster Ions: Rates of Formation and Decomposition of Hydrates of the Hydronium Ion. *J. Chem. Phys.* 1970, *53*, 4295–4302; also Ref. 1, p. 69.
4. Clemmer, D. E.; Jarrold, M. F. Ion Mobility Measurements and Their Applications to Clusters and Biomolecules. *J. Mass. Spectrom.* 1997, *92*, 577–592.
5. Gillig, K. J.; Ruotolo, B. T.; Stone, E. G.; Russell, D. H.; Fuhrer, K.; Gonin, M.; Schultz. J. A. Coupling High-Pressure MALDI with Ion Mobility/Orthogonal Time-of-Flight Mass Spectrometry. *Anal. Chem.* 2000, *72*, 3965–3971.
6. Wu, C.; Siems, W. F.; Klasmeier, J.; Hill, H. H. Separation of Isomeric Peptides Using Electrospray Ionization/High Resolution Ion Mobility Spectrometry. *Anal. Chem.* 2000, *72*, 391–395.
7. Wyttenbach, T.; Kemper, P. R.; Bowers. M. T. Design of a New Electrospray Ion Mobility Mass Spectrometer. *Int. J. Mass Spectrom.* 2001, *212*, 13–23.
8. Gidden, J.; Bowers, M. T. Gas Phase Conformations of Deprotonated and Protonated Mononucleotides Determined by Ion Mobility and Theoretical Modeling. *J. Phys. Chem. B* 2003, *107*, 12829–12837.
9. Ruotolo, B. T.; Gillig, K. J.; Woods, A. S.; Egan, T. F.; Ugarov, M. V.; Schultz, J. A.; Russell, D. H. Analysis of Phosphorylated Peptides by Ion Mobility-Mass Spectrometry. *Anal. Chem.* 2004, *76*, 6727–6733.
10. Woods, A. S.; Ugarov, M.; Egan, T.; Koomen, J.; Gillig, K. J.; Fuhrer, K.; Gonin, M.; Schultz J. A. Lipid/Peptide/Nucleotide Separation with MALDI-Ion Mobility-TOF MS. *Anal. Chem.* 2004, *76*, 2187–2195.
11. Baker, E. S.; Bernstein, S. L.; Bowers M. T. Structural Characterization of G-Quadruplexes in Deoxyguanosine Clusters Using Ion Mobility Mass Spectrometry. *J. Am. Soc. Mass Spectrom.* 2005, *16*, 989–997.
12. Bernstein, S. L.; Wyttenbach, T.; Baumketner, A.; Shea, J.-E.; Bitan, G.; Teplow, D. B.; Bowers M. T. Amyloid ß-Protein: Monomer Structure and Early Aggregation States of Aß42 and its Pro[19] Alloform. *J. Am. Chem. Soc.* 2005, *127*, 2075–2084.
13. Dwivedi, P.; Wu, C.; Matz, L. M.; Clowers, B. H.; Siems, W. F.; Hill, H. H. Gas Phase Chiral Separations by Ion Mobility Spectrometry. *Anal. Chem.* 2006, *78*, 8200–8206.
14. Tang, K.; Shvartsburg, A. A.; Lee, H.-N.; Prior, D. C.; Buschbach, M. A.; Li, F.; Tolmachev, A. V.; Anderson, G. A.; Smith. R. D. High-Sensitivity Ion Mobility Spectrometry/Mass Spectrometry Using Electrodynamic Ion Funnel Interfaces. *Anal. Chem.* 2005, *77*, 3330–3339.
15. Baker, E. S.; Tang, K.; Danielson III, W. F.; Prior, D. C.; Smith, R. D. Simultaneous Fragmentation of Multiple Ions Using IMS Drift Time Dependent Collision Energies. *J. Am. Soc. Mass Spectrom.* 2008, *19*, 411–419.
16. Kemper, P. R.; Dupuis, N. F.; Bowers. M. T. A New, Higher Resolution Ion Mobility Mass Spectrometer. *Int. J. Mass Spectrom.* 2009, *287*, 46–57.
17. Fernandez De La Mora, J.; Ude, S.; Thomson B. A. The Potential of Differential Mobility Analysis Coupled to MS for the Study of Very Large Singly and Multiply Charged Proteins and Protein Complexes in the Gas Phase. *Biotechnol. J.* 2006, *1*, 988–997.
18. Agbonkonkon, N. Counterflow Ion Mobility Analysis: Design, Instrumentation and Characterization, 2007, dissertation, Brigham Young University, Department of Chemistry and Biochemistry.

19. Kim, T.; Tolmachev, A. V.; Harkewicz, R.; Prior, D. C.; Anderson, G.; Udseth, H. R.; Smith, R. D.; Bailey, T. H.; Rakov, S.; Futrell J. H. Design and Implementation of a New Electrodynamic Ion Funnel. *Anal. Chem.* 2000, *72*, 2247–2255.

20. Baykut, G.; von Halem, O.; Raether, O. Applying a Dynamic Method to the Measurement of Ion Mobility. *J. Am. Soc. Mass Spectrom.* 2009, *20*, 2070–2081.

21. Page, J. S.; Tolmachev, A. V.; Tang, K.; Smith R. D. Variable Low Mass Filtering Using an Electrodynamic Ion Funnel. *J. Mass Spectrom.* 2005, *40*, 1215–1222.

22. Savitzky, A.; Golay M. J. E. Smoothing and Differentiation of Data by Simplified Least Squares Procedures. *Anal. Chem.* 1964, *36*, 1627–1639.

23. Baker, E. S. Biological Sciences Division, Pacific Northwest National Laboratory, Richland, WA, personal communication.

24. von Helden, G.; Wyttenbach, T.; Bowers M. T. Inclusion of a MALDI Ion Source in the Ion Chromatography Technique: Conformational Information on Polymer and Biomolecular Ions. *Int. J. Mass Spectrom. Ion Proc.* 1995, *146/147*, 349–364.

25. Julian R. R. Department of Chemistry, University of California Riverside, Riverside, CA, personal communication.

Instrumentation

4 Development of an Ion-Mobility-Capable Quadrupole Time-of-Flight Mass Spectrometer to Examine Protein Conformation in the Gas Phase

Bryan J. McCullough, Peter A. Faull, and Perdita E. Barran

CONTENTS

4.1 INTRODUCTION

In recent years mass spectrometry has established itself as a powerful method with which to determine the stoichiometry and conformations of proteins and their complexes.[1–3] Pioneering work in the groups of Robinson, Heck, and Loo, among others, has now established that with careful use of electrospray ionization and gentle transfer into the gas phase, bioactive complexes of significant size can be transported intact into the controlled environment of a mass spectrometer for subsequent analysis.[4] Mass spectra of very large protein complexes,[5] whole ribosomes,[6] and intact viruses[7] have been recorded, as well as several studies on the dynamics of large complex assembly or disassembly.[8–10]

There is, of course, controversy about whether a solution phase structure is retained in its entirety in the solvent-free environment of a mass spectrometer,[4,11] but for large macromolecular systems such as those referred to above, bound by many noncovalent interactions, there is evidence to suggest that macroscopic features of solution and even *in vivo* structures are retained. The growth and success of studies of macromolecular complexes by mass spectrometry increasingly places biological mass spectrometry as the first step in the structural analysis of unknown or as yet unquantified protein:protein architectures; in short, it now has a role as a predictive tool.

Alongside the developments in biological mass spectrometry the use of ion mobility coupled with mass spectrometry (IM-MS) has become a particularly important tool for structural analysis. After developments in soft ionization methods, IM-MS studies of biologically relevant species started in the mid-1990s on homebuilt instruments that coupled these two well-known analytical techniques. Bowers, Jarrold, Clemmer, and Hill performed some of the most influential work in this period, and their investigations both paved the way for others and prompted the development of commercially available mobility devices as the power of this technique for biological analysis became apparent.[12–16] Efforts in commercial development have concentrated on improving the low duty cycle that is inherent to linear ion mobility experiments, and also on increasing the resolution available. Initial studies have been focused as much on structural measurements as on mixture separation, reflecting the growing power of mass spectrometry as a tool to provide detailed conformational information on biological moieties.[17] Waters MS Technologies recently introduced the first commercially available integrated IM-MS instrument, the Synapt HDS.[18] The radio frequency (RF) applied to consecutive electrodes in the stacked ring ion guide within the ion mobility separator provides a potential well that keeps the ions radially confined within the device. In order to propel the ions through the device, a traveling wave comprising a series of transient DC voltages is superimposed on top of the RF voltage, and hence this device is sometimes referred to as a Traveling Wave

Ion Guide (TWIG). This voltage is applied sequentially to pairs of ring electrodes, providing a potential that can push ions through the device. Since it contains three ion guides the mobility component of the Synapt HDS is known as the Triwave.

Commercially available devices have already been used to good effect. Borysik and coworkers have employed a FAIMS (field asymmetric waveform ion mobility spectrometry) mobility device coupled to the front end of a quadrupole time-of-flight (QToF) instrument to examine conformers of β-microglobulin.[19] Using a Synapt HDS system, Ruotolo et al. have assessed conformations of multimeric proteins[20] and also the disassembly of complexes viewing the partial unfolding of monomer units while still retaining some of the integrity of the complex.[21] We recently built an ion mobility mass spectrometer that is capable of performing temperature variable measurements of collision cross sections and applied it to examine the conformations of biological molecules and complexes.[22,23] Our instrument is built from a Waters quadrupole ToF instrument and uses the same software as the Synapt device to obtain selected ion arrival time distributions from ToF spectra. This chapter details the construction of this instrument, with reference to design criteria and its application to determine protein conformations in the gas phase.[22]

4.2   INSTRUMENT DESIGN

The starting point for the design of the apparatus was a QToF I mass spectrometer (Waters Manchester, UK). This instrument, launched in the late 1990s as part of the first wave of commercial tandem mass spectrometers, possesses three major components: a z-spray ion source; MS 1, a quadrupole mass analyzer; and MS 2, a ToF mass analyzer.

4.2.1   INSTRUMENT LAYOUT

The literature states that an ion mobility cell can be placed in a number of positions within a mass spectrometer: before[24] and after[25] the mass analyzer, between mass analyzers on tandem MS instruments,[26] or before[27] or after[28] a collision cell. The positioning of the cell and the type of mass analyzer used partially determines the types of experiments that can be performed on an instrument. In a QToF there are two positions in which a drift cell can be easily accommodated: (i) in place of the hexapole collision cell, or (ii) in a new vacuum chamber between the source hexapole and the quadrupole.

Each of the locations has certain advantages over the other, e.g., placing the cell after the quadrupole mass analyzer would allow mass selection prior to ion mobility spectrometry (IMS) separation, but would require substantial extra pumping proximal to the ToF analyzer chamber. Ultimately we decided that the cell should be housed in a new chamber between the source hexapole and the quadrupole. The reasons for this were manifold but included:

1. The pre-cell hexapole can be used to store ions prior to pulsing them into the IMS cell.

2. Housing the cell separately from the quadrupole (and away from the ToF) will allow the instrument to operate at near normal pressures.
3. The hexapole collision cell and hence MS/MS capability can be retained.
4. Designing a new vacuum chamber rather than modifying an existing one means the dimensions are less constrained.

In order to ease the transition back into the original instrument, a short hexapole was also included in the new chamber to transmit and focus the ions exiting the drift cell and also provide a vacuum seal via the top hat. Figure 4.1 shows a schematic diagram of the mobility QToF (MoQToF) without the time-of-flight region, showing where the drift cell and its housing chamber are located within the instrument.

4.2.2 THE CELL

Ion mobility is a measure of how quickly a gas-phase ion moves through a buffer gas under the influence of an electric field, and this depends on two factors: the rotationally averaged collision cross section of the ion (Ω) and the charge present on it (z). By measuring the drift time of an ion through a known distance it is possible to determine its collision cross section with some degree of accuracy. Of the three currently available methods of recording ion mobility (FAIMS, Travelling Wave, and Linear), only linear ion mobility, where the drift field is a DC potential, provides absolute conformational information from measurements of the average drift time (tD) of mass selected ions as described by Equation (4.1):

$$\Omega = \frac{(18\pi)^{\frac{1}{2}}}{16} \frac{ze}{(k_B T)^{\frac{1}{2}}} \left[\frac{1}{m_I} + \frac{1}{m_B} \right]^{\frac{1}{2}} \frac{t_D E}{L} \frac{760}{P} \frac{T}{273.2} \frac{1}{N} \tag{4.1}$$

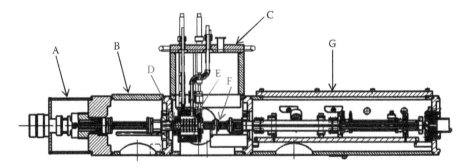

FIGURE 4.1 Schematic diagram of the MoQToF. A, z-spray ESI ion source; B, vacuum chamber 1 housing pre-cell hexapole; C, vacuum chamber 2: new chamber housing pre-cell Einzel lens (D), drift cell (E), and post-cell hexapole (F). This chamber also contains all gas and electric feedthroughs for the drift cell. G, vacuum chamber 3 housing quadrupole mass analyzer and hexapole collision cell leading to orthogonal ToF mass analyzer (not shown). (McCullough, B. J.; Kalapothakis, J.; Eastwood, H.; Kemper, P.; MacMillan, D.; Taylor, K.; Dorin, J.; Barran, P. E., Development of an ion mobility quadrupole time of flight mass spectrometer. *Analytical Chemistry* 2008, 80, (16), 6336–6344. With permission from ACS.)

where z is the charge state of the ion, m_I is the mass of the ion, m_B is the mass of the buffer gas, E is the electric field, L the length of the drift tube, P and T are the pressure and temperature of the buffer gas, and N is the neutral number density. This version of the equation contains the expression for the reduced mobility.

Our linear mobility cell was based on the design used by the Bowers group.[24] In order to incorporate the cell into the apparatus some significant redesigning of the ion optics (see Section 4.2.3), cell mounting, and feed troughs (power and gas supplies, etc.) was required. Figure 4.2 shows the final cell design. The figure comprises two cross section views of the cell (I and II) and the front and rear elevations of the cell.

The cell was made from a copper block and a copper end cap separated by a ceramic ring giving a cell of the following dimensions: $65.5 \times 88.9 \times 88.9$ mm (L × W × H). The cell contained five drift rings (30.5-mm outer diameter [o.d.] × 15.2-mm inner diameter [i.d.] × 3.2 mm) spaced 5.2 mm apart, separated by ceramic spacers. The drift field was achieved by applying a potential difference between the cell body

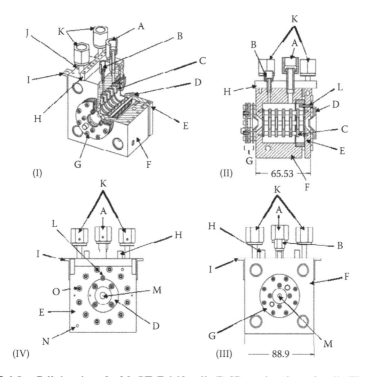

FIGURE 4.2   Cell drawings for MoQToF drift cell. (I) 3D section through cell; (II) section through cell viewed from side; (III) front elevation; (IV) rear elevation. Parts are labelled as follows: A, Baratron connection; B, gas in; C, drift rings; D, exit lens (L4); E, end cap (C2); F, cell body (C1); G, Einzel lens (L1, L2, and L3); H, heater terminal block; I, mounting brackets; J, heaters; K, cooling line inlets; L, feedthrough to drift rings; M, molybdenum orifice; N, thermocouple mounting; O, cell screws. (McCullough, B. J.; Kalapothakis, J.; Eastwood, H.; Kemper, P.; MacMillan, D.; Taylor, K.; Dorin, J.; Barran, P. E., Development of an ion mobility quadrupole time of flight mass spectrometer. *Analytical Chemistry* 2008, 80, (16), 6336–6344. With permission from ACS.)

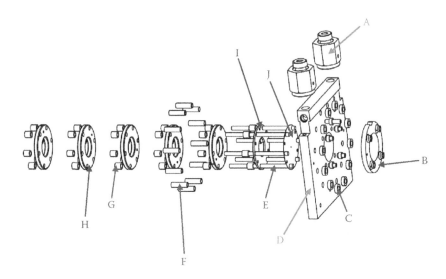

FIGURE 4.3 Cell end cap and drift ring stack. A, end cap cooling lines (1/4-inch Swagelok VCR fitting); B, exit lens (L4); C, cell screw external insulator; D, end cap; E, drift ring mounting ceramics; F, cell screw internal insulator; G, drift ring separating ceramics; H, drift ring; I, orifice retaining ring; J, molybdenum orifice.

and the end cap; the field was kept linear using six 1-MΩ resistors connected in series along the drift rings (the first resistor between the cell body and the first drift ring, the second resistor between the first and second drift rings, and so on). The cell orifices were cut into molybdenum discs (30.5-mm o.d. × 1.27 mm); the orifice size can be chosen at the discretion of the user but is typically 1 mm in diameter for the work shown here. The two orifices were held in place by copper retainer rings (30.5-mm o.d. × 15.2-mm i.d. × 1.6 mm). The drift ring stack was built up from the end cap on a set of six ceramic rods (Figure 4.3). The cell exit lens (L4) mounted from the exterior of the end cap.

Figure 4.4 shows the cell body construction. A retainer ring (D) was used to keep the front orifice in place; once this is in place the end cap and drift ring stack can be inserted through the ceramic spacer ring (C) and located in the retainer ring (D). The end cap was then kept in position by a set of 12 hex screws insulated from the end cap by a set of ceramics (Figure 4.4, labels C and F). The cell and end cap can be heated via a set of ceramic heaters. These consist of 71-mm ceramic rods threaded with Tantalum wire. The cell body contains 10 of these rods (five on either side) while the end cap contains two shorter ceramic rods (50 mm). Applying a potential to the heaters causes the wire to resistively heat and hence heat the cell. The cell body and end cap can be cooled using a stream of nitrogen gas, which can be passed though cooling channels in the copper. The cell temperature can be monitored via a set of three thermocouples—one on the cell body and two on the end cap.

Each ceramic heater is attached in series to the heater terminals (H)—the end cap and cell heater voltages are supplied separately.

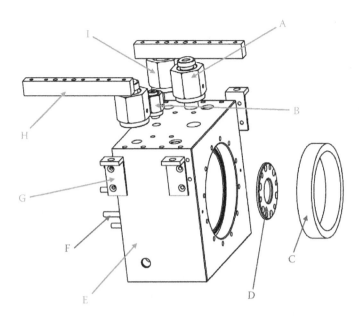

FIGURE 4.4 Cell body (C1). A, feedthrough for Baratron (1/4-inch Swagelok VCR); B, gas inlet (1/8-inch Swagelok VCR); C, ceramic spacer ring; D, orifice retainer ring (orifice not shown); E, cell body; F, ceramic mounting rods; G, cell mounting brackets; H, heater terminals; I, cell cooling line feedthroughs (1/4-inch Swagelok VCR).

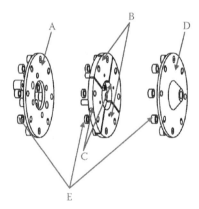

FIGURE 4.5 Cell entrance lens stack. A, L1; B, L2Y; C, L2X; D, L3; E, ceramic spacers.

4.2.3 THE LENS SYSTEMS

The cell entrance lens was mounted on a set of ceramic rods from the front of the cell body. The three lenses (L1 – A, L2 – B and C, and L3 – D) are stainless steel discs with the following dimensions: 40.6-mm o.d. × 22.4-mm i.d. × 3.8 mm (Figure 4.5). L2 (B and C) was split into four sections to allow x-y steering of the ions—voltages were supplied separately to each lens, and each pair of lenses (x and y) can be biased relative to each other using a variable resistor. The final lens (L3 – D) is tapered to a final i.d. of 7.6 mm to aid focusing into the cell.

4.2.4 LOCATING THE CELL IN THE MASS SPECTROMETER

In the Bowers instruments,[24,29] the cells are mounted on a set of horizontal bars; employing a similar arrangement here would have made removal of the cell an extremely time-consuming activity requiring initial detachment of the source and first vacuum chamber to gain access to the cell (see Figure 4.1). Our cell was therefore designed such that it could be inserted and removed vertically.

The cell was mounted on four vertical stainless steel rods attached to brackets on each corner of the cell body and to two plastic supports at the other end; the plastic supports were in turn attached to the top of an aluminum vacuum flange. Figure 4.6 shows the fully constructed cell with the plastic supports (A) and mounting rods (B) attached. This figure also shows the hose attachments for the various gas feedthroughs. Figure 4.7 shows the cell vacuum chamber (E and F), cell (I, J, K, L) mounted from the top flange (E), and post-cell hexapole (H2 – H, TH2 – G). The chamber has an iso-160 flange to the bottom allowing the mounting of a 500 l s⁻¹ Pfeiffer TMH1J20 turbomolecular pump (Pfeiffer Vacuum Ltd., Newport Pagnell, UK).

The voltages applied to the cell and the pre-cell hexapole (H1) and top hat (TH1) are supplied via the voltage feedthroughs (D). The top flange contains three voltage feedthroughs: one for applying voltage to the cell components, one for supplying voltage to the heaters, and one for connecting thermocouples.

4.2.6 POWER SUPPLY DESIGN

A bespoke DC power supply was designed and constructed in house. The elements that require a voltage supply are as follows: H1, TH1, L1, L2 (four inputs), L3, C1, C2, and L4. The post-cell hexapole and top hat (H2 and TH2) and the collision voltage (CV) are supplied via the instrument. The required voltages and their grounds are shown in Figure 4.8.

As can be seen from the figure, each voltage in the supply is set with reference to the collision voltage (the voltage applied to the hexapole collision cell with respect to ground); all other voltages float on top of this value, i.e., increasing the collision voltage by 5 V will cause all other voltages to increase by 5 V. Further to this L4 is referenced to the end cap voltage (C2) and all voltages to the source side of the cell body are referenced to C1. Table 4.1 shows the voltage ranges of each supply and the total voltage applied.

The power supply was constructed such that these voltages could be supplied in both negative and positive polarity. This was achieved by using two different printed circuit boards (PCBs): one for supplying negative voltages and the other for supplying positive voltages.

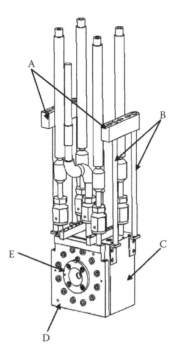

FIGURE 4.6   Cell with mounting rods and gas feedthroughs attached. A, plastic supports; B, mounting rods; C, cell body; D, end cap; E, L4.

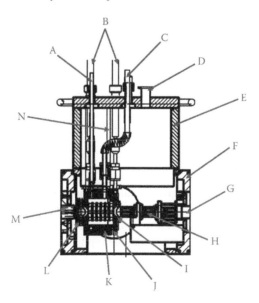

FIGURE 4.7   Cell chamber, cell, and post-cell hexapole. A, gas inlet; B, cooling lines; C, Baratron feedthrough; D, voltage feedthroughs; E, cell chamber top flange; F, cell chamber bottom flange; G, post-cell hexapole top hat (TH2); H, post-cell hexapole (H2); I, L4; J, end cap; K, cell body; L, Einzel lens; M, pre-cell top hat (TH1); N, cell mounting rods.

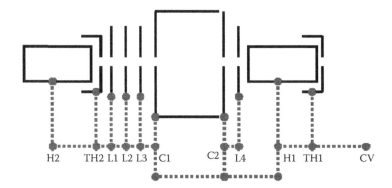

FIGURE 4.8   Schematic of voltages. Horizontal lines indicate the ground for each voltage, i.e., for TH1 the ground voltage is CV (the collision voltage). Voltages indicated in blue are supplied from the instrument and controlled via the software.

TABLE 4.1
Voltage Requirements for Drift Cell Power Supply

Element	Voltage Range	Output Voltage
CV	0 to 200 V	CV
TH2	0 to –5 V	CV + TH2
H2	0 to –5 V	CV + H2
L4	+50 to –50 V	CV + C2 + L4
C2	0 to 75 V	CV + C2
C1	0 to 200 V	CV + C1
L3	+50 to –50 V	CV + C1 + L3
L2	0 to –350 V	CV + C1 + L2
L1	0 to –350 V	CV + C1 + L1
TH1	0 to 200 V	CV + C1 + TH1
H1	0 to 200 V	CV + C1 + H1

4.2.7   PULSER DESIGN

An ion mobility experiment requires a pulse of ions, so when using a continuous ion source (such as electrospray) it is necessary to trap ions and pulse them into the cell. Here we employed the pre-cell transfer hexapole H1 (see Figure 4.1). This is achieved by placing a stopping voltage on the top hat lens (TH1). To pulse ions out of the hexapole it is necessary to apply a pulsed voltage to TH1, which will allow the stopping potential to be rapidly lowered for short periods of time at a set frequency.

A pulser unit was therefore designed and built in house to allow the TH1 to be pulsed. The unit requires two inputs: TH1 from the lens power supply described above and a transistor–transistor logic (TTL) pulse from a signal generator. A signal generator is used to define the pulse frequency and width; the amplitude of the stopping potential is controlled via the pulser unit.

4.2.8 GAS INLET SYSTEM, PRESSURE AND TEMPERATURE MEASUREMENT

The gas inlet system includes a gas filter used to remove oxygen, water, and hydrocarbons from the buffer gas. It was designed for use with one or two buffer gases (e.g., He or He and H_2O). The pressure in the drift cell is measured using an MKS Baratron attached to the cell via a ¼-inch Swagelok VCR fitting. The temperature of the cell is monitored using three K-type thermocouples: one on the cell body and two on the cell end cap. The thermocouples are read on an Omega CN1001TC thermocouple controller.

4.2.9 QUADRUPOLE UPGRADE

On a standard QToF I the quadrupole has a maximum isolation mass-to-charge ratio (m/z) of 4000. While this is sufficient for most applications, it may not be suitable for large biomolecular complexes, such as those studied by Rostom and coworkers,[31] which can often appear at above 4000 m/z on a mass spectrum. With this in mind, the quadrupole was modified to have a mass range of up to 32,000 m/z. This was achieved by replacing the quadrupole RF driver with one capable of lower frequency operation.

4.3 OPERATION OF THE MOQTOF

In order to take mobility data (i.e., measure arrival times) on the Synapt HDS system,[18] the Masslynx software was redesigned by Waters Micromass Technologies (Manchester, UK). The new software is equally applicable for our instrument. It allows for two modes of operation: (i) MS(/MS) mode—collection of mass spectra only (on point detector or micro channel plates [MCPs]); (ii) IMS mode—collection of mass spectra with mobility data (MCPs only).

4.3.1 MS MODE

The basic operation for MS mode has not changed greatly; the only significant software change was to two of the source voltages, those applied to the extractor cone and the sample cone.

The extractor cone voltage is referenced to the collision voltage, and prior to the instrument modifications it was typically held at 0–5 V above the collision voltage and crucially at a higher DC voltage than H1. In the modified instrument H1 is held at several tens of volts higher than the collision voltage (typically ~110 V); the extractor must therefore be held at 0–5 V higher than this value. To achieve this the software was altered to allow a greater range of voltages to be applied.

The sample cone is referenced to the extractor cone and therefore the software required some modification to also allow the desired voltage to be applied to it. The maximum voltage that can be applied to the sample cone (sample cone voltage + extractor cone voltage) is 204 V + CV; this can prove to be a limiting factor in examining high m/z species and some further modification is still required.

MS mode is typically used when tuning for signal; the drift cell is filled with buffer gas to the desired pressure, the drift voltage and injection energy (defined as the difference between the voltages applied to C1 and H1) are set, and the source voltages are tuned until the best possible signal is obtained.

4.3.2 IMS MODE

For mobility operation significant changes were made to the software so it can record mobility data (arrival times) along with mass data. In ToF-MS mode the instrument measures the time taken for ions to pass from the pusher to the detector. These times are then converted to *m/z* values. In IMS mode each ion has an associated arrival time (the time it takes to pass through the drift cell + dead time) that must also be recorded.

In the MoQToF software the ion arrival time is the time taken to pass from TH1 to the pusher. This is measured as follows:

1. TH1 is pulsed to allow a packet of ions into the drift cell (10–40 µs wide, 50–100 Hz). The signal generator sends a simultaneous pulse to the instrument time to digital converter (TDC) card, which signals the software to start the IMS experiment.
2. The first pulse of the pusher (push) after this start signal is scan 1, the second push is scan 2, and so on up to 200 scans. The width of each scan is equivalent to the pusher frequency being used.
3. The software records a mass spectrum for each scan to build up a 200-scan total ion chromatogram (TIC) count. The 200 scans can then be summed to give a mass spectrum for the experiment.
4. The software allows any single *m/z* or *m/z* range to be selected. It then builds up a chromatogram for the selected *m/z* with a peak after *n* scans corresponding to that ion's arrival at the pusher.
5. Multiplying the scan number of the peak by the pusher frequency gives the arrival time.

In practice a single set of 200 scans provides insufficient data and the software is set to sum over a number of seconds (10–25 s) to obtain a single set of 200 averaged scans. Further to this it is necessary to sum the data over a number of scan sets to obtain good arrival time data. For every set of 200 scans there can be only one pulse of ions into the drift cell; the frequency of the period of this pulse must therefore be greater than 200 × pusher period.

Figure 4.9 shows the pulse sequences of the pusher and IMS pulse. The upper trace shows the pulse sequence used to control the ion mobility pulse while the lower trace shows the ToF pusher pulse. The pulses in red indicate the first push after the IMS pulse, i.e., the start of the experiment. Although Figure 4.9 indicates that the IMS pulse and pusher pulse are synchronized, they are in fact asynchronous (the consequences of this on the measurements will be discussed later). In order for the software to start an IMS experiment, it must see an input pulse while it is pushing; for this reason the input pulse to the TDC card must be at least one pusher period in width (thus ensuring the input pulse and pusher pulse are concurrent).

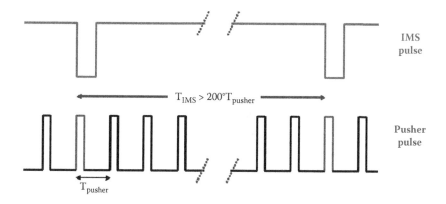

FIGURE 4.9   Pulse sequence for IMS pulser and ToF pusher.

4.3.3   FROM ARRIVAL TIME DISTRIBUTIONS TO DRIFT TIMES TO MOBILITIES

Figure 4.10 is a schematic representation of how to proceed from the total ion chromatograms (TIC) to the selected ion chromatograms (SIC) from which the mobilities of a given m/z species can be obtained. Arrival time distributions (ATDs) obtained from SICs give the time ions take to pass from TH1 to the pusher, i.e., the drift time (t_d time spent in the cell) + dead time (t_{dead} time spent outside the cell) (Equation (4.2)). To calculate the drift time accurately, it is necessary to determine the dead time or manipulate the data so that the dead time is not needed to calculate the mobility. In reality, both of these conditions can be satisfied by plotting the arrival time (t_a) of an ion against P/V as shown by Equation (4.3) where L is cell length, T_0 is 273.15 K, K_0 is reduced mobility, P_0 is 7.6×10^{-4} Torr, T is cell temperature in kelvin, P is cell pressure in Torr, and V is drift voltage (the difference between C1 and C2).

$$t_a = t_d + t_{dead} \tag{4.2}$$

$$t_d = \frac{L^2 T_0}{K_0 P_0 T}\frac{P}{V} \tag{4.3}$$

By substituting Equation (4.2) into Equation (4.3) it is clear that a plot of arrival time as a function of P/V for any one ion should give a straight line of gradient inversely proportional to the reduced mobility. For any experiment the values of T and L are well known, so by performing the measurement at several different P/V values, K_0 can be easily calculated from such a plot. The y-intercept of this plot will give the dead time for the ion of interest. In our group we only accept mobility data where the R^2 values from the straight line fits of these plots are accurate to 0.999 or better (Figure 4.10c).

To obtain different values of P/V it is possible to change one or both of the parameters; however, the cell pressure can take several seconds to stabilize when it is changed, while the drift voltage stabilizes almost instantly. It is therefore more

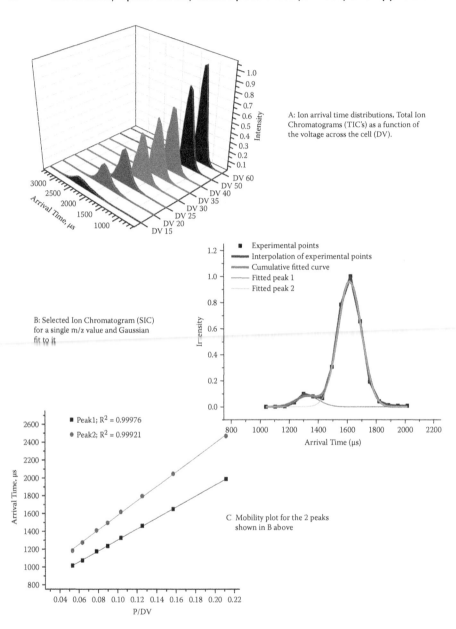

FIGURE 4.10 Schematic representation of the workflow IMS data obtained from Masslynx. A set of TIC (total ion count) arrival time distributions (ATDs) is obtained at a range of discrete drift voltages (a); the data is combined to give a mass spectrum; from the mass spectrum individual ions are selected to give SIC (selected ion chromatograms) and a corresponding ATD is created, which can be viewed in terms of time by multiplying the scan number by the pusher time (b). The mobility plots of *P/V* versus arrival time are used to obtain the drift time and dead time for each conformationally resolved species, and hence the mobilities (c).

practical to perform the measurements at a range of drift voltages with constant cell pressure.

4.3.4   Obtaining Good ATDs

As the software only allows the ions to be sampled once every pusher cycle, the resolution of the data is relatively poor, e.g., for data where the maximum $m/z = 1000$, the pusher period is set to 63 μs; the ions are therefore only sampled once every 63 μs. To try and counteract this low resolution, it is necessary to sum a number of distributions ($n \geq 10$) and fit a Gaussian distribution to the peak in order to obtain an accurate arrival time (see Figure 4.10b). Summing the data in this manner also allows low abundance peaks to be better sampled.

4.4   MOBILITY EXPERIMENTS ON PROTEINS— NON-NATIVE CONDITIONS

With the instrument tuned as described above, a set of test experiments were carried out to check the accuracy and consistency of mobility measurements made using the MoQToF. The experimental cross sections of the proteins cytochrome c,[12] ubiquitin,[31,32] and lysozyme[33] have been widely reported in the literature and are now frequently used to calibrate Synapt HDS data.[34] As these compounds are readily available they make ideal test/calibration compounds. Lyophilized, equine heart cytochrome c, ubiquitin, and hen's egg lysozyme were obtained from Sigma Aldrich and used without further purification. Samples were prepared for nano-ESI (electrospray ionization) in 50:50 methanol/water with 0.1% formic acid at concentrations ranging from 20 to 40 μM.

4.4.1   Cytochrome c

Cytochrome c is probably the protein most widely studied by IMS. It has been widely used as a test compound for new instruments[12,35,36] and as a calibrant molecule for some mobility measurements.[34,37,38] It is a 104-residue heme containing protein of mass 12,360 Da. Figure 4.11a shows ATDs for $[M+8H]^{8+}$, $[M+12H]^{12+}$, and $[M+17H]^{17+}$ charge states of cytochrome c. The two higher charge state species can be seen to yield single peaks in the ATD; the $[M+8H]^{8+}$ ATD, however, shows evidence of a shoulder indicative of a second conformation. Shelimov et al.[12] and others[39] report the presence of multiple resolvable confirmations for the +5 to +8 ions of Cyt C. The data obtained here shows only single resolvable conformers for each charge state. There was, however, some evidence of a more extended conformation for the +8 charge state of Cyt C. Unfortunately, this more unfolded species was only observable at this drift voltage; it was of too low intensity at all other voltages to obtain accurate arrival times. The mobility of this species cannot therefore be determined via the previously outlined method; however, as the more compact species was observable at all drift voltages, the experiment dead time for this charge state can

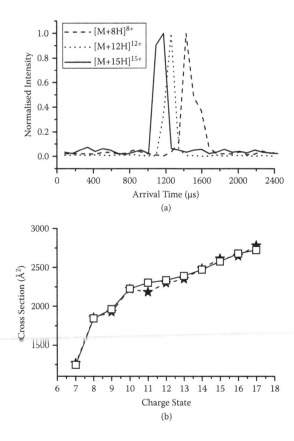

FIGURE 4.11 (a) Arrival time distributions for [M+8H]8+ (dashed line), [M+12H]12+ (dotted line), and [M+15H]15+ (solid line) cytochrome C at 3.5 Torr, Vdrift = 60 V. (b) Experimental cross sections of different charge states of the protein cytochrome C obtained on the MoQToF (black stars) and from literature (open squares; (From Shelimov, K. B.; Clemmer, D. E.; Hudgins, R. R.; Jarrold, M. F., Protein structure in vacuo: Gas-phase confirmations of BPTI and cytochrome c. *Journal of the American Chemical Society* 1997, 119, (9), 2240–2248.)).

be accurately determined. It is reasonable to assume that the dead time for any one charge state should be independent of collision cross section, and therefore the drift time for the more extended species at $V_d = 60$ V could be determined. The arrival time of the lower mobility species is 1504 μs and the dead time for the +8 species was determined to be 654 μs. This gives a drift time of 850 μs for this species. Using Equation (4.1) the mobility was therefore determined to be 1.977 cm²V⁻¹s⁻¹ (cf. 2.285 cm²V⁻¹s⁻¹ for the more abundant conformer) corresponding to a collision cross section of 2144 Å². This compares well—within 5%—with the cross section of 2061 Å² reported by Shelimov et al.[12] This agreement is especially pleasing as the mobility was only obtained using one arrival time; this suggests that the assumption made about the dead time is indeed a valid one. The mobility—and hence collision cross section—of this species can therefore be determined without the need for measurement at multiple drift voltages.

Figure 4.11b shows a plot of measured cross section against charge state along with published cross sections for cytochrome c. It can be seen that there is excellent agreement between the published data and data obtained from the MoQToF; all values agree within 6% with most in agreement within 2%. This provides a very nice example of the effect of protonation on the conformation of a gas-phase protein; as pointed out in previous work the collision cross sections increase as a function of the number of protons that charge the protein due to coulombic repulsion.

4.4.2   Ubiquitin and Lysozyme

Ubiquitin is a small protein (76 residues, 8564 Da) found in all eukaryotic cells. It has been widely studied by IM-MS. Figure 4.12a is a plot of cross sections obtained from the protein ubiquitin measured on the MoQToF against charge state along with published cross sections. Once more there is good agreement between the MoQToF data and previously published data (within 4%).

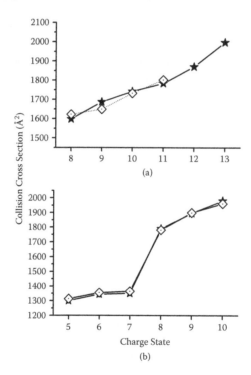

FIGURE 4.12 Experimental cross sections for the proteins ubiquitin (a) and lysozyme (b) obtained from the MoQToF (black stars) and literature (open squares; for ubiquitin, Li, J. W.; Taraszka, J. A.; Counterman, A. E.; Clemmer, D. E., Influence of solvent composition and capillary temperature on the conformations of electrosprayed ions: unfolding of compact ubiquitin conformers from pseudonative and denatured solutions. *International Journal of Mass Spectrometry* 1999, 187, 37–47, and for lysozyme, Valentine, S. J.; Anderson, J. G.; Ellington, A. D.; Clemmer, D. E., Disulfide-intact and -reduced lysozyme in the gas phase: conformations and pathways of folding and unfolding. *Journal of Physical Chemistry B* 1997, 101, (19), 3891–3900).

Lysozyme is a 129 amino acid protein (~14 kDa) that contains four intermolecular disulfide linkages. The presence of these disulfide bonds has been shown to have a stabilizing effect on the gas-phase conformations adopted by the protein. Here it was studied in its fully disulfide intact (oxidized) form. Figure 4.12b shows a plot of cross sections measured for the observed charge states ([M+5H]$^{5+}$ to [M+10H]$^{10+}$). Once more only single resolvable conformations were observed here for each charge state. Valentine et al.,[33] however, have reported the presence of a number of conformations for [M+7H]$^{7+}$ to [M+11H]$^{11+}$. They categorize these conformations as either highly folded, partially unfolded, or unfolded (with the latter category only evident for fully disulfide reduced lysozyme); the [M+5H]$^{5+}$ and [M+6H]$^{6+}$ charge states of lysozyme exist as only single conformers categorized as highly folded. The cross sections observed in this work for the three lowest charged species ([M+5H]$^{5+}$, [M+6H]$^{6+}$, and [M+7H]$^{7+}$) agree well with the values from Valentine et al. assigned as highly folded. The cross sections measured here for the [M+8H]$^{8+}$, [M+9H]$^{9+}$, and [M+10H]$^{10+}$ charge states also agree well with the reported values for partiallyfolded lysozyme. The conformational change can be clearly seen in Figure 4.12b.

It is worth noting that Valentine et al.[33] only report the partially unfolded species at high injection voltage (120 V) and not at the 30 V used here. The reason the partially unfolded conformation is sampled here is not clear, but it may be due to RF heating of the ions while they are being stored in the pre-cell hexapole.

4.5   USE OF THE MOQTOF ON PROTEIN COMPLEXES

Interrogating biological molecules in the gas phase using ion mobility mass spectrometry (IM-MS) techniques yields information not available using mass spectrometry alone, opening a new avenue to structural biology,[40–42] in particular when coupled with the use of nanoelectrospray ionization (nano-ESI) to probe near-native environments from solutions of biomolecules for analysis by mass spectrometry.[43–45] Recent studies systematically adjusting source hexapole pressure conditions[17,46,47] provide useful methodology for analyzing biomolecules composed of multiple constituent subunits or macromolecular assemblies.[48] Essentially the change in pressure from atmosphere to high vacuum must be achieved slowly to maintain noncovalent interactions and successfully transfer multicomponent structures into the mass spectrometer for subsequent analysis.

Despite the undoubted benefits of modified QTof instrumentation in retaining macromolecular species into the gas phase,[49,50] there is not yet a commercially available instrument that records absolute ion mobilities along with mass spectrometry data. Furthermore, interpretation of relative mobility data obtained on FAIMS or Synapt devices relies on data taken on biomolecules on homebuilt linear IM-MS instrumentation.[22,24,28,51,52] Given that it is possible to study extremely large assemblies that differ greatly from the standard proteins used for calibration (as reported in Section 4.4 above) both in terms of their mass and the charge supported on them, there would be a benefit in the provision of absolute mobilities for larger macromolecular species for calibration purposes. With this in mind, and also being mindful of the general insights into the biophysics of protein unfolding and assembly that can be obtained from gas-phase studies, we have considered two systems larger than those described in Section 4.4 that might be good models for further work.[53]

4.5.1 CYTOCHROME C AGGREGATES

The first macromolecular system is that of aggregates formed from the heme protein cytochrome c, as described in Section 4.4.1 above. These complexes have been reported in several studies, and although not biologically relevant, they have provided useful insights to dissociation in the gas phase.[54–56] They are readily observable, following nanoelectrospray of the protein from aqueous solution (pH 5.65) that has been left for a long time (over a year in some cases). The protein (again obtained from Sigma Aldrich) was stored for over three months in distilled water at a concentration of 200 μM and diluted to 62 μM for experimental work with a measured solution pH of 5.65. Under these conditions, cytochrome c exhibits a range of multimeric species (Figure 4.13) with the monomer species $[M+nH^+]^{z+}$ for $z = 5$ to 12 dominating. This is wider than we would expect. The solution is nativelike and ought to display a narrower distribution of charge indicative of a folded protein, which indeed it does at lower concentrations. Higher charge states may be the result of charge partitioning and dissociation from a higher order multimer.

Dimer ($z = 11^+$ to 19^+) and trimer ($z = 16^+$, 17^+, and 19^+) signals are observed as favorable pumping, and temperature conditions in the source hexapole region allowed aggregates formed in solution to be transferred to the gas phase. Signals for uneven

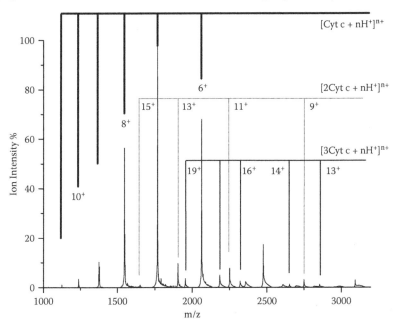

FIGURE 4.13 Mass spectrum of the protein cytochrome c, at a concentration of 62 μM, sprayed from aqueous solution (pH 5.65) taken on the MoQToF showing monomer, dimmer, and trimer peaks. The cross sections from all of these species are presented in Table 4.2.

TABLE 4.2
Collision Cross Sections (Å²) of Aqueous Cytochrome c (62 μM Concentration)

Charge	Monomer	Dimer	Trimer
5	1217	—	—
6	1243	—	—
7	1546	—	—
8	1865	—	—
9	2066	—	—
10	2228	—	—
11	2255	2586	—
12	2407	—	—
13	—	3248	—
14	—	—	—
15	—	3389	—
16	—	—	3812
17	—	3953	4014
18	—	—	—
19	—	4157	4430

Note: Ion mobility data was taken at 300 K temperature, 3.5 Torr helium, with an elevated source hexapole pressure of ~0.4 mbar for transfer of larger multimeric species.

charged dimers could be assigned only in mass spectrometry experiments as evenly charged dimer m/z values are concomitant with monomers with half the charge, i.e., a doubly charged dimer will have the same m/z as a singly charged monomer. Two higher order multimers, 17⁺ tetramer (2909 m/z) and 19⁺ pentamer (3254 m/z), were repeatedly observed when aqueous samples were sprayed. Monomer collision cross sections of cytochrome c ranged from 1217 to 2407 Å² for 5⁺ to 12⁺ charge states consistent with the work of Clemmer et al.[57] and as we have previously reported[22] and shown above (Table 4.2) are consistent with coulombically driven unfolding.

Dimer (D) cross sections were obtained for five species, and three trimer (T) collision cross sections were elucidated (Table 4.2). As z increases so does the cross section for both dimer and trimer. Collision cross sections of dimer species reveal interesting behavior, which we attribute to a gas-phase unfolding mechanism, and the collision cross sections increase rapidly with z, despite retaining the noncovalent dimeric interaction. Using the protein data bank (PBD) file 1CRC,[58] two cytochrome c monomers have been coarsely docked together to give an approximate dimer arrangement to calculate an effective hard spheres scattering (EHSS) collision cross section for this species of 2374 Å². This value is smaller than the lowest experimental values obtained for the lowest z, but provides an approximate value for comparison. Dimer unfolding has been shown[59,60] to be accompanied by dissociation to monomers whose charge is dependent upon the initial charge carried by the dimer molecule. Relatively low charged species (e.g., 11⁺) dissociate into monomers carrying asymmetrical charge, with one monomer carrying much more of the charge than the other. An intermediate conformation for this would contain one monomer unfolding to a greater extent than the other. As the initial dimer charge increases, this charge asymmetry lessens to a point whereby both monomers dissociate with close to symmetrical or symmetrical charge as they have unfolded to similar extents and exposed available ionizable sites for protonation. Our measured collision cross sections for the dimers increase while retaining a dimeric interface over a fairly substantial charge state range that is significantly greater than that observed with the hemoglobin dimers (see below) or the cytochrome c trimers. We only resolve single conformations, suggesting that dimers are stable in this partially unfolded state over the timescale of our experiment.

Protons within the trimer species will be distributed throughout the three subunits. We see a smaller number of trimers and a much smaller change in cross section with z (Table 4.2). Unfolding to larger cross sections will probably not occur until a large number of charges are sequestered on the assembly or, more likely, dissociation to (i) monomer subunits or (ii) a dimer and a monomer subunit will occur before unfolding. This would explain the small observed population of trimer species, and their gas-phase stability could be probed in future experiments using temperature and collision-induced dissociation (CID).

4.5.2 HEMOGLOBIN

Several gas-phase investigations on the abundant oxygen transport protein hemoglobin have been reported over the past decade with intense focus on interactions between the subunits and within the heme-globin monomer.[61–63] Measuring gas-phase stability using dissociative techniques (e.g., collision-induced dissociation) provides evidence for the retention of noncovalent interactions into the gas phase, and the globin monomeric chains (a and b) as well as the homodimeric (ab) species and the biologically active heterotetramer $(ab)_2$ have been interrogated in this way to determine information on both assembly and disassembly pathways.[64,65]

The hemoglobin used for these experiments was obtained from Sigma Aldrich. Prior to experimental work it was dissolved in 50 mM ammonium acetate and passed through a PD-10 column (GE Healthcare Bio-Sciences AB, Uppsala, Sweden) to remove any salts prior to analysis. The concentration was measured using a UV/Vis spectrometer (Cecil 1000 Series, Progen Scientific Ltd., Nuneaton, UK) at $1 = 274$ nm and $\varepsilon oxy = 138{,}048$ M-1cm-1 and established to be 120 μM.

Mass spectra for the hemoglobin protein were taken under a range of conditions,[53] but the tetrameric complex was best preserved by increasing the ammonium acetate buffer pH using sodium hydroxide, to pH 9.5, consistent with previous work.[66] Three tetramer (Q) signals are clearly apparent, $[Q+nH^+]^{z+}$ for $z = 16$, 17, and 18 with m/z values of 4030.0, 3793.0, and 3582.3, respectively (Figure 4.14). This mass spectrum also shows two holoheterodimers (a*b) with charge states 11+ and 12+, along with very weak signals for three a* monomer species with 6+ to 8+ charge states.

A single arrival time peak was observed for all species under nativelike conditions from IM-MS experimental data indicating compact structure—not necessarily indicative of a singular gas-phase conformation but perhaps a number of conformations with similar mobilities and cross sections.

Four a_{holo} monomers were distinguished under these experimental conditions for $[a_{holo}+nH^+]^{z+}$ $z = 6 – 9$. The collision cross section increases with increasing charge

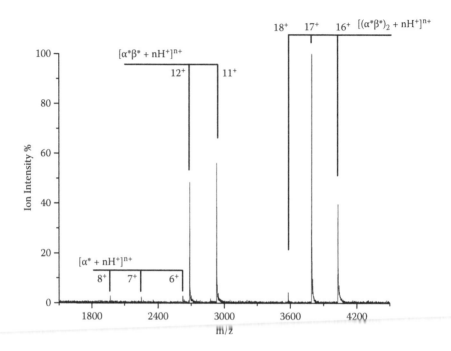

FIGURE 4.14   Mass spectrum of hemoglobin in buffered ammonium acetate at pH 9.5 from 1500 to 4500 m/z. All observed peaks are annotated. The elevated pH stabilizes tetramer signal with the $[Q+17H^+]^{17+}$ tetramer dominating the spectrum. Holoheterodimer and α^* are also present.

state (Table 4.1) with a near linear increase from $z = 7$ to 9, which is attributable to coulombic repulsion as noted above for cyctochrome c, ubiquitin, and lysozyme. There is a larger change in cross section from $z = 6$ to $z = 7$ indicating a protein unfolding event.

Ten b_{apo} globin monomers were distinguished: $[b_{apo}+nH^+]^{z+}$ for $z = 8–17$. The collision cross sections are reported in Table 4.3. Between $z = 8$ and 11, there is a near linear increase in collision cross section resulting in a total increase of 410 Å^2, a significant increase in the size of the protein. This accommodates the acquired positive charges. In the region $z = 11–14$ the cross section changes by less than 75 Å^2, which indicates the positive charge is being accepted at sites that are more remote and a large conformational change is not necessary to reduce the repulsive force. As charge increases further ($z = 15$ and 16) a dramatic size increase of 596 Å^2 is observed. This large increase could be attributed to the loss of a secondary fold in a significant portion of the monomeric protein.

These large cross section increases support the suggestion that the β-globin monomer has an intrinsically unstructured nature. Small regions of the protein are potentially structured, but addition of charge causes large structural changes indicating few conserved interactions within the molecule. Under physiological conditions, b_{apo} monomers use a_{holo} monomers as templates to order themselves into a structure capable of binding a heme group subsequently forming the a*b* holoheterodimer

TABLE 4.3

Collision Cross Sections (Å²) of α and β Monomer Chains under Native (10 mM Ammonium Acetate Buffer) and Denatured (10 mM Ammonium Acetate + 2% Formic Acid) Conditions

Charge	Holo-α (Native)	Apo-α (Denatured)	Apo-β (Native)	Apo-β (Denatured)
6	1115	—	—	—
7	1420	1381	—	1421
8	1525	1515	1461	1532
9	1608	1625	1626	1679
10	—	1700	1733	1864
11	—	1863	1871	1941
12	—	2041	1886	2068
13	—	2252	1899	2156
14	—	2441	1945	2267
15	—	2590	2080	2398
16	—	2614	2541	2595
17	—	2772	2622	2679
18	—	3200	—	3057
19	—	3227	—	3147
20	—	3430	—	—
21	—	3465	—	—
22	—	3498	—	—

of which were distinguished: $[a*b*+nH^+]^{z+}$ for $z = 10–12$. Collision cross sections for all three dimers (Table 4.4) imply minimal protein unfolding with increasing protonation. Also of interest is the fact that the dimer cross section is less than twice that of the monomers that could produce it, indicative of reordering by the monomer subunits to adopt a more compact dimer.

The a*b semihemoglobin dimer has been proposed as a crucial intermediate in the assembly pathway for the biologically active tetramer.[67–69] Three semihemoglobin dimers were observed here: $[a*b+nH^+]^{z+}$ for $z = 10–12$ (Table 4.4). The collision cross sections of the semihemoglobin and holoheterodimers are within 5% of one another of the same z value, indicating that for these gas-phase conformations, loss of a heme group through structural rearrangement and loss of noncovalent bonding has altered the collision cross section only to a small extent.

The $(a*b*)_2$ tetramer was observed at 3394 and 3520 m/z relating to $[(a*b*)_2 + 16H^+]^{16+}$ and $[(a*b*)_2 + 17H^+]^{17+}$, respectively. A narrow charge state distribution here is indicative of a well-folded tetramer showing that preservation of noncovalent interactions is sustained in the gas phase. There is a small (~4%) increase in collision cross section between $z = 16–17$, suggesting charge sequestering at a site that does not induce a large conformational change to overcome unfavorable interactions.

Using the Effective Hard Spheres Scattering (EHSS) method within MOBCAL,[70,71] theoretical cross sections were produced using the published structures for comparison

TABLE 4.4

Collision Cross Sections (Å^2) of $\alpha^*\beta$ Dimer, $\alpha^*\beta^*$ Dimer, and $(\alpha^*\beta^*)_2$ Tetramer Hemoglobin Species

Charge	$\alpha^*\beta$ Dimer	$\alpha^*\beta^*$ Dimer	$(\alpha^*\beta^*)_2$ Tetramer
10	2225	2174	—
11	2249	2355	—
12	2469	2484	—
13	—	—	3051
14	—	—	3215
15	—	—	3408
16	—	—	3460
17	—	—	3649

Note: Dimer species data was collected under native conditions (10 mM ammonium acetate).
For all five tetramer species, data was collected upon elevating the buffer pH to 9.5.

with the experimental work described above. PDB file 1GZX[72] was used to calculate the collision cross section for the hemoglobin tetramer (4272 Å^2); 2DN1[73] was modified to determine the collision cross sections of both the holoheterodimer, $a^m b^+$ (2099 Å^2) and the a_{holo} species (1587 Å^2). By comparing these numbers with experimentally derived values, we find that the dimer and tetramer experimental collision cross sections are smaller than those calculated theoretically. The tetramer experimental values are smaller by 15–30% compared to the calculated collision cross section. Holoheterodimer experimental collision cross sections are ~10–20% smaller than the calculated values. This decrease in collision cross section is attributed to a contraction of the gas-phase structure in the mass spectrometer. We can speculate that the protein has shrunk on transfer to a solvent (and buffer-free) environment. It is likely that the solvated crystal structures, taken in the presence of high salt concentrations, are likely to be larger than those of desolvated ions in the gas phase. This does not mean that the gas-phase form does not preserve the native contacts needed to form the biologically active complex but it does show a distinct conformational change from the crystal structure.

4.6 CONCLUSIONS AND OUTLOOK

The change from the commercial instrument to the modified ion mobility instrument has brought about many developments not only in the physical instrument, but in its performance as a mass spectrometer. Table 4.5 summarizes the performance of the instrument before and after the modifications. Our new ion mobility mass spectrometer, the MoQToF, has been shown to produce reliable ion mobility data in good agreement with previously published data. It is also capable of producing mobility measurements over a wide temperature range. The coupling of a high *m/z* capable QToF that possesses an ion source well suited to native electrospray ionization for proteins means that this IM-MS instrument has a role in obtaining accurate cross sections of large native proteins and their complexes, as demonstrated here with the model systems cytochrome c and hemoglobin.

TABLE 4.5
Instrument Specifications before and after Modifications

	QToF I	MoQToF
Cell chamber pressure (no gas in cell)	n/a	1×10^{-6} mBar
Cell chamber pressure (3.5 Torr He in cell)	n/a	1×10^{-3} mBar
Analyzer chamber pressure (no gas)	6×10^{-6} mBar	1×10^{-6} mBar
Analyzer chamber pressure (3.5 Torr He)	n/a	1×10^{-5} mBar
ToF chamber pressure (no gas)	3×10^{-7} mBar	2×10^{-7} mBar
ToF chamber pressure (3.5 Torr He)	n/a	5×10^{-7} mBar
Quadrupole mass range	2 to 4500 m/z	2 to 32,000 m/z
Quadrupole resolution	~500	~250
ToF mass range	1 to 5000 m/z	1 to 100,000 m/z
ToF resolution	8000	8000
Transmission efficiency (MS 3.5 Torr He)[a]	n/a	5%
Transmission efficiency (IMS, 3.5 Torr He)[a]	n/a	8%
IMS resolution[b]	n/a	~20
IMS pressure range[c]	n/a	≤4.5 Torr
IMS temperature range	n/a	100 to 600 K

[a] As compared to ion transmission prior to upgrade.
[b] Calculated for $[M+8H]^{8+}$ cytochrome c at $V_d = 60$ V, P = 3.5 T, T = 305.
[c] With orifices of 0.9 mm.

ACKNOWLEDGMENTS

This research was supported by the EPSRC grants GR/S77639/01 and EP/C541561/1 and in particular via the award of an Advanced Research Fellowship to PEB and also studentships to BJM and PAF. The entire Barran group cohort, 2002–2009, is thanked for their continued input. The construction of the MoQToF would not have been possible without Paul Kemper and Terry Hart (UCSB), and Douglas Munro (U of E). We also had support from the Royal Society, the Royal Society of Chemistry, the British Mass Spectrometry Society, and Waters MS Technologies Centre, and in particular we thank Steven Pringle, Kevin Giles, Jason Wildgoose, and Robert Bateman. We are also grateful for the continuing support from the School of Chemistry at the University of Edinburgh. Finally, we would like to acknowledge inspiration from the originators of the use of ion mobility together with mass spectrometry to look at biologically relevant species, in particular Martin Jarrold, David Clemmer, and Mike Bowers.

REFERENCES

1. Benesch, J. L.; Ruotolo, B. T.; Simmons, D. A.; Robinson, C. V., Protein complexes in the gas phase: technology for structural genomics and proteomics. *Chemical Reviews* 2007, 107, (8), 3544–3567.

2. Jarrold, M. F., Peptides and proteins in the vapor phase. *Annual Review of Physical Chemistry* 2000, 51, 179–207.

3. van den Heuvel, R. H.; Heck, A. J. R., Native protein mass spectrometry: from intact oligomers to functional machineries. *Current Opinion in Chemical Biology* 2004, 8, (5), 519–526.

4. Loo, J. A., Studying noncovalent protein complexes by electrospray ionization mass spectrometry. *Mass Spectrometry Reviews* 1997, 16, (1), 1–23.

5. Loo, J. A.; Berhane, B.; Kaddis, C. S.; Wooding, K. M.; Xie, Y. M.; Kaufman, S. L.; Chernushevich, I. V., Electrospray ionization mass spectrometry and ion mobility analysis of the 20S proteasome complex. *Journal of the American Society for Mass Spectrometry* 2005, 16, (7), 998–1008.

6. Ilag, L. L.; Videler, H.; McKay, A. R.; Sobott, F.; Fucini, P.; Nierhaus, K. H.; Robinson, C. V., Heptameric (L12)(6)/L10 rather than canonical pentameric complexes are found by tandem MS of intact ribosomes from thermophilic bacteria. *Proceedings of the National Academy of Sciences of the United States of America* 2005, 102, (23), 8192–8197.

7. Fuerstenau, S. D.; Benner, W. H.; Thomas, J. J.; Brugidou, C.; Bothner, B.; Siuzdak, G., Mass spectrometry of an intact virus. *Angewandte Chemie-International Edition* 2001, 40, (3), 542–544.

8. Sobott, F.; Benesch, J.; Vierling, E.; Robinson, C., Subunit exchange of multimeric protein complexes—real-time monitoring of subunit exchange between small heat shock proteins by using electrospray mass spectrometry. *Journal of Biological Chemistry* 2002, 277, (41), 38921–38929.

9. Sharon, M.; Robinson, C. V., The role of mass spectrometry in structure elucidation of dynamic protein complexes. *Annual Review of Biochemistry* 2007, 76, (1), 167–193.

10. van Duijn, E.; Bakkes, P.; Heeren, R.; van den Heuvell, R.; van Heerikhuizen, H.; van der Vies, S.; Heck, A., Monitoring macromolecular complexes involved in the chaperonin-assisted protein folding cycle by mass spectrometry. *Nature Methods* 2005, 2, (5), 371–376.

11. Breuker, K., The study of protein–ligand interactions by mass spectrometry—a personal view. *International Journal of Mass Spectrometry* 2004, 239, (1), 33–41.

12. Shelimov, K. B.; Clemmer, D. E.; Hudgins, R. R.; Jarrold, M. F., Protein structure in vacuo: gas-phase confirmations of BPTI and cytochrome c. *Journal of the American Chemical Society* 1997, 119, (9), 2240–2248.

13. Badman, E. R.; Hoaglund-Hyzer, C. S.; Clemmer, D. E., Monitoring structural changes of proteins in an ion trap over similar to 10–200 ms: Unfolding transitions in cytochrome c ions. *Analytical Chemistry* 2001, 73, (24), 6000–6007.

14. Valentine, S. J.; Clemmer, D. E., Temperature-dependent H/D exchange of compact and elongated cytochrome c ions in the gas phase. *Journal of the American Society for Mass Spectrometry* 2002, 13, (5), 506–517.

15. Gidden, J.; Ferzoco, A.; Baker, E. S.; Bowers, M. T., Duplex formation and the onset of helicity in poly d(CG)(n) oligonucleotides in a solvent-free environment. *Journal of the American Chemical Society* 2004, 126, (46), 15132–15140.

16. Bernstein, S. L.; Liu, D. F.; Wyttenbach, T.; Bowers, M. T.; Lee, J. C.; Gray, H. B.; Winkler, J. R., Alpha-synuclein: stable compact and extended monomeric structures and pH dependence of dimer formation. *Journal of the American Society for Mass Spectrometry* 2004, 15, (10), 1435–1443.

17. Benesch, J. L. P.; Ruotolo, B. T.; Simmons, D. A.; Robinson, C. V., Protein complexes in the gas phase: technology for structural genomics and proteomics. *Chemical Reviews* 2007, 107, (8), 3544–3567.

18. Pringle, S. D.; Giles, K.; Wildgoose, J. L.; Williams, J. P.; Slade, S. E.; Thalassinos, K.; Bateman, R. H.; Bowers, M. T.; Scrivens, J. H., An investigation of the mobility separation of some peptide and protein ions using a new hybrid quadrupole/travelling wave IMS/oa-ToF instrument. *International Journal of Mass Spectrometry* 2007, 261, (1), 1–12.

19. Borysik, A. J. H.; Read, P.; Little, D. R.; Bateman, R. H.; Radford, S. E.; Ashcroft, A. E., Separation of b-microglobulin conformers by high-field asymmetric waveform ion mobility spectrometry (FAIMS) coupled to electrospray ionisation mass spectrometry. *Rapid Communications in Mass Spectrometry* 2004, 18, (19), 2229–2234.

20. Ruotolo, B. T.; Giles, K.; Campuzano, I.; Sandercock, A. M.; Bateman, R. H.; Robinson, C. V., Evidence for macromolecular protein rings in the absence of bulk water. *Science* 2005, 310, (5754), 1658–1661.

21. Ruotolo, B. T.; Hyung, S. J.; Robinson, P. M.; Giles, K.; Bateman, R. H.; Robinson, C. V., Ion mobility-mass spectrometry reveals long-lived, unfolded intermediates in the dissociation of protein complexes. *Angewandte Chemie International Edition England* 2007, 46, (42), 8001–8004.

22. McCullough, B. J.; Kalapothakis, J.; Eastwood, H.; Kemper, P.; MacMillan, D.; Taylor, K.; Dorin, J.; Barran, P. E., Development of an ion mobility quadrupole time of flight mass spectrometer. *Analytical Chemistry* 2008, 80, 6336–6344.

23. Faull, P. A.; Korkeila, K. E.; Kalapothakis, J. M. D.; Gray, A. P.; McCullough, B. J.; Barran, P. E., Gas-phase metalloprotein complexes interrogated by ion mobility mass spectrometry. *International Journal of Mass Spectrometry.* 2009, 283, (1–3), 140–148.

24. Wyttenbach, T.; Kemper, P. R.; Bowers, M. T., Design of a new electrospray ion mobility mass spectrometer. *International Journal of Mass Spectrometry* 2001, 212, (1–3), 13–23.

25. Bluhm, B. K.; Gillig, K. J.; Russell, D. H., Development of a Fourier-transform ion cyclotron resonance mass spectrometer-ion mobility spectrometer. *Review of Scientific Instruments* 2000, 71, (11), 4078–4086.

26. Guo, Y. Z.; Wang, J. X.; Javahery, G.; Thomson, B. A.; Siu, K. W. M., Ion mobility spectrometer with radial collisional focusing. *Analytical Chemistry* 2005, 77, (1), 266–275.

27. Lee, Y. J.; Hoaglund-Hyzera, C. S.; Barnes, C. A. S.; Hilderbrand, A. E.; Valentine, S. J.; Clemmer, D. E., Development of high-throughput liquid chromatography injected ion mobility quadrupole time-of-flight techniques for analysis of complex peptide mixtures. *Journal of Chromatography B-Analytical Technologies in the Biomedical and Life Sciences* 2002, 782, (1–2), 343–351.

28. Koeniger, S. L.; Merenbloom, S. I.; Valentine, S. J.; Jarrold, M. F.; Udseth, H. R.; Smith, R. D.; Clemmer, D. E., An IMS-IMS analogue of MS-MS. *Analytical Chemistry* 2006, 78, (12), 4161–4174.

29. Kemper, P. R.; Bowers, M. T., A hybrid double-focusing mass-spectrometer—high-pressure drift reaction cell to study thermal-energy reactions of mass-selected ions. *Journal of the American Society for Mass Spectrometry* 1990, 1, (3), 197–207.

30. Rostom, A. A.; Sunde, M.; Richardson, S. J.; Schreiber, G.; Jarvis, S.; Bateman, R.; Dobson, C. M.; Robinson, C. V., Dissection of multi-protein complexes using mass spectrometry: subunit interactions in transthyretin and retinol-binding protein complexes. *Proteins–Structure Function and Genetics* 1998, 3–11.

31. Valentine, S. J.; Counterman, A. E.; Clemmer, D. E., Conformer-dependent proton-transfer reactions of ubiquitin ions. *Journal of the American Society for Mass Spectrometry* 1997, 8, (9), 954–961.

32. Li, J. W.; Taraszka, J. A.; Counterman, A. E.; Clemmer, D. E., Influence of solvent composition and capillary temperature on the conformations of electrosprayed ions: unfolding of compact ubiquitin conformers from pseudonative and denatured solutions. *International Journal of Mass Spectrometry* 1999, 187, 37–47.

33. Valentine, S. J.; Anderson, J. G.; Ellington, A. D.; Clemmer, D. E., Disulfide-intact and -reduced lysozyme in the gas phase: conformations and pathways of folding and unfolding. *Journal of Physical Chemistry B* 1997, 101, (19), 3891–3900.

34. Ruotolo, B. T.; Benesch, J. L.; Sandercock, A. M.; Hyung, S. J.; Robinson, C. V., Ion mobility-mass spectrometry analysis of large protein complexes. *Nat Protoc* 2008, 3, (7), 1139–1152.

35. Guevremont, R.; Barnett, D. A.; Purves, R. W.; Viehland, L. A., Calculation of ion mobilities from electrospray ionization high-field asymmetric waveform ion mobility spectrometry mass spectrometry. *Journal of Chemical Physics* 2001, 114, (23), 10270–10277.

36. Wu, C.; Siems, W. F.; Asbury, G. R.; Hill, H. H., Electrospray ionization high-resolution ion mobility spectrometry–mass spectrometry. *Analytical Chemistry* 1998, 70, (23), 4929–4938.

37. Gill, A. C.; Jennings, K. R.; Wyttenbach, T.; Bowers, M. T., Conformations of biopolymers in the gas phase: a new mass spectrometric method. *International Journal of Mass Spectrometry* 2000, 195, (196), 685–697.

38. Thalassinos, K.; Slade, S. E.; Jennings, K. R.; Scrivens, J. H.; Giles, K.; Wildgoose, J.; Hoyes, J.; Bateman, R. H.; Bowers, M. T., Ion mobility mass spectrometry of proteins in a modified commercial mass spectrometer. *International Journal of Mass Spectrometry* 2004, 236, (1–3), 55–63.

39. Clemmer, D. E.; Hudgins, R. R.; Jarrold, M. F., Naked protein conformations—cytochrome-c in the gas-phase. *Journal of the American Chemical Society* 1995, 117, (40), 10141–10142.

40. Barran, P. E.; Polfer, N. C.; Campopiano, D. J.; Clarke, D. J.; Langridge-Smith, P. R. R.; Langley, R. J.; Govan, J. R. W.; Maxwell, A.; Dorin, J. R.; Millar, R. P.; Bowers, M. T., Is it biologically relevant to measure the structures of small peptides in the gas-phase? *International Journal of Mass Spectrometry* 2005, 240, 273–284.

41. Wyttenbach, T.; von Helden, G.; Bowers, M. T., Gas-phase conformation of biological molecules: bradykinin. *Journal of the American Chemical Society* 1996, 118, (35), 8355–8364.

42. Clemmer, D. E.; Jarrold, M. F., Ion mobility measurements and their applications to clusters and biomolecules. *Journal of Mass Spectrometry* 1997, 32, (6), 577–592.

43. Fenn, J. B.; Mann, M.; Meng, C. K.; Wong, S. F.; Whitehouse, C. M., Electrospray ionization for mass spectrometry of large biomolecules. *Science* 1989, 246, 64–71.

44. Cech, N. B.; Enke, C. G., Practical implications of some recent studies in electrospray ionization fundamentals. *Mass Spectrometry Reviews* 2001, 20, (6), 362–387.

45. Wilm, M.; Mann, M., Analytical properties of the nanoelectrospray ion source. *Analytical Chemistry* 1996, 68, (1), 1–8.

46. Heck, A. J. R.; van den Heuvel, R. H. H., Investigation of intact protein complexes by mass spectrometry. *Mass Spectrometry Reviews* 2004, 23, (5), 368–389.

47. van den Heuvel, R. H. H.; vanDuijn, E.; Mazon, H.; Synowsky, S. A.; Lorenzen, K.; Versluis, C.; Brouns, S. J. J.; Langridge, D.; vanderOost, J.; Hoyes, J.; Heck, A. J. R., Improving the performance of a quadrupole time-of-flight instrument for macromolecular mass spectrometry. *Analytical Chemistry* 2006, 78, (21), 7473–7483.

48. Benesch, J. L. P.; Robinson, C. V., Mass spectrometry of macromolecular assemblies: preservation and dissociation. *Current Opinioin in Structural Biology* 2006, 16, (2), 245–251.

49. van den Heuvel, R. H.; van Duijn, E.; Mazon, H.; Synowsky, S. A.; Lorenzen, K.; Versluis, C.; Brouns, S. J.; Langridge, D.; van der Oost, J.; Hoyes, J.; Heck, A. J., Improving the performance of a quadrupole time-of-flight instrument for macromolecular mass spectrometry. *Analytical Chemistry* 2006, 78, (21), 7473–7483.

50. Sobott, F.; Hernandez, H.; McCammon, M.; Tito, M.; Robinson, C., A tandem mass spectrometer for improved transmission and analysis of large macromolecular assemblies. *Analytical Chemistry* 2002, 74, (6), 1402–1407.

51. Mao, Y.; Woenckhaus, J.; Kolafa, J.; Ratner, M. A.; Jarrold, M. F., Thermal unfolding of unsolvated cytochrome c: experiment and molecular dynamics simulations. *Journal of the American Chemical Society* 1999, 121, (12), 2712–2721.

52. Kanu, A. B.; Dwivedi, P.; Tam, M.; Matz, L.; Hill, H. H., Ion mobility-mass spectrometry. *Journal of Mass Spectrometry* 2008, 43, (1), 1–22.

53. Faull, P. A.; Korkeila, K. E.; Kalapothakis, J. M.; Gray, A.; McCullough, B. J.; Barran, P. E., Gas-phase metalloprotein complexes interrogated by ion mobility-mass spectrometry. *International Journal of Mass Spectrometry* 2009, 283, (1–3), 140–148.

54. Breuker, K.; McLafferty, F. W., The thermal unfolding of native cytochrome c in the transition from solution to gas phase probed by native electron capture dissociation. *Angewandte Chemie International Edition England* 2005, 44, (31), 4911–4914.

55. Jurchen, J. C.; Garcia, D. E.; Williams, E. R., Further studies on the origins of asymmetric charge partitioning in protein homodimers. *Journal of the American Society for Mass Spectrometry* 2004, 15, (10), 1408–1415.

56. Jurchen, J. C.; Williams, E. R., Origin of asymmetric charge partitioning in the dissociation of gas-phase protein homodimers. *Journal of the American Chemical Society* 2003, 125, (9), 2817–2826.

57. Clemmer, D. E.; Hudgins, R. R.; Jarrold, M. F., Naked protein conformations: cytochrome c in the gas phase. *Journal of the American Chemical Society* 1995, 117, (40), 10141–10142.

58. Sanishvili, R.; Volz, K. W.; Westbrook, E. M.; Margoliash, E., The low ionic strength crystal structure of horse cytochrome c at 2.1 A resolution and comparison with its high ionic strength counterpart. *Structure* 1995, 3, (7), 707–716.

59. Jurchen, J. C.; Williams, E. R., Origin of asymmetric charge partitioning in the dissociation of gas-phase protein homodimers. *Journal of the American Chemical Society* 2003, 125, (9), 2817–2826.

60. Jurchen, J. C.; Garcia, D. E.; Williams, E. R., Further studies on the origins of asymmetric charge partitioning in protein homodimers. *Journal of the American Society for Mass Spectrometry* 2004, 15, (10), 1408–1415.

61. Griffith, W. P.; Kaltashov, I. A., Mass spectrometry in the study of hemoglobin: from covalent structure to higher order assembly. *Current Organic Chemistry* 2006, 10, (5), 535–553.

62. Schmidt, A.; Karas, M., The influence of electrostatic interactions on the detection of hemeglobin complexes in ESI-MS. *Journal of the American Society for Mass Spectrometry* 2001, 12, (10), 1092–1098.

63. Gross, D. S.; Zhao, Y.; Williams, E. R., Dissociation of heme-globin complexes by blackbody infrared radiative dissociation: molecular specificity in the gas phase *Journal of the American Society for Mass Spectrometry* 1997, 8, (5), 519–524.

64. Vasudevan, G.; McDonald, M. J., Spectral demonstration of semihemoglobin formation during CN-hemin incorporation into human apohemoglobins. *Journal of Biological Chemistry* 1997, 272, (1), 517–524.

65. Vasudevan, G.; McDonald, M. J., Ordered heme binding ensures the assembly of fully functional hemoglobin: a hypothesis. *Current Protein & Peptide Science* 2002, 3, (4), 461–466.

66. Versluis, C.; Heck, A. J. R., Gas-phase dissociation of hemoglobin. *International Journal of Mass Spectrometry* 2001, 210, (1–3), 637–649.

67. Griffith, W. P.; Kaltashov, I. A., Protein conformational heterogeneity as a binding catalyst: ESI-MS study of hemoglobin H formation. *Biochemistry* 2007, 46, (7), 2020–2026.

68. Griffith, W. P.; Kaltashov, I. A., Highly asymmetric interactions between globin chains during hemoglobin assembly revealed by electrospray ionization mass spectrometry. *Biochemistry* 2003, 42, (33), 10024–10033.

69. Komar, A. A.; Kommer, A.; Krasheninnikov, I. A.; Spirin, A. S., Cotranslational folding of globin. *Journal of Biological Chemistry* 1997, 272, (16), 10646–10651.

70. Shvartsburg, A. A.; Jarrold, M. F., An exact hard-spheres scattering model for the mobilities of polyatomic ions. *Chemical Physics Letters* 1996, 261, (1–2), 86–91.

71. Mesleh, M. F.; Hunter, J. M.; Shvartsburg, A. A.; Schatz, G. C.; Jarrold, M. F., Structural information from ion mobility measurements: effects of the long-range potential. *Journal of Physical Chemistry* 1996, 100, (40), 16082–16086.

72. Paoli, M.; Liddington, R.; Tame, J.; Wilkinson, A.; Dodson, G., Crystal structure of T state haemoglobin with oxygen bound at all four haems. *Journal of Molecular Biology* 1996, 256, (4), 775–792.

73. Park, S.-Y.; Yokoyama, T.; Shibayama, N.; Shiro, Y.; Tame, J. R. H., 1.25 Å resolution crystal structures of human haemoglobin in the oxy, deoxy and carbonmonoxy forms. *Journal of Molecular Biology* 2006, 360, (3), 690–701.

5 The Differential Mobility Analyzer (DMA)

Adding a True Mobility Dimension to a Preexisting API-MS

Juan Fernandez de la Mora

CONTENTS

5.1   INTRODUCTION: THE CHALLENGES TO THE WIDE ADOPTION OF LINEAR IMS-MS

Numerous impressive studies on ion mobility spectrometry (IMS) in tandem with mass spectrometry (MS) have proven beyond any doubt the outstanding ability offered by this combination to (i) widen the range of species identifiable in complex biological mixtures, and (ii) provide structural information on complex ions.[1–5] The great potential impact of IMS-MS is so amply recognized that many laboratories that have followed its development are eager to incorporate IMS into their MS portfolios. This eagerness is particularly evident in the success of the only existing commercial IMS-MS system (Waters Synapt[6]), based on ion separation in a drift gas caused by a succession of travelling electrical waves (T waves). This solution, however, comes at some cost. First, the measured ion drift time is related to mobility in a complex and incompletely understood fashion, which limits the precision and reliability of the structural information obtained.[7] This is due to the fact that the travelling voltage wave pushing ions forward rises in each cycle as much as it falls, so its net effect is zero within the linear approximation. Only nonlinear effects survive on average, whose complex dependence on many details of the waveform and other circumstances is not easy to predict. Hence an extensive calibration effort is required, which is compounded by the lack of calibration standards of fixed mobility in the range of masses and mobilities typical of large biological ions. Second, the mobility resolution of the T-wave system, although extremely useful,[8] has been considerably smaller than that for the best pulsed IMS. This remains the case, although recent developments show T-wave resolving power of 45 for cross section.[9] Third, the capital investment required is high, as an entirely new tandem instrument needs to be acquired, rather than simply adding the mobility dimension to the many existing MS systems. Given the outstanding proven accomplishments of pulsed linear IMS-MS, its continuous development, and the importance of the structural information it unambiguously yields, it is surprising that it has not been successfully commercialized and widely adopted by clinical and research laboratories. This contrast between what the experts and the wider community can do is to a great measure attributable to the difficulty of incorporating the IMS component into the many well-developed existing MS systems.

Most high-resolution linear IMS-MS studies have relied on the separation of gated ion packets in time. But high resolution requires long mobility cells (~1 m) with associated substantial radial beam broadening by diffusion, calling for the use of reconcentrating ion funnels. Also, gated IMS is typically implemented within the vacuum system of the MS, which calls for costly and complex fully integrated IMS-MS instruments rather than the simple addition of a mobility cell within existing mass spectrometers. The integration is somewhat simpler in vacuum matrix-assisted laser desorption/ionization (MALDI) sources, which naturally produce ion pulses, and for which some commercial experience is already available.[5,10] Instruments with an electrospray ion source have to deal with the additional difficulty of coupling *pulsed* IMS efficiently with mass spectrometers making use of *steady* atmospheric ionization sources (atmospheric pressure ionization [API]-MS). The difficulty of coupling IMS with electrospray (ES)-MS is drastically diminished in the case of devices achieving mobility separation in space rather than in time, particularly when this is done at atmospheric pressure. Such *spatial mobility filters* can simply be placed between the atmospheric pressure ion source and an existing API-MS, adding to it the IMS dimension with minimal modifications to the MS and a modest capital investment. The most familiar device that separates ions in space is Field Asymmetric IMS (FAIMS[11], also called DMS). Its weak points are a modest resolving power; the requirement of a high-voltage, high-frequency power supply consuming considerable power; and the inability to measure true mobility. Its peculiar advantage is the possibility to scan over the amplitude of the field causing the nonlinear effects, adding separation power. In this chapter we focus on an alternative *spatial* mobility filter, the Differential Mobility Analyzer (DMA), which naturally overcomes some of these limitations, while being readily integrated upstream from any API-MS adding to it a true mobility dimension.

5.2 THE DIFFERENTIAL MOBILITY ANALYZER (DMA) AND THE MEASUREMENT OF TRUE MOBILITY

The DMA functions by combining an electric field E and a large flow field u of clean gas to separate a small sample flow of ions in space into a fan, as illustrated in Figure 5.1. Unlike drift-time IMS instruments carrying out their mobility separation in time, the DMA can supply a steady beam of mobility-selected ions for subsequent analysis, analogously to FAIMS and DMS. And like drift-time IMS but unlike FAIMS, the DMA can achieve resolving powers of 50–100 and can separate according to true mobility. The DMA therefore offers characteristics somewhere in between those of these two more familiar techniques. For this reason, some terminological precisions will be used in this chapter. The term IMS will be reserved to refer generically to mobility separation methods capable of measuring true linear mobility, including, among others, drift time IMS and the DMA. Other mobility separation methods incapable of measuring true linear mobility, like the T wave and FAIMS, will be referred to as *nonlinear IMS*. For greater precision they should be called *intrinsically nonlinear IMS*, since both drift time IMS and the DMA can operate under the nonlinear mobility regime and can therefore determine both linear as well as nonlinear terms in the mobility versus electric field relation. The importance of the distinction between instruments capable and incapable of measuring

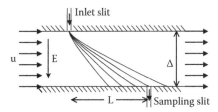

FIGURE 5.1 Sketch of a planar DMA. Ions are injected through an inlet slit (top) into a stream of ion-free sheath gas moving towards the right with velocity **u** between two parallel electrodes separated by a distance Δ. An electric field **E** pushes ions of mobility Z downwards at velocity ZE, so that the trajectories separate into a fan with a slope proportional to the electrical mobility Z. The sampling slit located at an axial distance L downstream from the inlet slit extracts ions for which ZE/U = Δ/L.

linear mobility follows evidently from the biological importance of structure, and the associated need for reliable methods to determine it. Linear mobility devices may in principle be used as primary standards for mobility measurement without the need for calibration. When faced with imperfectly known parameters (precise geometrical dimensions in gated IMS and the DMA, or sheath gas flow rates in the DMA), one single calibration is generally sufficient to provide a precise linear mobility scale. The issue of relating linear mobility to structure is, of course, different and by no means trivial.

Historically, the combined use of fluid flow and electric fields to separate ions goes back at least to the early twentieth century.[12,13] Some of the many variants tried have been discussed by Tammet[14] and more recently by Flagan.[15] The modern form of the DMA is due to Hewitt,[16] but its wide use for separation of submicron aerosols followed from the design of Knutson and Whitby,[17] commercialized by TSI. The use of DMAs for the analysis of ions smaller than some 10 nm was initially limited by poor resolution and transmission resulting from Brownian diffusion.[14,18,19] These difficulties were eventually overcome by short cylindrical geometries (ratio of width over length $\Delta/L \sim 1$ in Figure 5.1) and by flow field improvements permitting high gas speeds without turbulent transition.[20,21] As a result, resolving powers in excess of 100 have been obtained.[22] However, cylindrical DMA geometries have poor ion transmission at their inlet and outlet.[23]

An important recent development spearheaded by the company SEADM has been the return from the cylindrical geometries characteristic of aerosol work to the flat plate geometry of Figure 5.1, assuring high ion transmission from the source to the DMA and from the DMA to the detector. This last advance has been particularly useful when interfacing a DMA between an electrospray ionization (ESI) source and an atmospheric pressure ionization mass spectrometer (APIMS) to achieve the ESI-DMA-API-MS combination to be discussed later. SEADM's design has already been described in some detail.[24,25] The problems encountered in prior planar DMA work[26] relating to deformation of the thin electrodes at high speeds, leakage of destabilizing jets, and discontinuity of materials on wetted surfaces leading to transition to turbulence are solved with the insulating box structure of Figure 5.2, closed above and below by a pair of rigid metal electrodes. Unlike the schematic of Figure 5.1, these electrodes are curved in the conventional shape of a wind tunnel. The initial convergence shown on the right of Figure 5.2 is meant to accelerate the gas and reduce the level of flow perturbations in order to moderate the natural tendency of high-speed flows towards turbulent transition. The divergence of the channel following the separation region (left) is a diffuser that converts the high flow speed into pressure and enables attaining relatively high velocities with a blower of modest power such as those used in vacuum cleaners. The typical gap Δ used in these two-dimensional DMAs has been of 1 cm.

5.3   ADVANTAGES AND LIMITATIONS OF DMAS

The most peculiar analytical characteristics of DMAs follow rather directly from the instrument's structure. They will be briefly listed here to enable an easy comparison with more familiar mobility separation schemes. Some of these *peculiarities* will be further developed and illustrated throughout the chapter.

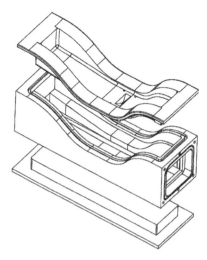

FIGURE 5.2 *Flat* DMA structure avoiding leaks and discontinuities on wetted surfaces based on an insulating box covered on its top and bottom with the two DMA electrodes. (From Rus J., Fernandez de la Mora J., Resolution improvement in the coupling of planar differential mobility analyzers with mass spectrometers or other analyzers and detectors. US patent application publication 20080251714, October 16, 2008. With permission.)

5.3.1 Resolution versus Transmission

The sharp ion trajectory lines depicted in Figure 5.1 have in reality a finite width that increases along the ion trajectory (top to bottom) due to Brownian diffusion. This diffusive spread of the ion beam orthogonally to the mean trajectory has two negative effects. First, it brings to the sampling slit ions with mobilities different from the one selected (resolution reduction). Second, it displaces away from the sampling slit some of the ions with the selected mobility (reduced transmission). As a result, resolution and transmission are strongly linked in DMAs to the point that the geometrical and fluid dynamical approaches already successfully used to increase the resolution[24] automatically achieve relatively high DMA transmission. The situation tends to be the opposite in gated IMS, where the long drift distances required for high resolution lead to lateral ion losses (which, at low operating pressures may be compensated via radio frequency [RF] focusing).

5.3.2 Duty Cycle

The DMA is a narrow band mobility filter that needs to be scanned to cover a desired mobility range. It therefore has a duty cycle limitation similar to that of a gated IMS instrument coupled to a steady ion source. For a pulsed ion source such as MALDI,[5,27] however, pulsed IMS would use the available ions more efficiently than the DMA, and the duty cycle would be determined by the repetition rate of the source. For a given mobility, however, the DMA has near unit transmission and is far more efficient than gated IMS with steady ion sources.

5.3.3 Single Ion Monitoring (SIM)

On the other hand, if a DMA is interfaced with a quadrupole MS for monitoring a target ion, it greatly increases the selectivity and reduces the noise with little reduction in sensitivity. In some instances of dilute species in complex matrices such as urine, the DMA provides an ultrafast alternative to high-performance liquid chromatography (HPLC).[28] Note further that the transit time for an ion in the DMA is typically 10^{-4} s (10^{-2}m/100 m/s), shorter than in usual gated IMS, which allows in principle synchronizing within a submillisecond scale the jump of the mobility and the mass (DMA and quadrupole voltages) when shifting from monitoring one ion to monitoring the next. As a result, multiple ions can be monitored via DMA-quadrupole MS with greatly improved resolution and reduced noise, yet without increasing the analysis time of the quadrupole alone.[29]

5.3.4 Interfacing to API-MS

The ability of the DMA to yield a steady mobility-selected ion beam provides one of its major advantages when interfacing it to an API-MS. As noted earlier, the DMA-API-MS combination is in fact the best known approach to add the mobility dimension to an existing MS without modifying its vacuum components, since the DMA is part of the atmospheric ion source. Other advantages of the DMA-MS pair depend considerably on the MS type. Single MS systems with an upstream DMA are turned into tandem instruments, where the DMA can select one parent ion whose mass can be measured. Fragmentation may then be introduced between the DMA and the MS (for instance, in the declustering region), and the mass of the fragments measured.[29] This approach has already been implemented in time-of-flight (TOF) mass spectrometers.[30] When full mass spectra of complex mixtures are investigated, the DMA-TOF combination is perhaps the most effective of those already tested (see below). But a DMA combined with an ion trap would be comparably useful when mobility-selected ions are accumulated in the trap over time periods larger than the time needed to measure their full mass spectrum. None of these combinations fully overcomes the duty cycle limitation following from the fact that only a small fraction of the ions (those within the selected mobility window) are analyzed at each instant. However, this selection is often desirable when analyzing complex mixtures, as it permits discerning individual peaks when the mass spectrum would otherwise be a featureless continuum.[23,28] Also, space charge limitations are greatly reduced in the ion guides and in trapping regions, enabling higher transmission and more effective accumulation of the selected ions.

5.3.5 Atmospheric Pressure Operation

The MS connectivity advantages just discussed come at a certain cost, since the DMA must run at relatively high pressure, typically with air or nitrogen near atmospheric pressure. Many gated IMS systems operate with helium gas, with two important advantages often assumed to facilitate the conversion of the measured mobility into the geometrical cross section of the ion. First, the ion-dipole interaction (polarization effect) is very weak in He, but sometimes needs to be included in air, and particularly in CO_2. Yet the necessary polarization corrections are not known precisely, particularly in the case of large, soft ions in molecular gases. Second, He is often presumed to bounce elastically rather than being absorbed and accommodating on the surface of the target ion, a fair assumption in the case of low energy collisions with atomic or small and rigid molecular targets. But whether this remains true for soft and flexible targets exhibiting vast numbers of internal degrees of freedom (such as proteins) remains to be proven. The elastic collision assumption is certainly far from true in the case of molecular gases (air, N_2, CO_2, ...), whose collisions with large ions involves an accommodation coefficient α whose value (not far from the purely inelastic limit $\alpha = 1$) has been well studied for relatively large particles, but only to a limited extent at nanometer dimensions.[31] One could of course run a DMA with He, but He cannot easily be used as electrospray gas due to sparking, nor sampled into the MS at atmospheric pressure due to the reduced capacity of turbopumps. Therefore, DMA-MS with He gas requires two interfaces with change of gas: one from the electrospray gas into He upstream of the DMA, and another from He to N_2 from the DMA to the MS. Consequently, most DMA-MS measurements have been performed in gases such as air, CO_2, or N_2, compatible with both turbopumps and electrosprays. Another complication of an atmospheric pressure drift gas (including He) is that the usual free-molecule expressions relating ion mobility to cross section require small "continuum" modifications that scale with the product of the background gas pressure times the size of the ion. At one atmosphere, such corrections must be taken into account for large ions such as megadalton proteins and viruses.[31–33] In spite of these perceived difficulties, careful calibration with liquid nanodrops show that the commonly used Millikan relation between ion mobility and diameter can be used reliably in nitrogen and air at ion diameters as small as 2 nm and above (see also Section 5.6 below), while the same verification above 2-nm diameters is still lacking for He.

In many IMS systems operating at low pressure it is possible to clean up the ions by inducing collisions with the background gas at modest energies. This so-called *declustering* is readily implemented at low pressures by application of modest electric fields, but much higher fields are required at atmospheric pressure for the approach to be effective. Consequently, all DMA measurements carried out to date have been performed without declustering prior to ion entry in the DMA. Declustering can, of course, be implemented following the DMA and the MS inlet, but then the ion whose mobility is measured is not necessarily the same subsequently mass analyzed. This mismatch has several consequences already explored to a certain extent.[30] The first is a shift of a few percent in ion mass. This is very useful for sensitivity enhancement and accurate mass determination in protein studies, since the low intensity of the original broad clustered mass peak is greatly enhanced as ions in various initial clustered states focus into the purely protonated mass. Although the declustering mass shift is not matched by a corresponding mobility shift, the resulting mismatch in the mobility-mass relation is often minor in relation to the typical width of a protein mobility peak.

5.3.6   Gas versus Liquid Phase Structure

One of the most useful features of true (linear) IMS is the ability to infer structural clues from a measured mobility. However, what is biologically relevant is the liquid rather than the gas-phase structure. Although the relation between both configurations remains unclear, it is apparent that the shorter the time and the milder the transfer from solution to gas, the closer these two structures will be. Recent studies on the evolution of native protein conformation in ES-MS show that the loss of the residual solvation typically taking place at the vacuum interface to the MS launches a very fast denaturing process, followed by a certain level of compaction.[34] Yet the residual solvation needed to maintain the native structure[34] could be retained through the electrospraying process, even in a relatively dry gas. If these ideas are correct, then an IMS method proceeding at atmospheric pressure above a minimal level of humidity would preserve and measure the native structure. In contrast, most prior IMS and all T-wave studies have provided substantial activation to the ions prior to mobility determination via transfer to a vacuum, strong RF fields for ion transfer, focusing or trapping, or by energetically injecting the ions from a vacuum into the drift tube. Once in the vacuum, these ions can be stored and the time evolution of their structure may be probed by timed injection into a drift tube,[35] either in a dry or a controlled humid background.[36] But the possibility offered by the DMA to delay the drying and denaturing of these ions until after mobility determination adds an important new dimension to these studies.

5.3.7   Mobility Range

The DMA has two characteristic dimensions, the gap Δ and the length L, whose ratio can be widely changed. The early DMAs used for submicron particle analysis were able to classify singly charged particles with diameters larger than 100 nm, which correspond to masses much larger than the largest ion of biological or industrial polymer interest. We have worked with parallel plate DMAs with Δ/L of 1, 2, and 4 and anticipate using values as large as 10 for analysis of particularly large masses, without significant loss in resolution. This means that, for all practical purposes, the DMA has an unlimited mobility range.

5.3.8   Miniaturization

Most DMAs coupled to an MS have had gaps Δ in the range of 1 cm, as a large gap favors resolution to a certain extent. However, DMA resolving powers of 50 have been obtained with cylindrical DMAs using a gap of 2 mm, and a resolving power of 50–100 is in principle possible at such dimensions. The DMA therefore has a high potential for miniaturization.

5.3.9   Limits to Resolution

There are many effects limiting DMA resolution. One is Brownian diffusion of the ions away from their mean trajectory. Another is the onset of flow instability, either directly in the form of turbulence in the separation region[37] or indirectly in the form of pressure fluctuations radiating upstream from the region of the pump moving the flow of clean gas.[22a] The finite sample flow rate extracted at the exit of the DMA and the finite widths of the inlet and outlet slits also set resolution limits. By a combination of proven methods it has been possible to overcome these various limitations to attain resolving powers beyond 100. Twice this value has been demonstrated in drift IMS, but would be rather difficult in a DMA.

5.3.10 LINEAR VERSUS NONLINEAR MOBILITY SEPARATION; TANDEM DMA

Very much as in FAIMS, the field across the two electrodes of a DMA can be made strong enough to bring the ions into a nonlinear mobility regime. In ambient air, this condition typically corresponds to ion velocities of hundreds of m/s, comparable to the sound velocity in the gas. Nonlinear ion separation can therefore be achieved in a DMA run at a Mach number of order unity, when the ratio Δ/L is close to unity or less. Both these conditions can be achieved in practice in a DMA.[22b] From the viewpoint of achieving a narrow band mobility filtration, operating in the nonlinear regime is not necessarily more useful than in the linear regime. However, two ions that cannot be separated in the linear regime will often be separable in the nonlinear regime. Also, if two DMAs are run in series at different levels of nonlinearity, they will separate according to different criteria. The resolving power of this tandem DMA (DMA^2) can therefore be much higher than 100. As already noted in the case of a DMA in tandem with a quadrupole mass filter, the use of several filters in series in single ion monitoring mode does not increase the measurement time. Hence, the DMA can in principle be used as a tandem instrument, even without modifying the ions analyzed in the second DMA with respect to those selected by the first.[29]

5.4 EXPERIENCE WITH TANDEM DMA-MS

The use of the DMA in tandem with a mass spectrometer is relatively recent, and only a few related studies have appeared. In what follows we will briefly review this limited experience and include some additional unpublished results. We have already mentioned several earlier DMA-MS studies based on cylindrical DMAs of limited transmission[23,38] and a parallel plate DMA prototype of excellent transmission but limited resolution.[26,39] Other results already noted have become available coupling SEADM's parallel plate DMAs with single and triple quadrupole instruments.[24a,28] However, more extensive studies have recently appeared based on combining one of these new DMAs enjoying both high resolution and high transmission with Sciex's Q-Star time-of-flight (quadrupole-TOF) MS having a mass range extending up to some 40,000 m/z (mass-to-charge ratio).[25,40,41] The results to be discussed subsequently in detail have all been obtained with two variants of this system: one installed at SEADM's laboratory (Boecillo, Spain) and another more recently installed at Yale's Keck Center. A schematic of the instruments is shown in Figure 5.3, with more detail provided by Rus et al.[25] The main variation involved in DMA-MS versus plain DMA operation is that the natural sample outlet geometry of a planar DMA is a linear slit, while the natural sample inlet to an API-MS

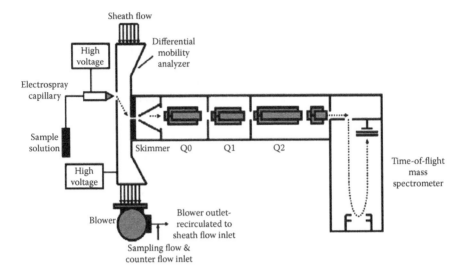

FIGURE 5.3 Schematic of the tandem arrangement of SEADM's P4 parallel plate DMA with MDS/Sciex's Q-Star quadrupole-TOF mass spectrometer. (Courtesy of Dr. C. Hogan.)

is a circular orifice. The mismatch has resolution disadvantages that are removed by a shaped outlet orifice on the lower DMA electrode, which starts as a slit on the DMA side and becomes a circle on the MS end.[24a] The sample flow is limited at the MS end, the only point in the line that operates critically. The electrospraying tip is brought very close (100–200 μm) to the entry of the sample inlet slit to the DMA. This short path is often used in nanospray to maximize ion transmission and suffices to enable complete drop evaporation and ion formation from the relatively conducting solutions used (typically 10–20 mM volatile ammonium salt in water or methanol). A gentle counterflow gas (typically 0.2–0.5 L/min) moving from the interior of the DMA through the slit into the electrospray chamber prevents entry of vapors, drops, and low mobility clusters from the ES source into the analyzer. Two-dimensional DMA-MS spectra are acquired by fixing a sequence of relatively close DMA voltages, at each of which a mass spectrum is recorded. This results in a set of ion intensities I as a function of m/z and DMA voltage V: $I(m/z, V)$. Although the variable V is discrete, the step in V is considerably smaller than the mobility peak width, whereby the representation of the DMA-MS spectrum $I(m/z, V)$ with I given by a color scale appears as continuous (Figure 5.4). The software used for this representation is kindly provided by SEADM and uses a logarithmic intensity scale in order to display peaks much weaker than the dominant one.

5.5   MEASUREMENTS WITH PROTEIN IONS

Figure 5.4 shows DMA-MS spectra from electrosprays of aqueous solution of lysozyme. The high concentration in Figure 5.4a (100 μM) is used here to artificially form clusters, to test the capacity of the tandem DMA-MS combination to distinguish large aggregates of n proteins (n-mers) in a wide range of charge states z: Lys_n^{+z}

(abbreviated n^z in Figure 5.4c). The lowest series of diagonally aligned peaks (joined with a line and marked 1^z in Figure 5.4c) corresponds to the lysozyme monomer carrying from 10 to 4 charges (Lys^{+4} with very little signal in Figure 5.4a). Ordered on a second diagonal line displaced upwards from the monomer line, one sees the series of dimer ions Lys_2^{+15}-Lys_2^{+7} (Lys_2^{+6} in Figure 5.4c). The analogous series for the trimers, tetramers, etc., are similarly distinguishable. The aqueous solutions used include 20 mM triethylammonium acetate (TEAA) in order to complement with lower charge states the usual higher ones produced from ammonium acetate buffers.[42] The ability of the system to also distinguish native multimers existing naturally in dilute solutions from those produced in the electrospraying process has been demonstrated by Hogan et al.[43] in the same instrument with proteins as small as insulin (m ~ 5 kDa, naturally forming hexamers), Aldolase (naturally forming tetramers with m ~ 150 kDa), and as large as GroEl (naturally forming tetradecamers with m ~ 800 kDa).

There is a substantial level of charge reduction at the entrance of the MS, revealed by the presence of pairs of vertically displaced neighboring charge states in the MS spectrum appearing at exactly the same DMA voltage. In each such pair, the ion with the larger m/z is a product of the lower peak resulting from a charge loss transition of the type $z + 1 \rightarrow z$. For instance, the most charged monomer ion seen in Figure 5.4a and c at $(z, m/z, V) = (10, 1430, 2500 \text{ V})$ has a shadow above it corresponding to $z = 9$, which appears as a tail of the real peak for the $z = 9$ state seen at 2620 V. The $z = 10$ peak appears as narrower in mobility space than the other charge states because it is unpolluted by the transition $z = 11 \rightarrow 10$ thanks to the almost complete absence of the $z = 11$ state in the spectrum. The $9 \rightarrow 8$ and $8 \rightarrow 7$ transitions are similarly clear in Figures 5.4a and b. The $7 \rightarrow 6$ transition is unambiguously present but is rather weak, while the lower ones, $6 \rightarrow 5$ or $4 \rightarrow 3$, are imperceptible, indicating the expected higher stability of lower charge states against charge loss. This phenomenon of charge loss at the entrance region of the MS is characteristic of all our measurements and provides a warning against the uncritical interpretation of MS peaks as ES ions naturally produced in the atmospheric pressure source. Note that the measurements of Figures 5.4 and 5.5 are taken with just 1 V of declustering potential. These transitions become substantially more probable at typical declustering voltages. The phenomenon also offers an opportunity to study the kinetics of protein charge reduction and illustrates one way in which the DMA-MS combination can be used similarly as in conventional MS-MS instruments.

A closer look at Figure 5.4b shows several main series of clustering peaks for the $z = 9$ and the $z = 8$ charge states, the lowest one being the purely protonated protein, and the two above it corresponding to substitutions of one or two protons by TEA^+. Each of these three broad peaks has its own substructure (seen better in the detail of Figure 5.5) involving substitution of one H^+ by either one Na^+ or one K^+. For instance, in Figure 5.5a for $z = 9$, the first trio includes the $9H^+$ (1590 m/z), $8H^+ + Na^+$, and $8H^+ + K^+$ ions. The second trio includes $8H^+ + TEA^+$ (1601 m/z) and two other substitutions of one of the eight protons by Na^+ or K^+. The third subseries of three peaks is a repetition of this pattern with two protons substituted by 2 TEA^+ ions, etc.

5.5.1 TRANSITIONS

The detailed products of the 10→9 transition are shown in Figure 5.5a to the left of the $z = 9$ peak (1590 m/z), at the mobility of the parent ion ($z = 10$; 2500 V). The products of the $z = 9$ and 8 peaks are similarly observed in Figures 5.5b and c, respectively. Note in Figure 5.5 that the adduct pattern of the product peaks differ from those of the parents. For instance, while the $z = 10$ peak is dominated by the combinations $2TEA^+ + 7H^+ + K^+$, $9H^+ + K^+$, and $TAA^+ + 8H^+ + K^+$, in decreasing order of abundance, its decay product in the $z = 9$ peak is dominated by the combination $8H^+ + K^+$, showing that the main transition is by loss of TAA^+. The anomalously small abundance of $TAA^+ + 8H^+ + K^+$ at $z = 10$ is in fact due to its almost complete conversion to $8H^+ + K^+$.

The 9→8 transition starts at $z = 9$ with a fairly uniform concentration of the various adducts, and ends at $z = 8$ with mostly the purely protonated and once sodiated peaks. The adducts with one and two TEA^+ have both decreased drastically, so, again, the decay mechanism is by loss of one and two TEA^+ units. The pattern in the $z = 8→7$ transition is similar. By careful analysis of these observable decay ratios in proteins of different radii, one could infer the dependence on charge state and radius of the rate constant for evaporation of TAA^+ ions. This charge loss seems to be an activated process with a high activation energy, which is nonetheless considerably

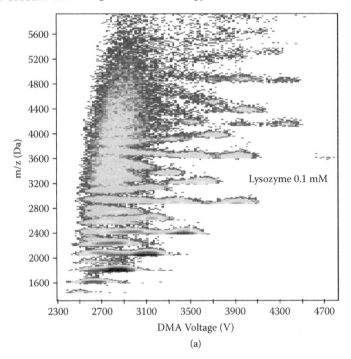

(a)

FIGURE 5.4 DMA-MS spectra in CO_2 of unpurified lysozyme in water with 20 mM triethyl-ammonium acetate. The protein concentration is 100 μM in (a) and (b), and 25 μM (c). DP = 1 V. (b) Is a close-up view of the high charge state region of (a).

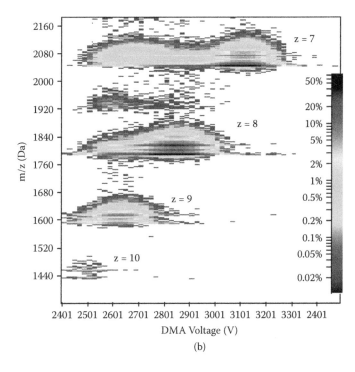

(b)

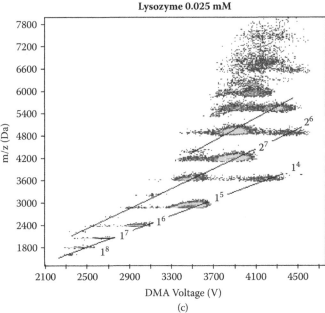

(c)

FIGURE 5.4 (CONTINUED) DMA-MS spectra in CO_2 of unpurified lysozyme in water with 20 mM triethylammonium acetate. The protein concentration is 100 µM in (a) and (b), and 25 µM (c). DP = 1 V. (b) Is a close-up view of the high charge state region of (a).

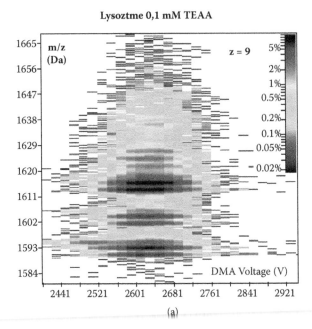

(a)

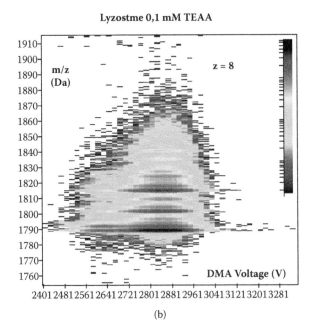

(b)

FIGURE 5.5 Details of charge states $z = 7$ to 9 from Figure 5.4b.

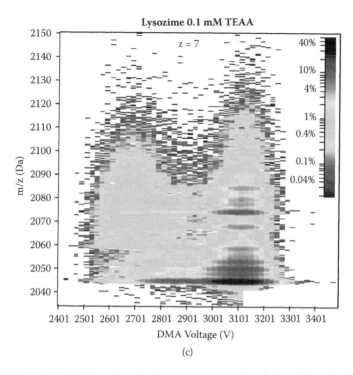

FIGURE 5.5 (CONTINUED) Details of charge states $z = 7$ to 9 from Figure 5.4b.

smaller than that for ammonium evaporation from water drops of size comparable to that of lysozyme. The alternative and often used notion that a proton on a charged protein can be stripped (without an energy barrier) by a gas-phase neutral amine may perhaps need to be revised once these rates and corresponding activation energies are measured through experiments similar to those just described. The fact that the low charge states are not discharged provides some evidence that the activation energy is very large, that it is reduced by the field on the protein surface, and that measurable rates take place only at sufficiently high protein charge or curvature, exactly as in the process of ion evaporation from charged drops discussed by Iribarne and Thomson. However, in some other lysozyme monomer experiments with 60 mM TEAA we see no trace of charge states below $z = 5$ and observe substantial $5 \rightarrow 4$ conversion, not only downstream of the DMA, but also within the DMA (clear from the abundant presence of $z = 4$ ions in the MS at mobilities intermediate between those of the $z = 5$ and 4 ions). The rather different conversion on the 20-mM and the 60-mM experiments may be suggesting that, perhaps, the limiting factor is not a kinetic barrier to charge stripping through the protonated amine, but merely complete loss of the amine between the DMA and the MS, as in the picture proposed by Kebarle. But then how could one explain the clear observation of ions carrying up to three TEA$^+$ units? Also, in the 60 mM TEAA experiments, the $z = 4 \rightarrow 3$ transition is very weak, apparently pointing again to an activated process. The issue will hopefully be resolved by more extensive and quantitative studies based on this DMA-MS tool.

One feature peculiar to the spectrum of Figure 5.4 relates to the fact already noted that the ions are not *declustered* upstream of the DMA (nor downstream in this particular case). This can be seen directly in the substantial vertical (m/z scale) width of the peaks. This is also seen in the fact that the series of cluster ions forming to the left of each monomer peak Lys_1^{+z} do not lie exactly on a horizontal line, but rise slightly towards the left. This series corresponds to aggregates of n monomers with a charge state nz, Lys_n^{+nz}, so all of them should appear at the same m/z. But they do not because the level of attachment of impurities (clustering) increases with the mass of the protein aggregate. Another undesirable problem associated to impurity attachment is that the ion signal is spread over a wide range of adduct masses, yielding a much smaller peak height than obtained when all the ions of a given charge z have exactly the same mass (as tends to be the case after the conventional declustering).

5.5.2 IONIZATION MECHANISM

The high lysozyme concentration (100 μM) in the measurements just discussed leads to an unusual level of adduction of Na and K. Figure 5.4c shows DMA-MS spectra for an electrospray of 25 μm lysozyme in 20 mM aqueous TEAF. One can see at first a greatly reduced contribution of Na and K adducts in charge states 9–6. The drastically simplified adduction pattern just noted applies only to the high charge states with $z \geq 6$. At $z = 5$ and 4 the pattern is so different that it is hard to imagine that the same ionization mechanism could be acting in both cases. This notion is further strengthened by the fact that the vast majority of the monomer ions carries 5 charges and are highly adducted. A similar pattern is seen in the case of the dimer ions. The trimers and higher states of aggregation also produce a narrow range of *dirty* low charge states, though without the corresponding clean highly charged ions (perhaps due to the decreasing abundance of these larger protein clusters). In conclusion, the major ionization mechanism acting is incapable of producing clean protein molecules and yields primarily $z = 5$ ions, with a modest contribution to $z = 4$. But there is a concurrent and completely different ionization mechanism producing clean ions over a broad range of charge states.

The case seems therefore to be quite strong for a charge residue mechanism yielding the abundant clustered ions and an ion evaporation mechanism producing less abundant clean ions. Indeed, if the ions originate from the dried drop, they must necessarily carry all the involatile impurities originally contained in the drop, which makes it almost impossible to produce clean ions as charged residues. But if clean ions are nonetheless produced, they must escape from the drop before it dries. This ion evaporation mechanism for proteins was proposed long ago by Fenn and colleagues,[44] but no direct experimental support for it had been available. In the picture they proposed, the protein would be spread over the drop surface prior to desorbtion, while our observations are more in line with the evaporation of the protein in its native state.

5.5.3 LARGER PROTEINS

Our early DMA-MS work with *unpurified* proteins (as received from Sigma Aldrich) confirmed the notion just noted that, at larger masses, their ES ions exhibit a growing level of clustering with increasingly broad mass peaks. In particular, the purely pro-tonated ion (manifested in Figure 5.4 by the flat-bottomed peaks of lysozyme mono-mers) becomes gradually lost in a sea of heterogeneity. Clustering must similarly widen the mobility distributions. For masses up to 45 kD (ovalbumin), this complex-ity does not preclude isolation of individual charge states by the DMA alone, as shown in Figure 5.6. One sees in the ammonium acetate buffer the same charge loss mechanism found for lysozyme in a TEAA buffer, though now it takes place at much larger charge states and without clear mass peaks to allow inference of the kinetics of the process. Larger proteins such as bovine albumin (m ~ 66.43 kDa; Figure 5.7) no longer show isolated mobility peaks (integrated over all masses). Figure 5.6b illus-trates the effect of declustering in unpurified ovalbumin. The bottom figure, at the maximum declustering voltage, shows narrower mass peaks, though not complete convergence to the bare protein. In exchange for this partial simplification of the mass spectrum, one sees a drastic increase in the number of phantom peaks due to charge loss between the DMA and the MS.

As illustrated in Figure 5.7b, some simplification of the level of clustering can be achieved without apparent denaturation by slightly increasing the acidity of the solu-tion electrosprayed. This results in a clear narrowing of the mobility distributions and an improvement in the accuracy with which mobility can be inferred.

In spite of these complications, it is possible to determine mobilities and masses for unpurified proteins up to at least 150 kDa, as shown in Figure 5.7c for immuno-globulin G (IgG). The full spectrum was accumulated over five minutes. Mass peaks are broad and barely recognizable in the pure mass spectrum. But DMA-MS peaks form distinct bands, probably because the variable level of clustering widening the mass and the mobility peaks does so in unison. The same slightly inclined bands of fixed z can be seen in the spectra of ovalbumin and bovine albumin. The appropri-ate mass assignment for the bare protein should therefore be based (for experiments without declustering) on the lower edge these bands. For the four higher charge states of IgG, this yields approximate m/z values of 5607, 5391, 5210, 5043, which can be assigned to $z = 27$–30 with the correct m ~ 151,000 Da.

These early findings made us wonder initially whether the DMA would be able to provide accurate structure information on large protein complexes. Fortunately, Yale's DMA-MS facility is open to visitors and has also benefited from collaborations with colleagues at the Keck Center. As a result, we have had recent access to purified protein samples, from which quite sharp DMA-MS spectra have been obtained up to masses of 800 kDa (GroEL tetradecamer), the largest we have tried so far.[43] Size exclusion chromatography (SEC) has proven to be particularly effective for this clean-ing process. The peaks are still sharpened by active declustering after mobility selec-tion, which also permits obtaining tandem information even in a single MS mode. Interestingly, the DMA-MS spectra of clean GroEL and rabbit Aldolase A (153 kDa) provide very well-defined peaks even with no declustering. For instance, the lowest charge state ($z = 27$) of Aldolase A shows a mobility width of 3% (full width at half maximum [FWHM]) based on the full mass peak (FWHM of 1% in mass). We are also making progress in the ion cleaning process by improving the electrospraying method. Several related studies are now in preparation in collaboration with C. Hogan (Yale Mechanical Engineering), J. Loo (UCLA), B. Ruotolo (now at Michigan), and

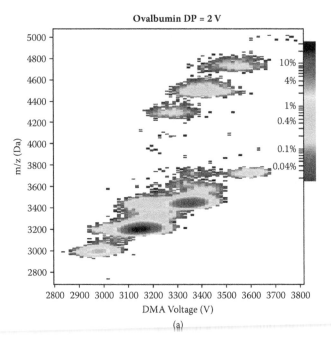

(a)

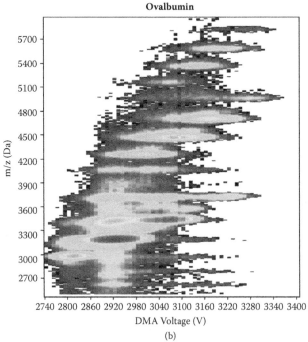

(b)

FIGURE 5.6 DMA-MS spectra in CO_2 for unpurified ovalbumin, showing the effect of declustering voltage. (a) DP = 2 V. (b) DP = 340 V. Declustering (b) partially narrows the mass peaks, but the complexity associated to charge loss transitions is greatly increased.

Cambridge colleagues E. Folta-Stogniev (Yale's Keck Center), B. Danrien (Alexion Pharmaceutical), and A. Heck and his colleagues (Utrecht). They demonstrate the ability of the DMA-MS to obtain accurate structural information for protein complexes, apparently with no upper mass limit beyond the range of the MS.

5.6 STUDIES WITH IONIC LIQUID NANODROPS

A DMA coupled to an MS with high mass range provides a splendid tool to unravel the charge and mass distributions of multiply charged cluster ions produced upon electrospraying concentrated solutions. The simplest aggregates result from solutions of a salt with anion A$^-$ and cation C$^+$ and have structures $(CA)_n(C^+)_z$. We have been particularly interested in so-called ionic liquids (room temperature molten salts) due to their ability to form ion beams when they are electrosprayed in a vacuum.[45] One important parameter when this process is used for electrical propulsion (or in other situations involving electrospray ionization) is the activation energy ΔG for ion evaporation. ΔG can be inferred from the z dependence of the radius of the smallest clusters that can hold a given number of charges, $R_{min}(z)$. Obtaining ΔG therefore requires the independent measurement of both charge and mass in relatively complex mixtures of highly aggregated and charged clusters. The DMA-MS spectrum of one such spray is shown in Figure 5.8 (adapted from Hogan et al.[40]), accumulated for 22 minutes for the ionic liquid EMI-Methide (Covalent Associates). The unusually well-ordered pattern of resolved peaks (reaching almost 200,000 Da) grouped in diagonal bands of fixed charge state (indicated in the figure) is due in part to the spherical geometry of the liquid nanodrops formed, and partly also to the very limited presence of the metastable transitions $z + 1 \rightarrow z$, which have complicated the appearance of prior related spectra.[25] The improved mobility resolution seen in these more recent data with respect to the other data shown in this chapter are not due to improvements in the DMA, but to a better stability of the sheath gas pump, enabling driving the DMA at higher speed. Of particular interest at the bottom of Figure 5.8 is the multiplicity of vertically displaced, singly charged peaks. They reveal that evaporation of one or several neutral pairs AC takes place after mobility measurement.[41]

Based on mobility data for EMI-Methide and other ionic liquid nanodrops, we have shown that the conventional Millikan relation between radius and mobility is accurately satisfied down to diameters smaller than 2 nm.[32,33] The radius is inferred from the measured ion mass and the bulk density of the ionic liquid, which is also found in these experiments to coincide with the density of the nanodrops down to diameters as small as 3 nm. We also find that polarization effects are negligible in air and N$_2$ at all the charge levels represented in Figure 5.8. These findings remove the ambiguities associated to the use of molecular gases as drift media and suggest the use of ionic liquid nanodrops as sharp mobility standards for nonlinear IMS calibration.[33] Hopefully these or other rigid standards of mobility and mass will help determine under what conditions T-wave drift times depend only on mobility, enabling reliable mobility determination after calibration. In general, the low pressure dynamics of an ion in a time-dependent field depends both on m/z and mobility. Because inertia (m/z dependence) is essential to produce the lateral confinement also

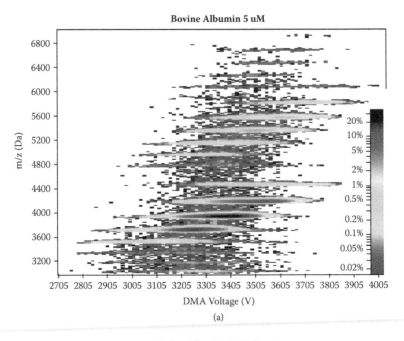

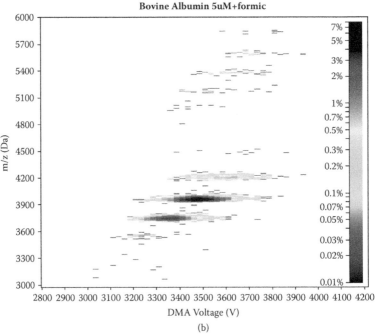

FIGURE 5.7 DMA-MS spectra for larger proteins in CO_2 gas. (a) (DP = 300 V) and (b) (DP = 200 V) are for 5 μM bovine albumin, with charge states $z = 15$–19 (~4500–3500 m/z) and 26–23 for monomers and dimers, respectively. In (b) addition of 370 μM formic acid sharpens the peaks. Ammonium acetate buffer for all.

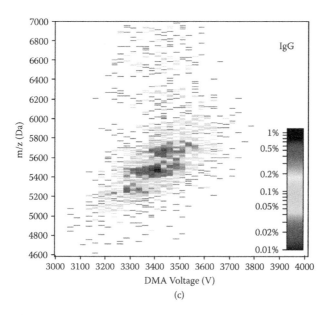

(c)

FIGURE 5.7 (CONTINUED)   DMA-MS spectra for larger proteins in CO_2 gas. (c) IgG 10 μM (DP = 340 V). Ammonium acetate buffer for all.

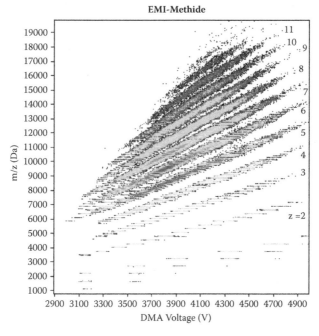

FIGURE 5.8   DMA-MS spectrum in N_2 of an electrospray of EMI-Methide 20 mM in acetonitrile. The various bands shown correspond to fixed charge states, as indicated. (Adapted from Hogan, C. J., Jr., Fernández de la Mora, J., *Chem. Phys. Phys. Chem.*, 2009 11, 8079–8090. With permission.)

taking place during the T-wave separation, it is by no means evident that inertial effects are negligible for the axial movement.

5.7   ELECTROSPRAYED POLYMER IONS

Two well-known kinds of difficulties relating to the determination of polymer mass distributions by electrospray ionization are (i) the two-dimensional complexity associated to the unknown distribution of both mass m and charge z and (ii) the fact that many industrially important polymers are soluble only in nonpolar liquids, which are difficult or impossible to electrospray into drops small enough to yield reasonably clean ions.

5.7.1   POLYMERS SOLUBLE IN POLAR SOLVENTS

Polar polymer chains generally yield electrospray ions attached to such a large number of charges[46] that individual mass peaks cannot be resolved in ESI-MS spectra past polymer masses of 20–40 kDa, even at resolving powers of 10,000. Unscrambling the two quasi-continuum variables z and m is possible in pure polymers via a pair of measurements yielding two pieces of information depending differently and independently on m and z. IMS-MS does this by determining mobility Z and m/z, both of which depend separately on z and m. However, in practice, highly charged ions take stretched conformations for which, to a first approximation, Z depends only on m/z. [38a] This phenomenon is shown in Figure 5.9a for a polyethylene glycol (PEG) sample with a relatively narrow distribution centered at 3.4 kDa. More than 90% of the signal of these naturally charged ions is in the approximately straight diagonal line starting on the low left and continuing to the right with the singly charged line (labeled 1). This colorful region is crowded with numerous charge states in close proximity. From it, various bands corresponding to just one charge state branch off at different positions, undergo various sharp bends, and end at large masses in a final monotonic line (labeled $z = 1, 2$ in Figure 5.10a, $z = 3$ in Figure 5.10b). The intense low-lying crowded region corresponds to fully stretched, highly charged polymers[38,47] in which determination of mass distributions is most laborious and requires very high resolution in both the IMS and the MS stages. The final monotonic lines correspond to nearly spherical shapes where the various charge states are widely separated (though crossed by higher z values), very much like the nanodrops bands of Figure 5.8. If most of the ion signal fell in this region of quasi-spherical ions and readily separated charge states, the determination of polymer mass distributions would be relatively straightforward.

The globular region and its first transition into a nonspherical shape has been well described up to 3000 m/z and $z = 5$.[38] This final kink occurs at m/z $= m^* z$, with $m^* \sim 500$ Da for PEG. The transition for $z = 2$ is clear in Figure 5.9a; the transitions for $z = 3$ and 5 in Figure 5.9b, while the transition for $z = 4$ can be guessed in Figure 5.9c. Figure 5.9d shows rather weakly what appear to be the $z = 6$ and 7 transitions. These regions were much more abundantly populated in the work of Ude et al.[38] because the natural charge states from their ammonium acetate buffer were reduced with a radioactive source. A stunning global view of the transitional kinks covering the mass range up to 2500 m/z has been reported with the formidable mobility resolution afforded by a 3-m drift tube.[47] The individual meandering charge state lines are well separated from each other in some parts of the transition region, except at occasional crossings where only two z values coexist and can be distinguished with ordinary MS resolution. However, it appears from Figure 5.9d that other pieces of the transition region contain so many bends and kinks that many charge states coexist. Also, the detailed position of these kinks is sensitive to the polymer terminal groups and the attached cation.[47,48] Therefore, inferring mass distributions from IMS-MS data away from the spherical asymptote will require a careful analysis. In contrast, when the ions lie in the spherical region, the ratio Z/z depends in a known fashion on the mass over bulk density ratio, m/ρ.[38] Therefore, once z is determined as in Figure 5.8, the complex two-dimensional DMA-MS spectra can be turned into mass distributions, even when the signal intensity or the MS resolution are insufficient for sharp mass peaks to be individually recognizable. This inversion depends only on the polymer density ρ, which is apparently quite independent of external circumstances (such as cation type). The criterion $m > m^* z^2$ already given for an ion of mass m and charge ze to lie in the spherical region is in reality a Rayleigh limit criterion: $m > z^2 e^2 \rho/(48\pi\gamma\varepsilon_o)$, fixed by only one polymer property, γ/ρ, where γ is the polymer *surface energy*, and ε_o is the electrical permittivity of vacuum. Although prior work had assumed that γ/ρ was fixed for a given polymer,[38] the spectra of Trimpin et al.[47] show several branches with a certain variability in the ion mobility at the critical mass, at least for $z = 3$. Another complication observed in Figure 5.9 is the appearance of several anomalous series of singly charged ions, the most prominent one clearly visible at their far left. Singly charged ions with these masses cannot have such large mobilities, so these oddly placed, singly charged ions must be products from the highly charged ions in the fully stretched region. Because they have exactly the same mass sequence $H(C_2H_4O)_nOH\text{-}(NH_4)^+$ as those in the main $z = 1$ series, they must originate from polymers present in the sample rather than from gas-phase breakup of a chain. The singly charged product ion must emerge initially from an ES drop attached to a multiply charged fully stretched chain and must travel with this partner through the DMA, but then they must part company somewhere between the DMA and the MS. This ion evaporation of a polymer chain had apparently never been observed before, but had been hypothesized by Fenn and colleagues[44] (though in their picture the evaporated ions would originate with multiple charges from a liquid drop rather than a rodlike solid).

A variety of schemes have been demonstrated for reducing the charge state of electrospray ions prior to DMA analysis, including the use of a radioactive source for slight discharging[38] or for elimination of all charge states but $z = 1$.[49,50] Charge reduction in an ion trap has also been found effective for PEG.[51] Adopting these approaches here, however, requires placing the ES needle at some distance from the DMA inlet, which reduces the transmitted signal. Charge reduction down to $z = 1$ also greatly limits the mass range at which a DMA can be coupled to an MS. For these reasons we have made preliminary experiments based not on charge reduction in the gas phase, but on the use of buffers naturally yielding reduced charge states from the liquid phase. As shown in Figure 5.10, shifting from ammonium acetate to triethylammonium formate in the buffer salt has far more drastic consequences on the charge of PEG ions than on that of proteins. Use of the same buffer has been previously reported (without detail) to produce charge reduction effects in PEG ions

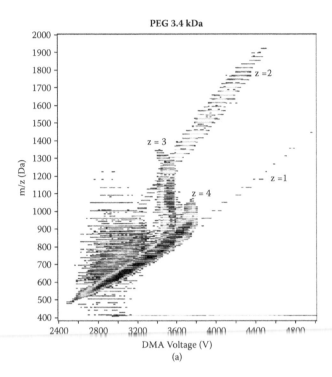

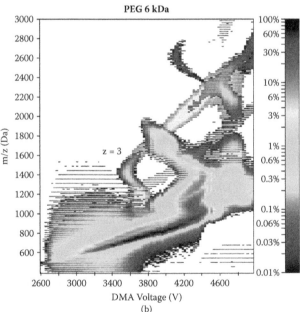

FIGURE 5.9   DMA-MS spectra of polyethylene glycol samples electrosprayed in CO_2 from water/methanol 50/50 (v) containing ammonium acetate. The average polymer mass increases from top to bottom. Several well-resolved bands of low charge states are indicated.

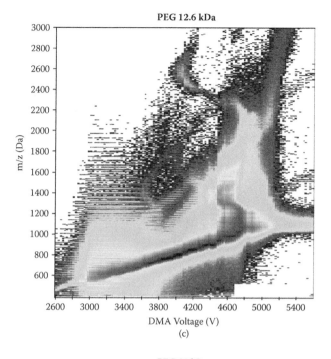

PEG 12.6 kDa

(c)

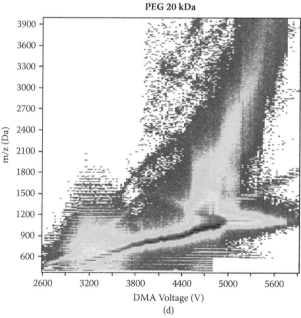

PEG 20 kDa

(d)

FIGURE 5.9 (CONTINUED) DMA-MS spectra of polyethylene glycol samples electro-sprayed in CO_2 from water/methanol 50/50 (v) containing ammonium acetate. The average polymer mass increases from left to right. Several well-resolved bands of low charge states are indicated.

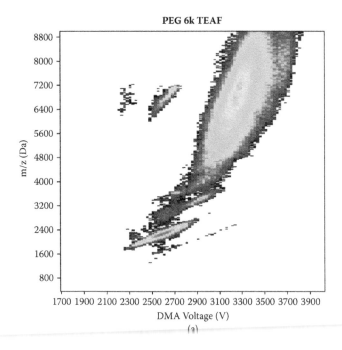

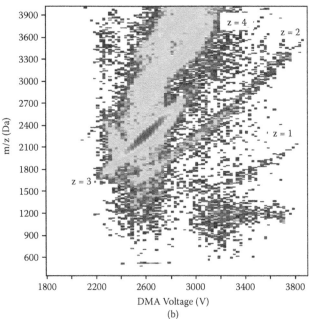

FIGURE 5.10 DMA-MS spectra of low charge states of PEG 6 kDa (150 μM) electro-sprayed from 50/50 (v) water/methanol with 10 mM triethylammonium formate. DP = 20 V. Note the absence of the fully stretched and transitional conformations, and the dominant contribution from spherical ions.

comparable to the addition of triethylamine in the gas phase (dominant charge state $z = 5$ for PEG 40 kDa).[52] As shown in Figure 5.10, the lower region of fully stretched and transitional ions is completely removed for PEG 6 kDa. Essentially all the signal appears in the ideal form of spherical ions, most of them carrying three charges, with small but clearly resolved contributions from $z = 1$, 2, and 4. Additional contributions from higher charge states seem to be present (especially $z = 5$), but the signal is too weak for determining z from the mass peaks. Notice in Figure 5.10a an anomalously mobile band of ions in the range of 6400–7200 m/z, directly above the triply charged line. They can be identified as singly charged members of the expected series $H(C_2H_4O)_nOH\text{-}TEA^+$, and are surely also product ions released between the DMA and the MS from a parent ion of the same mobility.

5.7.2 WATER-INSOLUBLE POLYMERS

We have previously shown that the solvent l-methyl-2-pyrrolidone (NMP) seeded with either formic acid[53] or (preferably) dimethylammonium formate (DMAF)[54] produces fine electrosprays. Because NMP is an excellent solvent for a variety of nonpolar industrial polymers, including polystyrene and polymethylmetacrylate, the combination of NMP with 1% DMAF was shown to be excellent for forming gas-phase ions of these polymers. Our initial work made minimal use of mass spectrometry, relying mostly on mobility measurements.[53,54] No measurements were done with naturally charged electrospray ions. Instead, we reduced to unity their initial charge by passage through a weakly ionized gas (formed with a radioactive source) and then determined the mobility spectrum. We have now used the DMA-MS setup to study the natural charge distribution of polystyrene ions formed from NMP-1%DMAF. Remarkably, as shown in Figure 5.11, most ions from a sample with narrow mass distribution centered at 9200 Da are singly charged. As in our earlier work, the level of involatile residues in commercial NMP (patent in Figure 5.11a, in spite of the high sample concentration) confirms the known need for a solvent of higher purity. Pending this improvement, the quality of published MALDI spectra for a similar polystyrene sample is much better than that afforded by the present ES spectra.[55] The coincidence seen between the mobilities of the singly and doubly charged ions suggests that the former are products from $z = 2 \rightarrow 1$ transitions arising between the DMA and the MS. This point is confirmed by the complete conversion of the smallest doubly charged ions (expected to be least stable with respect to charge loss) in the voltage range from 3.2 to 3.4 kV.

5.8 CONCLUSIONS

This chapter introduced the DMA, particularly for tandem IMS-MS, with an emphasis on the peculiar characteristics distinguishing it from other mobility separation methods. Some of these idiosyncrasies were illustrated from examples based mostly on unpublished studies carried out by the author during 2007–2008, in brief visits to the only DMA-MS facility then available (Boecillo, Valladolid). The performance of that prototype instrument has since improved considerably, as a number of articles now in preparation, under review, or in press will hopefully soon show. The range of applications is also growing rapidly, thanks in part to the hosting of

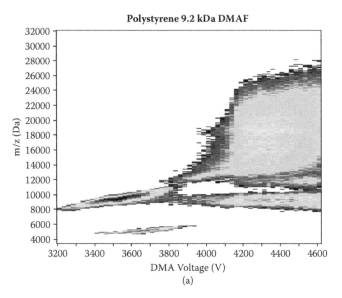

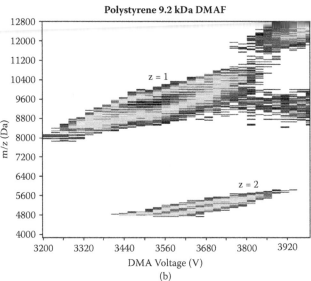

FIGURE 5.11 DMA-MS spectrum in CO_2 gas of 2.4 mM polystyrene electrosprayed from a solution of 1% dimethylammonium formate in NMP. (a) Broad spectrum showing the high contamination level of the solvent. (b) Detail of the lower left corner of (a) showing resolved singly and doubly charged peaks.

visitors by the new DMA-MS facility at Yale (since early 2009), to cover various facets of the analysis of high-molecular-weight ions. Of particular interest are several discoveries that were enabled by the ability to isolate in the DMA many previously undetected metastable ions, which normally decay into other ions prior to MS. We observed a diversity of ion evaporation events, some reducing the charge of initially electrosprayed protein, polymer, and solid or liquid cluster ions, and others detaching relatively long, singly charged polymer chains initially entangled within a multiply charged rodlike polymer ion. We have also found neutral evaporation events of the form $(AC)_nC^+ \rightarrow (AC)_{n-1}C^+ + (AC)$. The use of the DMA as a tool for tandem studies is not restricted to unstable clusters. Virtually any ion selected by the DMA (with no apparent m/z limit) may be activated prior to MS analysis. This possibility has been found useful for simultaneous determination of protein structure (prior to declustering) and mass (after declustering). It is also useful for fragmentation studies of mobility-selected ions of known mass in TOF-MS instruments that either do not have a quadrupole or have one with insufficiently high mass range.

ACKNOWLEDGMENTS

I am grateful to many colleagues at SEADM, who turned the complicated idea of high transmission parallel plate DMAs into a reality, have coupled it to a growing number of mass spectrometers, and have developed increasingly efficient hardware and software tools to run the tandem instrument. Special thanks are due to Gonzalo Fernandez de la Mora for starting and maintaining the venture, and to Juan Rus and Alejandro Casado for engineering the DMA-MS combination and improving the software tools to acquire and analyze DMA-MS spectra. My thanks also go to D. Gostick of Applied Biosystems, B. Thomson (Sciex), SEADM, Yale's Deputy Provost and Dean of Engineering for their many contributions to bring about Yale's DMA-MS facility, and to Chris Hogan, Carlos Larriba, and Juan Fernandez-Garcia for their pioneering use of it. Our study of ionic liquids was supported in part by AFOSR grant FA9550-06-1-0104.

REFERENCES

1. N. G. Gotts, G. Von Helden, and M. T. Bowers, Carbon cluster anions. Structure and growth from C5- to C62-, *Int. J. Mass Spectrom. Ion Proc.* 150, 217–229, 1995.
2. B. C. Bohrer, S. I. Merenbloom, S. L. Koeniger, A. E. Hilderbrand, D. E. Clemmer, Biomolecule analysis by ion mobility spectrometry, *Ann. Rev. Anal. Chem.* 1, 293–327, 2008.
3. (a) M. F. Jarrold, Peptides and proteins in the vapor phase, *Ann. Rev. Phys. Chem.* 51, 179, 2000. (b) S. L. Koeniger, et al., Evidence for many resolvable structures within conformation types of electrosprayed ubiquitin ions, *J. Phys. Chem. B* 110 (13), 7017, 2006.
4. A. A. Shvartsburg, et al., Characterizing the structures and folding of free proteins using 2-D gas-phase separations: Observation of multiple unfolded conformers, *Anal. Chem.* 78 (10), 3304–3315, 2006.
5. C. Becker, K. Qian, and D. H. Russell, Molecular weight distributions of asphaltenes and deasphaltened oils studied by laser desorption ionization and ion mobility mass spectrometry, *Anal. Chem.* 80, 8592–8597, 2008.
6. (a) J. Wildgoose, T. McKenna, C. Hughes, K. Giles, S. Pringle, I. Campuzano, J. Langridge, and R. H. Bateman, Using a novel travelling wave ion mobility device coupled with a time-of-flight mass spectrometer for the analysis of intact proteins, *Mol. Cell. Proteomics*, 5 (10), S14–S14, Supplement, Meeting Abstract: 78 , 2006. (b) S. D. Pringle, K. Giles, J. L. Wildgoose, J. P. Williams, S. E. Slade, K. Thalassinos, R. H. Bateman, M. T. Bowers, and J. H. Scrivens, An investigation of the mobility separation

of some peptide and protein ions using a new hybrid quadrupole/travelling wave IMS/oa-ToF instrument, *Int. J. Mass Spectrom* 261 (1),1–12, 2007.

7. (a) A. A. Shvartsburg, R. D. Smith, Fundamentals of traveling wave ion mobility spectrometry, *Anal. Chem.* 80 (24), 9689–9699, 2008. (b) D. P. Smith, T. W. Knapman, I. Campuzano, R. W. Malham, J. T. Berryman, S. E. Radford, and A. E. Ashcroft, Deciphering drift time measurements from travelling wave ion mobility spectrometry-mass spectrometry studies, *Europ. J. Mass Spectom.* 15 (2), Special Issue, 113–130, 2009.

8. J. L. P. Benesch, Collisional activation of protein complexes: Picking up the pieces, *J. Am. Soc. Mass Spectr.* 20 (3), 341–348, 2009.

9. K. Giles, T. Gilbert, M. Green, and G. Scott, Enhancements to the ion mobility performance of a travelling wave separation device, Paper presented at the 2009 ASMS conference, Philadelphia, PA, June 2009; Session TPH, poster 227.

10. R. Mukhopadhyay, IMS/MS: Its time has come, *Anal. Chem.* 80, 7918–7920, 2008.

11. (a) R. W. Purves, R. Guevremont, S. Day, C. W. Pipich, and M. S. Matyjaszczyk, Mass spectrometric characterization of a high-field asymmetric waveform ion mobility spectrometer, *Rev. Sci. Instrum.* 69, 4094–4104, 1998. (b) A. A. Shvartsburg, K. Tang, and R. D. Smith, FAIMS operation for realistic gas flow profile and asymmetric waveforms including electronic noise and ripple, *J. Am. Soc. Mass Spectrom.* 16, 1447–1455, 2005.

12. (a) H. A. Erikson, The change of mobility of the positive ions in air with age, *Phys. Rev.* 18, 100–101, 1921. (b) H. A. Erikson, On the effect of the medium on gas ion mobility, *Phys. Rev.* 30, 339–348, 1927.

13. J. Zeleny, The distribution of mobilities of ions in moist air, *Phys. Rev.* 34, 310–334, 1929.

14. H. F. Tammet, The aspiration method for the determination of atmospheric-ion spectra, Israel Program for Scientific Translations, Jerusalem, 1970 (original work in Russian from 1967).

15. R. C. Flagan, History of electrical aerosol measurements, *Aerosol Sci. Technol.* 28, 301–380, 1998.

16. G. W. Hewitt, The charging of small particles for electrostatic precipitation, *Commun. Electron.* 31, 300–306, 1957.

17. (a) E. O. Knutson and K. T. Whitby, Aerosol classification by electric mobility: Apparatus, theory and applications, *J. Aerosol Sci.* 6, 443, 1975. (b) Y. H. Liu and D. Y. H. Pui, *J. Colloid Interface Sci.* 47, 155, 1974.

18. Y. Kousaka, K. Okuyama, and T. Mimura, Diffusion on electrical classification of ultrafine aerosol-particles in differential mobility analyzer, *J. Chem. Eng. Japan* 19, 401–407, 1986.

19. M. R. Stolzenburg, An ultrafine aerosol size distribution measuring system. PhD thesis, University of Minnesota, St. Paul, MN, 1988.

20. J. Rosell, I. G. Loscertales, D. Bingham, and J. Fernández de la Mora, Sizing nanoparticles and ions with a short differential mobility analyzer. *J. Aerosol Sci.* 27, 695–719, 1996.

21. L. de Juan and J. Fernandez de la Mora, Size analysis of nanoparticles and ions: Running a Vienna DMA of near optimal length at Reynolds numbers up to 5000, *J. Aerosol Sci.* 29, 617–626, 1998.

22. (a) P. Martínez-Lozano and J. Fernández de la Mora, Effect of acoustic radiation on DMA resolution, *Aerosol Sci. Technol.* 39 (9), 866–870, 2005. (b) P. Martínez-Lozano and J. Fernández de la Mora, Resolution improvements of a nano-DMA operating transonically, *J. Aerosol Sci.*, 37, 500–512, 2006.

23. J. Fernandez de la Mora, B. A. Thomson, and M. Gamero-Castaño, Tandem mobility mass spectrometry study of electrosprayed Heptyl4N+Br- clusters, *J. Am. Soc. Mass Spectrom.* 16 (5), 717–732, 2005.

24. (a) J. Rus and J. Fernandez de la Mora, Resolution improvement in the coupling of planar differential mobility analyzers with mass spectrometers or other analyzers and detectors. US patent application publication 20080251714, October 16, 2008. (b) J. Rus,

F. Estévez, and J. Fernández de la Mora, A planar DMA coupled to a MS for tandem IMS-MS separation at high transmission, with IMS resolution approaching 100. Poster 105, ASMS conference, Indianapolis, 3–7 June 2007. (c) J. Rus, D. Moro, J. A. Sillero, J. Freixa, and J. Fernández de la Mora, A high flow rate DMA with high transmission and resolution designed for new API instruments, Poster 042, Annual ASMS conference, 1–6 June, Denver, Colorado, 2008.

25. J. Rus, D. Moro, J. A. Sillero, J. Royuela, A. Casado, and J. Fernández de la Mora, IMS-MS studies based on coupling a differential mobility analyzer (DMA) to commercial API-MS systems, submitted to *Int. J. Mass Spectrom.* In press 2010, doi:10.1016/j.ijms.2010.05.008.

26. J. Fernández de la Mora, S. Ude, and B. A. Thomson, The potential of differential mobility analysis coupled to mass spectrometry for the study of very large singly and multiply charged proteins and protein complexes in the gas phase, *Biotech. J.* 1, 988–997, 2006.

27. E. K. Lewis, T. Egan, K. Waters, S. Yates, J. F. Moore, C. Kittrell, S. R. Ripley, K. S. Ho, V. Womack, R. H. Hauge, V. N. Khabashesku, A. S. Woods, and J. A. Schultz, Vacuum ultraviolet post-ionization combined with ion-mobility for the characterization and application of functionalized nanomaterials as MALDI matrices, Paper presented at the 2009 ASMS conference, Philadelphia, PA, June 2009; Session TPH, poster 205.

28. H. Javaheri, Y. Le Blanc, B. A. Thomson, J. Fernandez de la Mora, J. Rus, and J. A. Sillero-Sepúlveda, Analytical characteristics of a differential mobility analyzer coupled to a triple quadrupole system (DMA-MSMS), Poster 061, Annual ASMS conference, 1–6 June, Denver, Colorado, 2008.

29. J. Fernandez de la Mora, A. Casado, and G. Fernandez de la Mora, Method and apparatus to accurately discriminate gas phase ions with several filtering devices in tandem, US patent Application Publication 20080203290, August 28, 2008.

30. C. J. Hogan Jr. and J. Fernández de la Mora, Ion mobility measurements of non-denatured 12-150 kDa proteins and protein multimers by tandem differential mobility analysis-mass spectrometry (DMA-MS), submitted to *J. Am. Soc. Mass Spectr.* 2009.

31. H. Tammet, Size and mobility of nanometer particles, clusters and ions. *J. Aerosol Sci.* 26, (3), 459–475, 1995.

32. B. K. Ku and J. Fernandez de la Mora, Relation between electrical mobility, mass, and size for nanodrops 1–6.5 nm in diameter in air, *Aerosol Sci. Technol.* 43 (3), 241–249, 2009.

33. C. Larriba, C.r J. Hogan Jr., M. Attoui, R. Borrajo, J. Fernandez Garcia, J. Fernandez de la Mora, The Mobility-Volume Relationship below 3.0 nm examined by Tandem Mobility-Mass Measurement, *Aerosol Sci. Tech*, accepted for publication, July 2010.

34. K. Breuker, F. W. McLafferty, Stepwise evolution of protein native structure with electrospray into the gas phase, $10^{-12} - 10^2$ s, *Proc. Nat. Acad. Sci. USA* 105, 18145–18152, 2008.

35. E. R. Badman, S. Myung, and D. E. Clemmer, Evidence for unfolding and refolding of gas-phase cytochrome c ions in a Paul trap, *J. Am. Soc. Mass Spectrom.* 16, 1493–1497, 2005.

36. J. Woenckhaus, Y. Mao, and M. F. Jarrold, Hydration of gas phase proteins: Folded +5 and unfolded +7 charge states of cytochrome c, *J. Phys. Chem. B* 101 (6), 847–851, 1997.

37. J. Fernández de la Mora, L. de Juan, R. Eichler, and J. Rosell, Differential mobility analysis of molecular ions and nanometer particles, *Trends Anal. Chem.* 17, 328–339, 1998.

38. (a) S. Ude, J. Fernandez de la Mora, and B. A. Thomson, Charge-induced unfolding of multiply charged polyethylene glycol ions, *J. Am. Chem. Soc.* 126, 12184–12190, 2004. (b) S. Ude, J. Fernandez de la Mora, J. N. Alexander IV, and D. A. Saucy, Aerosol size standards in the nanometer size range: II, Narrow size distributions of polystyrene 3–11 nm in diameter, *J. Coll. Interface Sci.* 293, 384–393, 2006.

39. S. Ude, Properties and measurement of nanometer particles in the gas phase, PhD thesis, Yale University, 2004.
40. C. J. Hogan Jr. and J. Fernández de la Mora, Tandem ion mobility-mass spectrometry (IMS-MS) study of ion evaporation from ionic liquid-acetonitrile nanodrops, *Phys. Chem. Chem. Phys.*, 2009, 11, 8079–8090, 2009.
41. C. J. Hogan Jr. and J. Fernández de la Mora, Ion-pair evaporation from ionic liquid clusters, *J. Am. Soc. Mass Spectr.*, http://dx.doi.org/10.1016/j.jasms.2010.03.044, in press, 2010.
42. M. I. Catalina, R. H. H. van den Heuvel, E. van Duijn, and A. J. R. Heck, Decharging of globular proteins and protein complexes in electrospray, *Chem. Eur. J.* 11, 960–968, 2005.
43. C. J. Hogan, E. Folta-Stogniev, B. Ruotolo, J. Fernandez de la Mora, IMS-MS study of native aggregates, from insulin to GroEL. To be submitted to *Anal. Chem.* 2009; Hogan, C., Ruotolo, B., Robinson, C., Fernandez de la Mora, J. Tandem differential mobility analysis-mass spectrometry of the GroEL complex: Structure compaction in the gas phase and inelastic air–protein interaction, Submitted to *J. Phys. Chem. B*, 2010.
44. (a) J. B. Fenn, Ion formation from charged droplets—Roles of geometry, energy, and time, *J. Am. Soc. Mass Spectrom.* 4 (7), 524–535, 1993; (b) J. B. Fenn, J. Rosell, C. K. Meng, In electrospray ionization, how much pull does an ion need to escape its droplet prison? *J. Am. Soc. Mass Spectr.* 8 (11), 1147–1157, 1997.
45. (a) I. Romero-Sanz, R. Bocanegra, J. Fernández de la Mora, and M. Gamero-Castaño, Source of heavy molecular ions based on Taylor cones of ionic liquids operating in the pure ion evaporation regime, *J. Appl. Phys.* 94 (5), 3599–3605, September 2003, (b) C. Larriba, S. Castro, J. Fernandez de la Mora, and P. Lozano, Monoenergetic source of kilodalton ions from Taylor cones of ionic liquids, *J. Applied Phys.* 101 (8), Art. No. 084303, 2007. See also Figure 10.3 in this volume.
46. S. F. Wong, C. K. Meng, and J. B. Fenn, Multiple charging in electrospray ionization of poly(ethylene glycols), *J. Phys. Chem.* 92, 546–550, 1988.
47. S. Trimpin and D. E. Clemmer, Ion mobility spectrometry/mass spectrometry snapshots for assessing the molecular compositions of complex polymeric systems, *Anal. Chem.* 80 (23), 9073–9083, 2008.
48. C. Larriba, C. Hogan, and J. Fernandez de la Mora, Measurement of the surface tension of single polymer molecules, in preparation, 2009.
49. S. L. Kaufman, J. W. Skogen, F. D. Dorman, F. Zarrin, and L. C. Lewis, Macromolecule analysis based on electrophoretic mobility in air: Globular proteins. *Anal. Chem.* 68, 1895–1904, 1996.
50. D. Saucy, S. Ude, W. Lenggoro, and J. Fernandez de la Mora, Mass analysis of water-soluble polymers by mobility measurement of charge-reduced electrosprays, *Anal. Chem.* 76, 1045–1053, 2004.
51. J. D. Lennon, III, S. P. Cole, and G. L. Glish, Ion/molecule reactions to chemically deconvolute the electrospray ionization mass spectra of synthetic polymers, *Anal. Chem.* 78, 8472–8476, 2006.
52. L. Huang, P. C. Gough, and M. R. DeFelippis, Characterization of poly(ethylene glycol) and PEGylated products by LC/MS with postcolumn addition of amines, *Anal. Chem.* 81, 567–577, 2009.
53. B. K. Ku, J. Fernandez de la Mora, D. A. Saucy, and J. N. Alexander, IV, Mass distribution measurement of water-insoluble polymers by charge-reduced electrospray mobility analysis, Anal. Chem., 76, 814–822, 2004.
54. S. Ude, J. Fernandez de la Mora, J. N. Alexander IV, and D. A. Saucy, Aerosol size standards in the nanometer size range: II, Narrow size distributions of polystyrene 3–11 nm in diameter, *J. Colloid Interface Sci.* 293, 384–393, 2006.
55. J. H. Scrivens and A. T. Jackson, Characterisation of synthetic polymer systems, *Int. J. Mass Spectr.* 200, 261–276, 2000.

6 A Cryogenic-Temperature Ion Mobility Mass Spectrometer for Improved Ion Mobility Resolution

Jody C. May and David H. Russell

CONTENTS

6.1 INTRODUCTION

Ion mobility spectrometry (IMS) has been used for a wide variety of analytical applications, ranging from applications related to forensic chemistry to security checkpoint screening and detection for small molecules.[1] More recently ion mobility–mass spectrometry (IM-MS) has emerged as an analytical technique for the characterization of petroleum crude fractions,[2] biophysical studies of biological molecules,[3] studies of protein complexes,[4] and studies of protein aggregation related to diseases such as Alzheimer's and Parkinson's diseases and diabetes.[5] Although IMS has changed very little since it began to emerge as an analytical technique in the 1970s,[6,7] the coupling of IMS with modern high-performance mass spectrometers equipped with electrospray ionization (ESI) and/or matrix-assisted laser desorption ionization (MALDI) sources has catalyzed new developments and applications of the technique. In addition, the growing numbers of problems now being addressed by using IM-MS has spawned new technological developments.[8] Although IMS was first combined with orthogonal time-of-flight mass spectrometry (TOFMS) instruments more than 40 years ago,[9] over the past decade there has been tremendous growth in IMS-TOFMS instrumentation. As the complexity of chemical problems being addressed by IMS has increased, the demand for higher resolution IMS techniques has also increased. As is typical in the field of technology development, researchers have attempted a variety of avenues to increase the resolution of IMS, such as increasing the pressure of the drift gas,[10] decreasing the temperature of the drift gas,[11,12] increasing the length of the IMS drift tube,[13,14] and seeking alternatives to uniform drift field IMS experiments.[15,16] This chapter focuses primarily on studies aimed at developing variable-temperature (VT)IMS with the goal of using low temperature to increase the resolution for the drift tube ion mobility method.

The mobility of an ion (K) through a neutral buffer gas is defined as the ratio of drift velocity (v_d) to the applied electric field (E_0) as shown by Equation (6.1), where the velocity is expressed in cm $\cdot$ s^{-1}, E is V $\cdot$ cm^{-1} to yield units of cm$^2 \cdot$ V$^{-1} \cdot$ s^{-1} for K.

$$K = \frac{v_d}{E_0} \tag{6.1}$$

Under a given set of conditions the experimentally measured quantity, the ion arrival-time distribution (ATD), in IMS is the time that is required for a packet of ions to traverse a drift cell of finite length and held at a constant electric potential and gas pressure; however, the experimental ATD of an ion in the drift tube will depend on several fundamental parameters of the experiment. These governing parameters are described by the Mason-Schamp generalized expression, which is exact in the limit of a vanishing electric field[17]:

$$K = \frac{3e}{16N} \sqrt{\frac{2}{\pi \mu k T_{ion}}} \frac{1}{\Omega(T_{ion})} \tag{6.2}$$

In the above expression, K is the measured ionic mobility, N is the gas number density, T_{ion} is the ion's effective temperature, μ is the ion-gas reduced mass term, and $\Omega(T_{ion})$ is the averaged ion collision cross section. The effective ion temperature is a composite of the drift gas temperature and the energy the ion gains from the electric field:

$$T_{ion} = T_{gas} + \frac{m_{gas} v_d^2}{3k} \tag{6.3}$$

In this expression, T_{gas} is the drift gas temperature, m_{gas} is the drift gas mass, and k is the Boltzmann constant. A consequence of Equation (6.3) is that the ion's temperature can never be less than or equal to the temperature of the drift gas in the IMS experiment.[18] In other words, the electric field always imparts some amount of heating to ions; however, for reasonably low field conditions (field divided by gas number density, E/N < 2 Td), the extent of ion heating results in no measurable consequence to the IMS experiment. As the size of the ion increases, this ion energy can be imparted into more degrees of freedom, resulting in a higher threshold for the onset of nonlinear field behavior (high field ion mobility) and eventual ion activation/fragmentation.[19] Thus it is not uncommon to find drift tube IMS experiments with ion temperatures 50 K or more in excess of the drift gas temperature, which for all practical purposes can be considered as low field conditions due to the ion's size.

The collision cross section term in Equation (6.2) also contains a dependence on ion temperature due to the attractive ion-neutral interaction potential. For sufficiently strong interactions, reduced temperatures result in increased ion clustering, which serves to increase the effective cross-sectional size of the ion. A more appropriate way of visualizing this effect is to consider an ion and neutral gas approaching one another prior to a collision. With too much energy, the two bodies collide and part ways (a glancing collision), but if they approach with sufficiently low velocities as might occur at low temperatures, then the ion-neutral may interact for a longer duration (a so-called orbiting collision).[12] One can visualize this temperature consequence by plotting a representative ion-neutral interaction potential that accounts for the centrifugal capture of an ion and neutral during a collision. Such a plot is contained in Figure 6.1a for a representative (2,6,4) interaction potential.[20]

The general shape of the interaction potential exhibits a deeper attractive well as the temperature is reduced due to the decreased steepness of the repulsive term. Another consequence of temperature can be seen in the repulsive portion of the curve, shown in Figure 6.1b. This energy barrier decreases as the temperature of the system is lowered, representing a decrease in the centrifugal energy associated with capturing the ion and neutral within the attractive well. While the depth of the energy well will be contingent on the ion-neutral system, the attractive interaction will be enhanced for all systems at reduced temperatures. Thus differences in

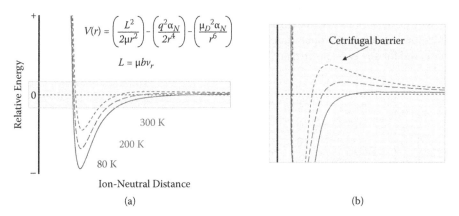

FIGURE 6.1 (a) A (2,4,6) ion-neutral interaction potential plotted at three temperatures. The potential chosen contains a centrifugal capture expression in the repulsive term (r^2) along with the attractive polarization (r^4) and charge-induced/London dispersion terms (r^6). The temperature dependence is contained within the relative velocity dependence of angular momentum, L. The general shape of the interaction potential is that of a deeper attractive well at decreased temperature, which predicts a stronger attractive interaction as the temperature of the ion and neutral are reduced. (b) Closer examination of the shape of the curves reveals a repulsive barrier at higher temperatures resulting from the associated higher centrifugal energies of the ion-neutral approach. This is a consequence of the ion and neutral approaching one another with too much energy to be captured in an orbiting collision.

the relative attraction of ions to the neutral drift gas can be enhanced and perhaps even exploited at low temperatures to effectuate a better IMS separation; however, as noted by Wyttenbach et al., accurately modeling such phenomena is difficult as the details of the interactions are dependent upon the chemical characteristics of the ion, e.g., ions that project hydrogen atoms (hydrocarbon polymer ions) to the neutral atom differ from those that project carbon atoms $(C_{60}{}^{+})$.[12] The complexities of such interactions were also noted by Dougherty.[21] Ultimately Equation (6.2) predicts that changes in the IMS drift gas temperature will have both direct and indirect influences on the measured ionic mobility: direct in terms of the extent of thermal diffusion and indirect as it relates to changes in the ion-gas interaction potential and effective collision cross section.[12] This complicated relationship between temperature and ionic mobility is perhaps one reason why few investigators choose to pursue (VT)IMS research; the other reason being that the design and construction of a variable temperature drift tube spectrometer requires a fair degree of engineering. Nonetheless, the analytical prospects of a low temperature IMS instrument are appealing enough to pursue its development.

There are several analytical reasons why one might be interested in varying the temperature in the IMS experiment. For practical reasons, VT experiments afford an added degree of flexibility when optimizing the IMS experiment for a particular separation of ions. For example, Figure 6.2a contains a plot of the reduced ionic mobilities for several atomic ions at two temperatures (80 and 300 K).[22–25] A straight line has been interpolated between the two datasets for reasons of clarity. As illustrated by the hypothetical mobility spectra in Figure 6.2b, the elution orders of these ions will change considerably from one temperature to another; thus there is a distinct operational temperature whereby an optimal separation is observed for any two ions in a mixture. The ability to vary the drift gas temperature

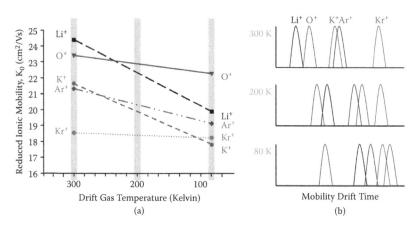

FIGURE 6.2   (a) A plot of reduced ionic mobility as a function of drift gas temperature, illustrating the temperature-dependent differences in ionic mobility for ions possessing different chemical characteristics. (b) Hypothetical mobility spectra for a mixture of ions from (a) at three different temperatures, illustrating how elution orders change relative to the temperature. (Data was obtained from Refs. 22–25.)

can have a powerful effect on the resulting separation abilities of the IMS experiment. Experimentally, the effects are much less dramatic than those depicted in Figure 6.2 since oftentimes ion mixtures are comprised of ions possessing very similar chemical characteristics.

Temperature also directly affects the observed peak widths in the IMS experiment. Specifically, as the temperature of the drift gas is lowered, the extent of ion diffusion is also diminished, resulting in both an increase in drift time and a decrease in the ATD (as per a Maxwell-Boltzmann distribution of speeds). An underlying motivation for cryogenic IMS is to increase the resolving power, R, as defined by the measured drift time divided by full width at half maximum (FWHM):

$$R = \frac{t_{drift}}{\Delta t_{FWHM}}$$ (6.4)

Because the ion-neutral interaction potential is influenced by temperature, low temperature IMS can be used to effectuate separations that would otherwise not be seen at elevated temperatures and electric fields. Some dramatic examples include the separation of ion electronic configurations[11] and electronic states,[26] the latter of which will be revisited briefly at the end of this chapter. Such systems represent true isomeric ions for which there are very few methods available that can differentiate the various isomeric forms. Finally, cooling the drift gas at temperatures below the condensation point of most atmospheric contaminants affords the added analytical advantage of drift gas purity. Gas impurities are a ubiquitous issue for practitioners of IMS as trace impurities give rise to undesirable reaction chemistry that broadens peak profiles and introduces added chemical noise to the experiment. Table 6.1 summarizes several commonly encountered gas impurities and their condensed phase change temperatures. A convenient cryogen to use is liquid nitrogen, which is chemically inert and has a stable phase change temperature near 80 K—if the IMS experiment is conducted at or near 80 K then virtually all other drift gas impurities condense from the system. This, of course, precludes the use of most all other drift gases except helium, though for low temperature IMS experiments helium is a preferred drift gas to use as it has the lowest propensity for clustering with analyte ions[1] and results in the least amount of ion heating, as per Equation (6.3). This, in addition to the large body of reported ion cross sections available in helium drift gas, provides the motivation for using helium in this work.

TABLE 6.1

Phase Change Temperatures for Several Commonly Encountered Gas Impurities

	H_2O	CO_2	Total HCs	O_2	Ar	CO	N_2
Phase change temperature (K)	273.2	194.7	CH_4: 112.0 C_2H_6: 184.5 C_3H_8: 231.1	90.2	87.3	81.0	77.4

6.2 THE VARIABLE-TEMPERATURE IM-MS INSTRUMENT DESIGN

The variable-temperature (VT) IM-MS instrument constructed at Texas A&M is shown schematically in Figure 6.3. For all the studies reported here the MS measurements were performed in the linear-TOF mode; however, for experiments that require high mass-to-charge ratio (m/z) resolution the instrument can also be operated as an orthogonal reflectron TOF instrument. Individual regions of the instrument are separated by conductance-limiting apertures, such that each component operates independently and can be readily serviced or replaced. This system of conductance limits allows the instrument to operate in pressure regimes, as illustrated on the lower bar in Figure 6.2. Custom ion optics are necessary to focus and transfer ions efficiently from one component to the next.

The instrument is designed to accommodate both MALDI and ESI ionization sources, but for the initial proof-of-concept experiments described here an electron ionization (EI) source is used. The EI source is adapted from an antiquated piece of equipment in the laboratory (Kratos AEI MS9/50 sector instrument) and is of a classic, Nier-type geometry (orthogonal electron beam to ion beam axis),[27] which generates a collimated beam of electrons having a well-defined kinetic energy. The EI source is mounted onto a vacuum flange and operated in a vacuum chamber at a base pressure of less than 10^{-6} torr. Sample neutrals are introduced to the ionization volume, either through a controlled leak of volatiles or through sublimation of a

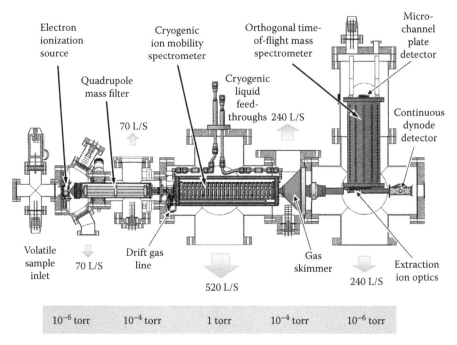

FIGURE 6.3 A schematic of the mass selected, cryogenic ion mobility–mass spectrometer with the important components labeled. Representative operational base pressures are provided in the lower bar. Components are to scale.

solid sample directly inserted into the source block. Ions formed from the electron ionization process are immediately extracted orthogonally to the electron beam through the use of parallel electrodes, the ion repeller, and ion extractor elements. The electron energy is defined as the potential difference between the emission filament and the center of the EI source directly between the repeller and extractor electrodes. Total ion current generation, including molecular ions and subsequent fragments, reaches a maximum somewhere between 50 and 70 eV electron energy for most small organic ions, which is the typical operational energy of the EI source. A wide range of small molecules can be studied with EI, given that they can be made sufficiently volatile to be introduced to the ionization region.

The instrument design also incorporates a quadrupole mass spectrometer between the ionization source and the IM-MS. For most experiments, the quadrupole operates as a broadband mass transmission ion guide (radio frequency only mode), efficiently transferring ions across the differential pressure region between the ionization source and the IMS. Switching the quadrupole to mass selective operation reduces ion transmission by as much as 50% for a particular m/z value, and so is used sparingly—most often the dispersive IM-MS analysis alone is sufficient to make confident ion assignments. The ability to mass select ions prior to IM-MS analysis is, however, essential for eliminating ambiguity in the analysis. Mass selective IM-MS experiments are perhaps the most effective means of elucidating ion chemistry, which occurs in the drift region. If, for example, a single m/z value is selected and introduced to the IM-MS, then the subsequent mass analysis (TOFMS) should verify the presence of a single m/z in the mass spectrum, if no additional ion chemistry is taking place. If product ions appear in the mass spectrum, then they can be unambiguously correlated to a precursor m/z value.

A schematic of the cryogenic drift tube assembly and vacuum system is contained in Figure 6.4. The IMS drift region is constructed using a conventional stacked ring electrode design. Twenty-three ring electrodes are stacked to a total drift length of 30.2 cm from ion entrance to exit. Electrodes are of a sufficiently large inner diameter (~3 cm) so as not to introduce any significant nonlinear fields within the drift corridor of ions. The ring electrodes are spaced with ceramics and chained together through a network of high precision resistors, effectively creating the linear electric field drop from end to end. The ring assembly is inserted within a cylindrical volume and compressed with end cap electrodes incorporating small conductance-limiting apertures (~500-μm diameter laser drilled orifices), creating a closed drift volume capable of supporting an elevated atmosphere of pressure relative to the surrounding vacuum system. With sufficient pumping, a closed drift tube design of this type can sustain tens of torr of pressure within the tube while operating the end regions below millitorr vacuum. The temperature envelope is fabricated by welding together two concentric tubes with end cap rings such that a closed volume is created within the annular region. This annular region serves as a Dewar jacket where liquid nitrogen, or another cooling fluid, is introduced. In this configuration the drift gas is in direct thermal contact with this temperature envelope. The temperature is monitored with two resistive temperature devices (RTDs), one inserted at either end of the drift tube and mounted in thermal contact with the drift gas and nothing else. The RTD is a thin platinum wire with well-characterized resistances across

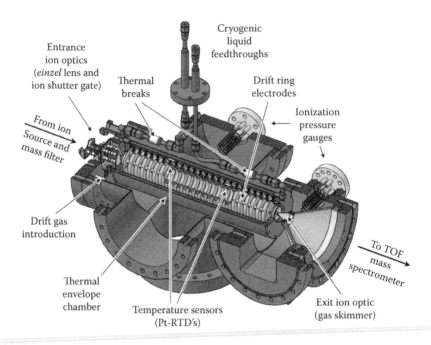

FIGURE 6.4   A partial cutaway schematic of the cryogenic drift tube and vacuum chamber assembly developed at Texas A&M.

a wide range of temperatures. Different drift gas temperatures can be accessed by filling the envelope with different levels of liquid nitrogen. Considerations for thermal expansion and contraction of materials must also be incorporated into the drift tube design. Since the drift tube physically moves as the temperature is altered, ion steering optics are incorporated into the ion lens assembly at the front of the IMS and must be adjusted to track the position of the entrance aperture to the drift tube. The cryogenic drift tube described here is constructed using only stainless steel and alumina ceramics, which have comparatively similar thermal expansion properties, thus minimizing mechanical failure problems associated with differential contraction of materials. Stainless steel is a particularly favorable material for constructing the cryogenic envelope as the poor thermal conduction properties result in very good temperature stability. In order to take full advantage of this temperature stability, the entire cryogenic drift tube assembly is thermally isolated through the use of ceramic standoff mounting and cryogenic feed-through breaks for gas and liquid introduction. This also allows the entire assembly to be electrically biased to minimize problems associated with electrical breakdown of the drift gas during elevated field operation. Placing the assembly within a vacuum system further improves thermal insulation and eliminates problems associated with ice buildup and current leakage.

The back end mass analysis is performed with an orthogonally configured time-of-flight mass spectrometer (TOFMS) built in house. Coupling these two temporally dispersive types of analyses, drift tube ion mobility and TOFMS, results in a very powerful analytical configuration since for every analysis cycle, a complete snapshot of mobility and mass information is obtained. The performance of the orthogonal TOFMS also benefits from the ion beam cooling effect of the IMS, that is, ions that elute from the IMS have kinetic energies that are essentially thermal, which greatly improves the subsequent TOFMS analysis.[28] As a result, a relatively simple linear geometry TOFMS and ion optics can be used, yielding adequate mass resolution and very good sensitivity. The TOFMS is constructed using a two-stage ion optical configuration essential in design to what was described by Guilhaus and coworkers for orthogonal ion extraction.[29] The flight tube is an electrically isolated stainless steel liner inserted within the vacuum chamber, permitting the entire TOFMS to be operated at a biased potential. Because the drift tube IMS experiment necessitates the use of a declining electric potential, either the front end or back end spectrometer components must be electrically biased—biasing the back end TOFMS to a negative potential is more convenient since it is desirable to maintain the ion source end at or near ground potential. Additionally, negative potentials are more resistant to gaseous electrical breakdown than are positive polarities, for reasons fundamental to the high mobility of free electrons. Ion detection is accomplished through the use of a fast response microchannel plate detector operated in an ion counting mode. For small mass ions, flight times through the 29.6-cm linear flight tube are on the order of tens of microseconds, such that the TOFMS can be operated at frequencies approaching 100 kHz or, put another way, as many as 100,000 spectra can be acquired each second with this instrument. Fast analysis times is one of the fundamental advantages necessary for conducting successful (VT)IMS experiments where temperature variability directly influences the resulting spectral quality. All spectra are acquired within a 2–3 degree temperature window to maintain the statistical precision of the measurement. A smaller temperature variation is possible using a thermostated system, which should further improve the IMS resolution obtainable on this kind of instrument.

6.3   EXPERIMENTAL RESULTS

The drift time of a given ion in the ion mobility experiment will have a temperature dependence owing to ion-gas interaction potential. The ion-neutral interaction becomes stronger as the temperature of the gas and ion is decreased, resulting from a decreased centrifugal barrier associated with the decrease in ion-gas relative velocities.[12,20] Additionally, the thermodynamically driven intermolecular forces that would otherwise destabilize an ion-gas interaction are dampened at lower temperature, enhancing the strength of the interaction. For strongly interacting neutrals such as water, lower temperatures result in very strong attractive forces, resulting in ion hydration (clustering), which increases the apparent ion cross section and ultimately the ion drift time in the drift tube.[30] The same general explanation can be applied to weakly interacting neutrals, such as in the case of helium. Low temperature helium-ion interactions do not yield detectable clusters in this experiment; however, the results of such interactions can be observed in some cases, as will be shown below for the separation of electronic states of Cr^+. In addition, the numerous ion-neutral interactions that occur over the course of the IMS experiment (tens of thousands of ion-gas collisions at 1 torr) can result in an increase in the ion's apparent (measured) cross section. Because the ion-neutral interaction potential will be unique for each ion and gas combination, different ions in the same drift gas may exhibit very different changes in drift time when the IMS gas temperature is changed. This is why the atomic ions in Figure 6.1a exhibit different mobility changes as the temperature is changed.

For the analytical chemist, temperature is an additional parameter that can be adjusted to improve the observed separations in the IMS experiment. Figure 6.5 illustrates one such example with the ions observed from the 70 eV electron ionization of acetone. Figure 6.5a contains the composite 2-dimensional (2D) ion mobility–mass spectrum taken with the drift gas (helium) at room temperature. Most ions in the spectrum can be unambiguously assigned, except ions at 28–31 m/z, which can represent more than one possible ion, i.e., mass isomers. As the temperature of the drift gas is lowered (Figure 6.5b), these nominal mass isomers begin to separate in the mobility dimension. At the lowest temperature in which the drift tube can operate (80 K; Figure 6.5c), the isobaric ions at 28–31 m/z are completely separated in the mobility dimension. Note also that these ions partition into chemical-

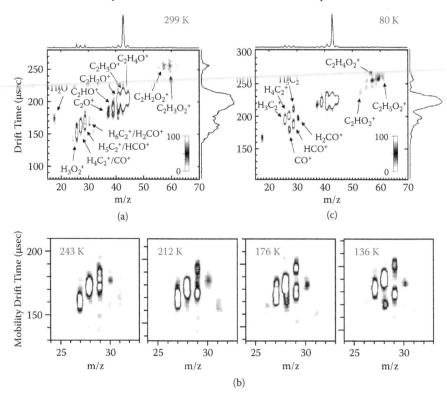

FIGURE 6.5 (a) A 2-dimensional ion mobility (y-axis) and mass-to-charge (x-axis) spectrum of the ions resulting from the 70 eV electron ionization of acetone. (b) A series of zoomed-in spectra for the ion signals at 27–30 m/z taken at decreasing temperatures from left to right, and (c) the 2D spectrum of acetone at 80 K, showing complete separation of the nominal isomeric ions 28–30 m/z. Some additional ions also appear in the low temperature spectrum due to the suppression of chemical noise.

specific trendlines in the 2D projection in Figure 6.5c, which can be used to make unambiguous identifications of these ions. This trendline partitioning is a powerful added piece of information that is unique to this particular kind of 2D mobility-mass analysis. An added observation in the spectrum in Figure 6.5c is the appearance of some additional ions, labeled in the upper right. These ions are not observed in sufficient abundance at room temperature due to chemical suppression resulting from reaction chemistry with trace impurities in the drift gas. By condensing out the drift gas impurities, the chemical noise resulting from reaction chemistry in the IMS is attenuated and these ion signals are enhanced.

Another consequence of low temperature drift tube operation is a decrease in ion and gas diffusion, which has a direct benefit to the analytical separation. The decreased diffusion leads to a narrower arrival time distribution in mobility space, which results in an improvement in resolving power and ultimately enhances the peak capacity of the IMS analysis. Figure 6.6 contains an example of the gains in resolving power that are typically observed in this experiment, here for the 70 eV electron ionization of carbon tetrachloride. The composite 2D spectrum is contained in Figure 6.6a, while Figure 6.6b contains a side-by-side comparison of mobility spectra for three selected ions taken at two different temperatures. For most ions, an improvement of over 50% in resolving power is observed when going from room temperature to 80 K ($1/T^{1/2}$). A few ions will exhibit either less of an improvement or

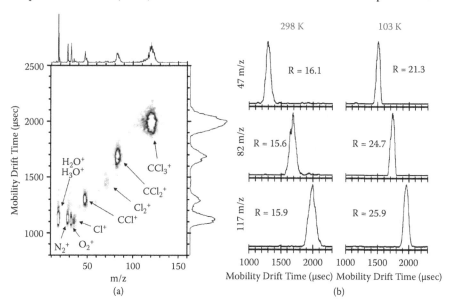

FIGURE 6.6 (a) The 2D mobility–mass spectrum of 70 eV electron ionized carbon tetrachloride. (b) Mobility spectra for three selected ions, CCl^+ (47 m/z), CCl_2^+ (82 m/z), and CCl_3^+ (117 m/z), taken at two different drift gas temperatures (298 and 103 K). At lower temperatures the mobility peaks narrow, leading to an increase in the analytical resolving power (denoted as R in b). Mobility spectra are extracted from the 2D data by integrating the ion current across an m/z window.

even a loss in resolving power as the temperature is lowered, as is the case with the ions in Figure 6.4 immediately prior to their separation.

One of the more dramatic examples of the separation abilities of cryogenic IMS is the separation of ion electronic states. In the early 1990s, Kemper and Bowers published a pioneering research article demonstrating the ability of IMS to separate the electronic states of transition metal cations in a technique they coined "electronic state chromatography."[26] A key aspect of the electronic state chromatography experiment was an IM-MS instrument fitted with a low temperature drift cell that allowed high-resolution separations to be obtained for the various electronic states. Figure 6.7 contains the experimental data of an electronic state system we recently revisited from Bowers' original work: the chromium cation generated from chromyl chloride (CrO_2Cl_2). Figure 6.7a contains the 2D mobility-mass spectrum of chromyl chloride electron ionized at 60 eV. A complete series of ligand ions appears in the spectrum, which gives rise to several resolved peaks in the composite mobility spectrum (y-axis). Of interest is the chromium cation (Cr^+) at 52 m/z, which exhibits a bimodal distribution in mobility space. Appearance energy experiments point to the higher mobility distribution as being represented by higher ionization energy, that is, at low ionization

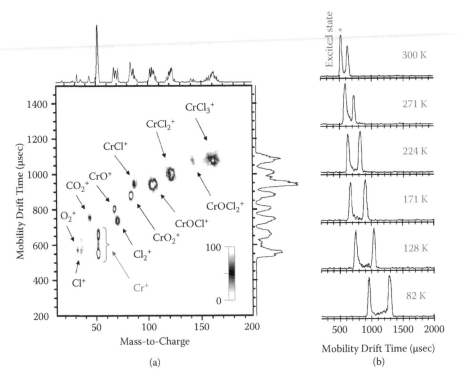

FIGURE 6.7   (a) The 2D mobility–mass spectrum of the ions resulting from the electron ionization of chromyl chloride (CrO_2Cl_2) at 60 eV. (b) Mobility spectra of Cr^+ taken at decreasing gas temperatures from 300 to 82 K. The separation of the electronic states is enhanced at lower temperatures. Some excited state depletion is also observed as the ion mobility residence time of ions increases in the experiment.

energy (<36 eV) only the lower mobility ion distribution appears in the spectrum. This suggests that the excited state or states are contained in the first arriving mobility peak and possess a higher ionic mobility than the ground electronic states. Because the electronic state system represents a case where the ion-neutral interaction is governing the IMS drift time, this interaction can be enhanced by reducing the drift gas temperature. This experiment is depicted in Figure 6.7b with several mobility spectra of Cr^+ taken at temperatures between 300 and 80 K. Note that as the temperature is reduced, the separation between the two ion distributions continues to increase. Another subtle observation is that the excited state mobility distribution decreases relative to the ground state as the temperature is decreased. Lowering the electric field at 82 K results in complete depletion of the excited state and a "filling in" between the two distributions (data not shown). This excited state depletion at low field and low temperature is likely a consequence of the increased frequency of "deactivating" collisions between Cr^+ and either a trace impurity in the drift gas or the result of long-lived ion-He interactions, as noted above. In the Ti^+ mobility spectrum (data not shown) we observe reaction chemistry and excited state depletion with trace amounts of an impurity and infer that a similar quenching mechanism exists in the Cr^+ system.

6.4 CONCLUSIONS

Cryogenic IM-MS is a powerful technique for enhancing the separation efficiency and spectral quality of the IMS experiment. We observe dramatic improvements in resolving powers upon cooling the helium-filled drift tube to ~80 K with no obvious detrimental consequences. This resolving power improvement is a net effect of decreased diffusional broadening of ion distributions and the condensing out of reactive impurities that would otherwise serve to broaden mobility distributions. In addition, the ion-neutral interaction is enhanced at low temperature and can be used to improve the resolution for select ion systems, such as was demonstrated with isobaric ion mixtures from acetone and the ground and excited ion state(s) of the chromium cation.

We are actively pursuing other variable temperature experiments, and an improved cryogenic IM-MS instrument platform is in development that will incorporate ESI, MALDI, and laser desorption/ionization (LDI) sources. Cryogenic IMS experiments for nanomaterials, polymers, and biological ion systems hold promise for obtaining thermodynamic structural information on reordering processes. In addition, variable temperature may provide answers to many longstanding questions regarding the nature of conformational interconversion, which may be occurring at room temperature for many of the biological ions that have already been investigated with ion mobility methods.

REFERENCES

1. Eiceman, G. A. and Z. Karpas. 1994. *Ion mobility spectrometry.* Boca Raton, FL: CRC Press.
2. Becker, C., K. Qian and D. H. Russell. 2008. Molecular weight distributions of asphaltenes and deasphaltened oils studied by laser desorption ionization and ion mobility mass spectrometry. *Anal. Chem.* 80:8592–7.

3. Bohrer, B. C., S. I. Merenbloom, S. L. Koeniger, A. E. Hilderbrand and D. E. Clemmer. 2008. Biomolecule analysis by ion mobility spectrometry. *Annu. Rev. Anal. Chem.* 1:293–327.

4. Ruotolo, B. T., S.-J. Hyung, P. M. Robinson, et al. 2007. Ion mobility-mass spectrometry reveals long-lived, unfolded intermediates in the dissociation of protein complexes13. *Angew. Chem. Int. Ed.* 46:8001–4.

5. Bernstein, S. L., N. F. Dupuis, N. D. Lazo, et al. 2009. Amyloid-β^2 protein oligomerization and the importance of tetramers and dodecamers in the aetiology of Alzheimer's disease. *Nat. Chem.* 1:326–31.

6. Griffin, G. W., I. Dzidic, D. I. Carroll, R. N. Stillwell and E. C. Horning. 2002. Ion mass assignments based on mobility measurements. Validity of plasma chromatographic mass mobility correlations. *Anal. Chem.* 45:1204–9.

7. Carr, T. W. 1984. *Plasma chromatography.* New York: Plenum Press.

8. Creaser, C. S., J. R. Griffiths, C. J. Bramwell, et al. 2004. Ion mobility spectrometry: A review. Part 1. Structural analysis by mobility measurement. *Analyst* 129:984–94.

9. Young, C. E., D. Edelson and W. E. Falconer. 1970. Water cluster ions: Rates of formation and decomposition of hydrates of the hydronium ion. *J. Chem. Phys.* 53:4295–302.

10. Davis, E. J., P. Dwivedi, M. Tam, W. F. Siems and H. H. Hill. 2009. High-pressure ion mobility spectrometry. *Anal. Chem.* 81:3270–5.

11. Verbeck, G. F., K. J. Gillig and D. H. Russell. 2003. Variable-temperature ion mobility time-of-flight mass spectrometry studies of electronic isomers of kr2+ and CH_3OH^{*+} radical cations. *Eur. J. Mass Spectrom.* 9:579–87.

12. Wyttenbach, T., G. v. Helden, J. J. J. Batka, D. Carlat and M. T. Bowers. 1997. Effect of the long-range potential on ion mobility measurements. *J. Amer. Soc. Mass Spectrom.* 8:275–82.

13. Srebalus, C. A., J. Li, W. S. Marshall and D. E. Clemmer. 1999. Gas-phase separations of electrosprayed peptide libraries. *Anal. Chem.* 71:3918–27.

14. Kemper, P. R., N. F. Dupuis and M. T. Bowers. 2009. A new, higher resolution, ion mobility mass spectrometer. *Int. J. Mass Spectrom.* 287:46–57.

15. Shvartsburg, A. A. 2009. *Differential mobility spectrometry: Nonlinear ion transport and fundamentals of FAIMS.* Boca Raton, FL: CRC Press.

16. Gillig, K. J., B. T. Ruotolo, E. G. Stone and D. H. Russell. 2004. An electrostatic focusing ion guide for ion mobility–mass spectrometry. *Int. J. Mass Spectrom.* 239:43–9.

17. Mason, E. A. and H. W. Schamp. 1958. Mobility of gaseous ions in weak electric fields. *Ann. Phys.* 4:233–70.

18. Fernandez-Lima, F. A., C. Becker, K. J. Gillig, et al. 2008. Ion mobility–mass spectrometer interface for collisional activation of mobility separated ions. *Anal. Chem.* 81:618–24.

19. Becker, C., F. A. Fernandez-Lima, K. J. Gillig, et al. 2009. A novel approach to collision-induced dissociation (CID) for ion mobility–mass spectrometry experiments. *J. Amer. Soc. Mass Spectrom.* 20:907–14.

20. Su, T., E. C. F. Su and M. T. Bowers. 1978. Theory of ion-molecule collisions: Effect of the induced dipole–induced dipole potential on ion-molecule capture rate constants. *Int. J. Mass Spectrom. Ion Phys.* 28:285–8.

21. Dougherty, R. C. 2001. Molecular orbital treatment of gas-phase ion molecule collision rates: Reactive and nonreactive collisions. *Mass Spectrom. Rev.* 20:142–52.

22. Ellis, H. W., R. Y. Pai, E. W. McDaniel, E. A. Mason and L. A. Viehland. 1976. Transport properties of gaseous ions over a wide energy range. *At. Data Nucl. Data Tables* 17:177–210.

23. Ellis, H. W., E. W. McDaniel, D. L. Albritton, et al. 1978. Transport properties of gaseous ions over a wide energy range. Part II. *At. Data Nucl. Data Tables* 22:179–217.

24. Ellis, H. W., M. G. Thackston, E. W. McDaniel and E. A. Mason. 1984. Transport properties of gaseous ions over a wide energy range. Part III. *At. Data Nucl. Data Tables* 31:113–51.

25. Viehland, L. A. and E. A. Mason. 1995. Transport properties of gaseous ions over a wide energy range, IV. *At. Data Nucl. Data Tables* 60:37–95.

26. Kemper, P. R. and M. T. Bowers. 1991. Electronic-state chromatography: Application to first-row transition-metal ions. *J. Phys. Chem.* 95:5134–46.

27. Nier, A. O. 1940. A mass spectrometer for routine isotope abundance measurements. *Rev. Sci. Inst.* 11:212–6.

28. Krutchinsky, A. N., I. V. Chernushevich, V. L. Spicer, W. Ens and K. G. Standing. 1998. Collisional damping interface for an electrospray ionization time-of-flight mass spectrometer. *J. Amer. Soc. Mass Spectrom.* 9:569–79.

29. Guilhaus, M., D. Selby and V. Mlynski. 2000. Orthogonal acceleration time-of-flight mass spectrometry. *Mass Spectrom. Rev.* 19:65–107.

30. We observe a complete hydration series for the Ti^{2+} system ($Ti^{2+} + [H_2O]_n$, where n = 1–4, unpublished data). These hydrated ions increase in relative abundance as the temperature is lowered, then disappear from the spectrum when the temperature of the drift gas is reduced below the freezing point of water, 273 K.

7 Multiplexed Ion Mobility Spectrometry and Ion Mobility– Mass Spectrometry

Glenn A. Harris, Mark Kwasnik,
and Facundo M. Fernández[*]

CONTENTS

7.1 INTRODUCTION

The drive to resolve and characterize samples of ever increasing complexity motivates the development and use of higher-throughput, more reproducible and sensitive analytical techniques. Chromatographic techniques such as capillary electrophoresis (CE), gas chromatography (GC), and high-performance liquid chromatography (HPLC) all have contributed to the continued evolution of separation methods for complex mixture analysis. More often than not, these separation techniques are coupled to a mass spectrometer for the sensitive quantitation and/or identification of chemical species of interest. In these cases, separation occurs pre-ionization in a time scale ranging from minutes to hours, which may lead to a throughput bottleneck. Speeding up the pre-ionization separation can result in loss of temporal separation of the eluted compounds. Conversely, increasing complexity of the mixture under study forces the analyst to choose separation or mass analysis approaches with higher peak capacities. In most scenarios, this choice is accompanied by an increase in the initial acquisition cost of the necessary instrumentation.

---

[*] These authors contributed equally to this work.

153

The formation of an intact gaseous ion from a neutral analyte eluting from a separation stream is a necessary step for mass spectrometric analysis. This is a delicate process that competes with fragmentation, ion-molecule reactions, ion-ion recombination, and/or neutralization. Pre-ionization separation simplifies these processes to some extent, limiting competitive reactions to a manageable degree. However, isomeric ions (species with the same molecular formula but different structure) and ions with accurate mass differences smaller than the mass spectrometric detector resolving power can be difficult, if not impossible, to resolve, even with highly selective stationary phases or by energy-resolved tandem mass spectrometry (MS/MS) experiments. Further separation of the formed ions in the gas phase prior to MS analysis has been shown to improve peak capacity without a noticeable effect on analysis throughput.

Ion mobility spectrometry (IMS) is a gas-phase ion separation technique that focuses on the resolution of ionized chemical species by their differential behavior or velocity in an electric field applied under atmospheric or reduced (1–10 Torr) pressure. Originally termed plasma chromatography, IMS and ion mobility–mass spectrometry (IM-MS) have made steady improvements, significantly contributing to the addressing of peak capacity and signal-to-noise ratio (SNR) challenges involved in the analysis of complex mixtures. Most ion mobility instruments utilize a pseudo-continuous or continuous ionization source such as corona discharge, electrospray ionization (ESI), or ⁶³Ni chemical ionization. Utilization of this stream is maximum in high-duty cycle ion mobility approaches, such as differential mobility spectrometry (DMS), also known as field-asymmetric ion mobility spectrometry (FAIMS).

Pulsed IMS approaches utilizing a continuous ion beam, such as drift tube IMS (DTIMS), require the creation of an ion packet. The most common approach for pulsing an ion beam in DTIMS is by the use of an interleaved Bradbury-Nielsen ion gate (BNG) or a Tyndall ion gate, creating discrete packets of ions into the drift region of the instrument where separation occurs. Unfortunately, DTIMS operated in the conventional signal averaging (SA) mode suffers from the drawback that only when the closing electric potential supplied to the ion gate is shut off (i.e., the ion gate is open) will ions be separated and detected. All other times when closing potential is applied to the gate (i.e., ion gate is closed), the continuous stream of ions is neutralized against the ion gate wires. The interval by which the ion gate is pulsed opened is called the injection time with typical values of 25–400 μs under atmospheric pressure (AP) conditions. The time it takes the species in the ion packet to travel down the drift tube under AP conditions (drift or arrival time) is typically in the range of 25–100 ms. Ion gating events are created in a periodic fashion, following a master clocking period larger than the maximum drift time to avoid overlap of ions in successive ion packets. Therefore, the theoretical duty cycle (the ratio of the injection time to the master period) for AP-SA DTIMS experiments is of the order of ~0.4–1%. In other words, 99% or more of ions reaching the ion gate are not analyzed. The low duty cycle of SA DTIMS experiments severely limits the overall sensitivity of the analysis, forcing the user to increase the averaging time. Effectively, this issue can translate to improper matching between the time scales of the pre- and post-separation processes, forcing the acquisition to slow down the former for improving detectability of trace analytes. Failure to do so would result in under-sampling of the stream eluting into the ion source and aliasing of its inherent chemical information. Increases in the ion injection time, although beneficial in terms of sensitivity, simultaneously decrease resolving power, and are thus not considered a satisfactory alternative in practical terms.[1]

Efforts to increase sensitivity and improve resolving power of DTIMS experiments have focused on instrumental improvements such as lengthening the drift distance, lowering drift gas temperature, increasing electric field homogeneity, ion trapping at reduced pressure, improving ion transmission into the drift cell, and increasing ionization efficiency prior to ion injection. These methods are discussed thoroughly in this book. An alternative approach to improve on traditional SA IMS and SA IM-MS analysis focuses on how duty cycle can be improved with only minor instrumental modifications to the ion pulsing and detection scheme, by using a multiplexed ion gating procedure.

In the context of analytical instrumentation, multiplexing is the process of overlaying multiple sources of information coherently along a single channel to increase the overall signal quality without a change in sampling time. Specifically, multiplexing methods in spectroscopy and spectrometry allow multiple experimental trials to be run *simultaneously*, thereby increasing throughput. In multiplexed DTIMS, multiple packets of ions are successively injected into the drift tube instead of just one packet per master clocking period. By knowing the ion injection time, the sequence of gating events during which the gate was opened or closed, and the frequency at which the ions' arrival times or mass-to-charge ratios were monitored, one can deconvolute the signals measured at the detector to generate mobility or mass spectra. As a result, a greater total percentage of ions are available to contribute to the measured signal resulting in an enhanced duty cycle.

This chapter aims to serve as an introductory tutorial to multiplexed IMS and IM-MS. A glimpse into weighing designs and mathematical transform-based methods of multiplexing is first provided. Subsequently, a synopsis of the implementation of multiplexing approaches in other physical and chemical analytical techniques serves us as a stepping stone for our look into multiplexed IMS. The duty cycle–resolving power conundrum in DTIMS will be discussed, including a description of reduced pressure IMS trapping methods as a promising avenue to solve it. Finally, current developments in the field of multiplexed DTIMS and DTIM-MS will be reviewed.

7.2 MULTIPLEXING OF ANALYTICAL INSTRUMENTATION: WEIGHING DESIGNS

A problem of all measurement-based fields is obtaining reliable, consistent, and accurate results of whatever variable is of interest. The goal is to perform reproducible measurements with as few interfering factors as possible. A common metaphor would be the weighing of apples at a grocery store using a single pan scale. Imagine wanting to know the weight of three apples. The most straightforward way to do this is to weigh the group of apples at once to obtain the weight of the group. Although this method is time efficient, there is only one measurement, thus it will not provide the most reliable of results since it is based on only one trial.

An improvement to this method would be to weigh each apple separately and sum the weights of all three. Although direct and simple, each step in this type of experiment could easily fall victim to any uncertainty like a fly landing on an apple during the measurement or leftover water found on the bottom of the measuring pan from a previous weighing of freshly cleaned produce. Because each apple is measured only once, this design has no fail-safe way of ensuring the data generated is not biased. Random noise, imprecision, instrument or environmental background, etc., all contribute to inaccurate measurement results.

How, then, can we improve the confidence in the obtained mass of the apples without spending any extra time at the grocery store? By utilizing a multiplexed method of placing two apples at a time into the scale in the three possible configurations (apples 1 and 2, apples 2 and 3, and apples 1 and 3) and subtracting the mass of one pair from the sum of the other two will result in the weight of the individual apple common to the second and third pairs. All three masses of apples can then be obtained in a similar way and summed to yield the total mass of the three apples. With this experiment no additional time is spent weighing the apples versus weighing each apple independently, since the total number of experiments is still three. Why is this experimental design more accurate? In comparison to our "one-apple-at-a-time" experiment, the calculated SNR gain obtained with this multiplexed approach increases with $(n + 1)/2n^{1/2}$, where n is the number of apples investigated.[2] By multiplexing the measurements, each apple is weighed twice, allowing for a realized gain in the SNR of 15% over weighing the apples separately.

The use of two weighing pans (equivalent to detectors in a spectrometer) could increase our SNR gains even further. Instead of weighing the apples in a single pan scale they could be weighed in a two-pan scale with two apples placed on one pan while one is placed on the other. A similar method is followed to solve for the masses of each apple; however, by using two pans each apple is weighed in every trial since there are three possible combinations of weighing experiments. Therefore, the realized SNR gain in a dual-pan scale is 40%.[2,3]

Making the leap from grocery shopping to precise analytical measurements may seem far-fetched, yet the principle of multiplexing remains the same. Multiplexing in spectroscopy involves encoding multiple independent signals so that combinations of them can be transmitted simultaneously in a controlled way, without increasing the total analysis time. Since the way the signals were initially combined (encoded) is known, the readout produced by the detector can be deconvoluted at the end of the experiment into the individual original signal components, each enhanced by SNR gains.

Both single detector (one pan scale) and multiple detector (two pan scale) multiplexed spectroscopic experiments are possible, all resulting in improved performance[4]; however, only single detector arrangements will be discussed here as they are more closely related to multiplexed IMS. One single detector approach for encoding spectroscopic information, known as frequency division multiplexing, allows for multiple wavelengths of light to strike the spectrometer's detector simultaneously. By modulating each wavelength of interest with a different "carrier" frequency, individual components of the signal can be interpreted by electronic means using tuned amplifiers. Interference from crosstalk of the modulating frequencies can be a problem if they are not sufficiently different from one another. Another approach, time-division multiplexing, successively arranges incoming wavelengths to be detected for a fraction of the total scan time.[4] In other words, a particular wavelength is always being detected at a given time during the scan. It is susceptible to instability in the spectral sweep caused by instrumental or sample-based noise that is compounded by the short scan time at any particular wavelength. Frequency-division multiplexing has been used with flame atomic fluorescence spectrometry,[5] fluorometry,[6] diode laser spectroscopy,[7,8] reflectometry,[9] radiography,[10] tomography,[11,12] Fourier transform profilometry,[13] and fiber-optic sensors.[14] Time-division multiplexing has been implemented in three-dimensional integral imaging,[15,16] x-ray microcalorimetry,[17] gas[18] and fiber-optic sensors,[19] various rapid scanning spectrometers,[20] and monochromators[21] for atomic spectrometry.

Transform-based multiplexing, also a single detector approach, allows multiple spectral elements to be recorded simultaneously without loss in selectivity. Monitoring multiplexed wavelength combinations instead of single channels can, in principle, yield improvements in SNR and spectral line shapes, aiding in increasing the spectral acquisition rate at constant SNR.[4,22] Based on Fourier transform (FT) and Hadamard transform (HT) mathematics, these methods rely less on physical instrument modifications than frequency- and time-division multiplexing approaches.

In FT methods, the detection of multiple frequencies is performed in the time domain, e.g., the free induction decay in a nuclear magnetic resonance (NMR) experiment. By simultaneously measuring several spectral elements, a multiplex advantage, also called Fellgett's advantage, is obtained.[23] The data digitized in time can be connected mathematically via the fast FT (FFT) to the frequency domain. This leads to the additional possibility of enhancing the analytical information by removing unwanted components in the frequency spectrum. Apodization, digital filtering, baseline smoothing, and resolution enhancement are all possible.[4,22] Examples of techniques that make use of FT multiplexing include ion cyclotron resonance MS,[24] IMS,[25] quadrupole resonance,[26] dielectric,[27] microwave,[28] NMR,[29] ultraviolet-visible spectroscopy (UV/Vis),[30] and infrared (IR) spectroscopy.[31] As with FT methods, transform methods based on Hadamard mathematics have also proven to be powerful, but because of their particular importance to the field of IMS we describe them in detail in the next section.

7.3   HADAMARD MULTIPLEXING AND THE FELLGETT ADVANTAGE

Traditionally, DTIMS is performed in the SA mode. The BNG is quickly pulsed open, delivering a small packet of ions to migrate within the drift region until they are detected (Figure 7.1). The total time for the experiment between gating pulses (i.e., the master period) is often referred to as the sweep or acquisition time. After a given sweep is completed, the BNG is pulsed again followed by detection during the drift time. This process is repeated many times until a sufficient signal is acquired, minimizing noise and maximizing signal intensity. The drawback of this operational mode is that each sweep suffers from low duty cycle. In the best-case scenario, the total ion utilization approaches 1%.

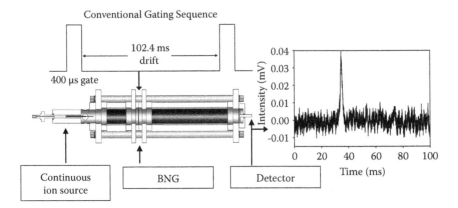

FIGURE 7.1 Conventional AP-SA DTIMS gating sequence and acquired spectrum. The gating sequence is shown with typical values, but these are generally user selectable.

As a way to improve the throughput of the experiment and SNR of SA IMS, an HT-based multiplexing experiment can be utilized. HT multiplexing uses a binary code arranged in a matrix to determine which combination of ions with different mobilities is present at any given time within the drift tube. In other words, the code used for gating represents successive opening and closing events of the BNG. A mobility spectrum (ion current versus drift time in a stand-alone DTIMS) is produced by the deconvolution of the acquired multiplexed data points by the application of the fast HT, often resulting in gains approaching the theoretical multiplex or Fellgett's advantage.[23] The binary code is derived from a Hadamard matrix (H_n) composed of ones (1) and minus ones (-1) of size n such that the dimensions of the matrix are $n \times n$. Hadamard matrices must have their scalar product of any two different rows equal to zero such that:

$$H_n H_n^T = n I_n \tag{7.1}$$

where the superscript "T" denotes the transpose, and I_n is the $n \times n$ identity matrix. [32] In HT-AP DTIMS, the spectral elements are either present or omitted, reflecting when the BNG is open or closed. Therefore, the experimental design is better represented by a Simplex matrix (S_{n-1}) created by replacing the ones of the Hadamard matrix with zeros (0) and minus ones with ones, correlating to the opening (1) and closing (0) events of the BNG. Further modification of the Hadamard matrix involves deleting the first row and column of the matrix (by definition all ones) for the creation of the S_{n-1} matrix of order $n - 1$. Simplex matrices must satisfy[32]:

$$S_n S_n^T = \frac{1}{4}(n+1)(I_n + J_n) \tag{7.2}$$

$$S_n J_n = J_n S_n = \frac{1}{2}(n+1)J_n \tag{7.3}$$

$$S_n^{-1} = \frac{2}{n+1}(2S_n^T - J_n) \tag{7.4}$$

where J_n is the $n \times n$ matrix of ones. Making a Simplex matrix practical for use involves making it cyclic. In replacement of several different n sequences to apply to the BNG, one long coding sequence can be generated with the reading frame of the experiment shifting down along the series following the length of $2n - 1$. For example, if we chose a reading frame for the conventional SA sweep to be five bits (units) long it would then resemble:

```
1 0 0 0 0
```

If there were five total runs in the entire SA experiment, this experiment would be represented as:

```
1 0 0 0 0
1 0 0 0 0
1 0 0 0 0
1 0 0 0 0
1 0 0 0 0
```

When applied to the BNG, the sequence would look like:

```
1 0 0 0 0 1 0 0 0 0 1 0 0 0 0 1 0 0 0 0 1 0 0 0 0
```

A corresponding 5×5 Simplex matrix used for an HT multiplexed experiment could resemble:

```
1 0 1 0 1
0 1 0 1 1
1 0 1 1 0
0 1 1 0 1
1 1 0 1 0
```

The applied sequence to the BNG has length $2n - 1 = 9$ so that the final zero in the gating sequence is omitted. The actual experiment is based on the following sequence of gating events:

```
1 0 1 0 1 0 1 0 1 1 1 0 1 1 0 0 1 1 0 1 1 1 0 1
```

One can now start to see the advantage of the HT multiplexing technique. The increase in the number of open gate events (the ones in the sequence) results in more ions gated into the drift tube (Figure 7.2). In SA mode, the only avenue to increase the number of ions is to increase the duration of the experiment (i.e., the averaging time). In the HT multiplexed experiment, ions overlap with one another within the drift tube while traveling towards the detector. At any given time, the detector does not "see" ions

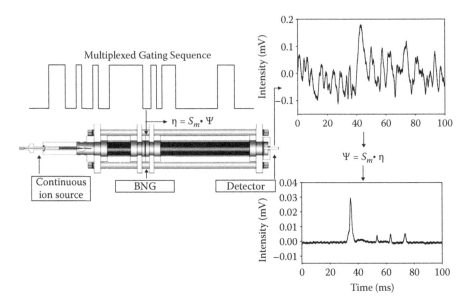

FIGURE 7.2   A representative multiplexed DTIMS gating sequence. The ion beam is modulated using a pseudo-random sequence (PRS) derived from a Simplex matrix, which is applied to the BNG. This is mathematically expressed as a multiplication of the original ion mobility spectra (Ψ) by the Simplex matrix (S_m). The multiplexed spectrum (η) is collected at the detector. The deconvoluted data ($\hat{\Psi}$) is recovered by multiplication of the multiplexed spectrum (η) with the inverse of the Simplex matrix (S_m^{-1}).

of a single ionic mobility, but predetermined combinations of ions of various mobilities. This is in general counterintuitive, but is equivalent to a multiplexed optical spectrometer where photons of several wavelengths strike the photomultiplier at the same time. The key point is that there is a priori knowledge of the encoded sequence for the ion injections that is used to deconvolute the acquired signal. Implementing a coding sequence with an equal number of ones and zeros, we realize a 50% duty cycle without added instrument complexity, such as an additional ion gate as in FT-DTIMS.[25,33]

It is important to remember that HT multiplexed instrumentation was initially conceived for optical spectroscopy where the detected photons travel orders of magnitude faster than the fastest modulation sequence. Unfortunately, ions traveling down a drift tube, especially at atmospheric pressure, do not travel anywhere near the speed of light. As a result, any processes that change the shape of the ion packet from a "perfect" square pulse to a more diffuse ion cloud during the time frame of the experiment will result in less-than-ideal multiplexing gains. Space-charge and thermal diffusion along the radial and longitudinal axes of the drift tube are known to occur[34] in addition to ringing of the square waveform applied to the ion gate[35] and ion depletion in the region between the BNG and the ion source.[1] Deviations from ideal square ion packets decrease SNR gains and create "ghost" or "echo" peaks, complicating spectral identification.[36]

In addition to DTIMS,[1,37,38] HT multiplexing has been applied to infrared spectrometry,[39] nuclear magnetic resonance,[40] Raman[41] and fluorescence microscopy,[42] capillary electrophoresis,[43] and gas and liquid chromatography MS.[44] Perhaps most related to HT-DTIMS is HT-TOFMS (time-of-flight MS). The increased attention towards broad-range molecular mass analysis has caused greater use of TOFMS due to its excellent resolving power, practically unlimited mass range, high ion transmission, and high spectral acquisition rate.[45] To further improve both sensitivity and speed of TOFMS the Zare group developed a linear extraction HT-TOFMS.[3,35,36,45–48] In its original implementation, duty cycle improvements of up to 50% and possible SNR gains of 45–64 for single detector configurations could be realized.[3,45] Further development of a dual anode detector setup saw improvements of increased SNR of 41% over a single detector HT-TOF system while reaching the theoretical 100% maximum duty cycle.[48]

A combination of the methodology employed in both HT-IMS and HT-TOFMS could be applied to develop an ion mobility/time-of-flight mass spectrometer equipped with Hadamard-multiplexed ion gates or pulsing elements within the IMS and TOFMS instruments. By simultaneously multiplexing both ion injection events, one could theoretically develop what could be described as an "imaging" IMS-TOF with full encoding of the signal (HT-IMS-HT-TOF MS). Figure 7.3 describes the

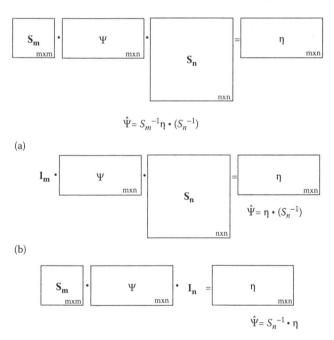

FIGURE 7.3 Example of the encoding and decoding matrix algebra that would be used for an imaging Hadamard spectrometer. Part (a) shows an experiment with a DTIMS operated in SA mode (represented by the $m \times m$ identity matrix, I_m) coupled to a Hadamard TOF (IMS-HT-TOF MS). Part (b) shows an experiment where Hadamard DTIMS is coupled to a conventionally pulsed TOF (HT-IMS-TOF MS, I_n represents an $n \times n$ identity matrix). S_n is the Hadamard matrix used for multiplexing the TOF ion pulsing.

encoding and decoding matrix algebra that would be used for an imaging Hadamard spectrometer and gives examples of both IMS-HT-TOF MS (Figure 7.3a) and HT-IMS-TOF MS (Figure 7.3b). The use of the term "imaging" derives from the nomenclature by Harwit and Sloane[2] used in optical spectroscopies and should not be confused with imaging mass spectrometric techniques used for collecting mass spectra in an x-y plane and assembling those into a molecular image.

7.4   MULTIPLEXING OF ION MOBILITY SPECTROMETRY AND ION MOBILITY–MASS SPECTROMETRY: EXAMPLES IN THE PEER-REVIEWED LITERATURE

In 1985, Knorr and coworkers[25] were the first to address the issue of low SNR in DTIMS caused by the intrinsically low duty cycle associated with SA data acquisition methods. By adding a second BNG mounted at the exit of the drift tube and before the detector, and applying a square wave pulse swept from low to high frequencies between the entrance and exit gates, a duty cycle of 25% was obtained. As the gates were swept, ion populations of a fixed mobility would appear in and out of phase with the gating events, resulting in a mobility "interferogram" that could be deconvoluted via application of an FFT. The SNR gain observed by FT multiplexing of DTIMS was approximately three fold.[33] In 2004, Tarver[10] performed FT-IMS without modification of the spectrometer through the use of an external "electronic" software-based gate to simulate the behavior of a physical gate. Utilizing this software-based approach to "filter" select ions allowed for an improvement in SNR gain of seven. Although FT-IMS did provide significant improvements in SNR, the drawbacks associated with this approach were three-fold: (1) it was sensitive to changes in the efficiency of the ion gate especially at high pulse frequencies and the use of an apodization function was needed to remove both baseline noise and the presence of spectral echoes; (2) the 25% duty cycle was only achieved for ion populations that were in phase with the gate pulses, and not all ions within the drift tube; and (3) in the original approach, an existing instrument had to be modified through the addition of an extra ion gate at the end of the drift tube.

A different multiplexing method was proposed by McLean and Russell[50] utilizing rapid injection of ions into the drift tube at a frequency much faster than could be processed by a stand-alone IMS using one-dimensional time correlation creating a pseudo-continuous ion beam within the drift tube. In this instrumental arrangement, a TOFMS was utilized as a rapid and sensitive detector since both IMS and TOFMS achieve separation based on time dispersion of ions. Correlated multiplexed data acquisition allowed the use of ion injection rates much faster than that predicted to achieve 100% duty cycle in a sequential pulse-and-wait experiment, and improved usage of the bidimensional separation space.

An alternative yet complementary approach to multiplexing the ion injection sequence is to trap and accumulate the ions prior to injection. When working in the low pressure regime (10^{-3}–10^{-1} Torr), ion manipulation methods utilizing both quadrupole and octopole ion traps have been successfully employed not only to accumulate ions prior to reduced pressure IMS analysis, but also to inject or gate the ions into the drift tube.[51–53] These types of low pressure ion trapping methods provide for increased sensitivity associated with higher ion utilization resulting in gains between 60% and 100% depending on the experimental setup. In 2001, Wyttenbach et al.[54] made advances in the manipulation of ions in a medium pressure regime by employing an electrodynamic ion funnel that could be operated in the 0.2-Torr range to efficiently accumulate and transfer ions into the drift tube of an IMS. Over the next several years, improvements in the design and construction of these ion funnel "traps" made possible ion manipulation in the 2–5 Torr pressure range while enhancing ion utilization efficiency by a factor of seven.[55,56] With this type of setup, the drift tube is positioned between two electrodynamic ion funnels. The first funnel accumulates ions for 50–100 ms and injects them in 50-μs pulses while the second funnel collects and focuses the now spatially disperse ion packets as they exit the drift tube. The second funnel then directs ions to an RF ion guide prior to mass analysis by an orthogonal TOF. Although ion efficiency was dramatically improved with this new dual ion funnel setup, the millisecond ion trapping and accumulation time was still disproportionally large when compared to the microsecond time scale of the ion injection events, and thus limited the overall ion utilization efficiency or duty cycle.

In 2007, Belov et al.[57] used the above-mentioned dual ion funnel trap setup and applied Hadamard-based pseudorandom pulsing sequences (PRS) to the ion trap for multiplexing of the ion injection events. PRS-based approaches utilize a binary, 50% duty cycle sequence generated from the first row elements of a Simplex matrix. Through the simultaneous use of ion funnel trapping and PRS-based multiplexing, they were able to increase the duty cycle of their instrument beyond the typical 50% associated with Hadamard-based multiplexing and realized improvements in SNR of up to 10-fold. The drawback of this method was that the ion funnel trap would accumulate ions with different efficiencies dependent on the intensity of the incoming ion beam. To correct for this phenomenon, a sample-specific weighing matrix was needed to accurately deconvolute or reconstruct the data and eliminate spectral echoes and, as such, this multiplexing method could not be universally applied. In 2008, it was proposed that by dynamically modifying the previously used PRS and trapping events to normalize the quantity of injected ions, the need for sample-specific weighing matrices could be eliminated. By employing fixed accumulation times in the ion traps prior to the individual gating events, they were able to improve the robustness of the multiplexing approach. These improvements made this method amenable to a wider range of samples while concurrently achieving a two-fold enhancement in the amount of ion injection events compared to the traditional HT-IMS experiment.[58] The latest advancements in the field of dynamic multiplexing now allow for the analysis of highly complex samples, such as proteolytic digests of blood plasma by IM-MS. Dynamically multiplexed IMS-TOFMS correlates analyzer performance with "brightness" of the ion source and ion funnel trapping function to minimize space charge-induced ion discrimination.[59] Through the use of automated feedback algorithms to select an optimum multiplexing sequence, peptides were detected at a concentration of 1 nM in complex mixtures with an accuracy better than 5 ppm, while providing a three orders of magnitude dynamic range.[60]

Other Hadamard transform-based multiplexing methods applied to stand-alone DTIMS were implemented in 2006 by Clowers et al.[1] and Szumlas et al.[38] using PRS gating, and in 2009 by Kwasnik et al. using arbitrary gating.[37] In all cases, the multiplexing sequence is applied to the two wire sets of a BNG causing the injection of multiple ion packets of varying sizes into the drift tube in rapid succession. These ion packets are separated in the drift tube based on their mobility and are detected as a pseudo-randomly distributed overlay of multiple arrival time distributions that is deconvoluted to extract a conventional spectrum. Deconvolution is achieved either via an FHT algorithm for PRS approaches or a modified transform involving a pseudo-inverse matrix multiplication for arbitrary sequences. As mentioned earlier, Hadamard multiplexing methods can suffer from the presence of spectral errors or echoes that appear as false peaks in the deconvoluted spectrum. Examples of these "echoes" are provided in Figure 7.4. Echo peaks are caused by imperfect modulation, or encoding, of the ion packets by the BNG and will be alternatively referred to as modulation defects. A detailed explanation as to the sources and manifestations of these defects can be found in articles by Hanley[61] and Kimmel et al.[35] In their work, Clowers et al. minimized the contribution of these defects on the deconvoluted spectrum by discarding the portion of the multiplexed ion signal that was most

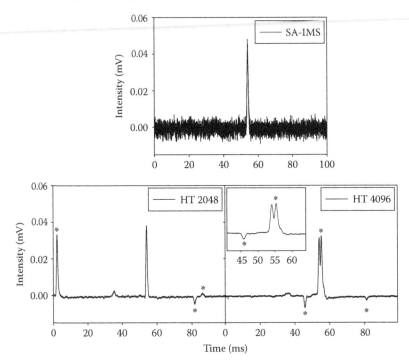

FIGURE 7.4 Examples of spectra collected in conventional SA-DTIMS (top panel) and HT-DTIMS encoded with sequence of varying length: $n = 2048$ (bottom left panel) and $n = 4096$ (bottom right panel). Although sensitivity is improved by 12- and 14-fold, respectively, the Hadamard mobility spectra suffer from spectral defects labeled with an (*) that are not present in SA-DTIMS.

heavily affected by imperfect performance of the BNG. SIMION simulations of the diffusive effects that wire grids have on ions under both atmospheric pressure and vacuum have further illustrated the deleterious influence of BNG ion optics on an incoming ion cloud before, during, and after a given gating event,[62] providing further evidence to understand modulation defects, especially under AP conditions.

"Digital multiplexing" is a term coined by our group that encompasses multiplexing with not only traditional Hadamard PRS, but also user-generated arbitrary binary ion injection waveforms with variable duty cycles ranging from 0.5% to 50%, and "extended" Hadamard sequences.[37] Extended Hadamard sequences are PRS that have had their duty cycle artificially reduced to values between 6.2% and 25% by appending additional "0", or gate-closed events, into the sequence. Digital multiplexing was developed to investigate the performance of both randomly generated nonconventional sequences and the effect of varying the amount of ions within the drift tube as an alternative approach to minimize or eliminate the appearance of spectral echoes in the deconvoluted spectra. By decreasing the frequency of ion injection events in these new digital sequences, the cumulative contributions of factors such as ion depletion prior to gating, columbic interaction of the individual ion packets within the drift tube, and imperfect gating events caused by non-ideal performance of the BNG were minimized. This in turn mitigated the overall impact of imperfect modulation on the deconvoluted spectra and reduced, or in some cases eliminated, spectral echoes. Arbitrary and extended Hadamard sequences produced improvements in SNR gain in the two- to seven-fold range, and conventional Hadamard sequences provided greater increases in SNR, with gains ranging from 9- to 12-fold; however, the mobility spectra suffered from echoes that appeared as false peaks in the deconvoluted spectra.

The above-mentioned multiplexing methods approach the sensitivity versus resolving power conundrum typical to DTIMS experiments from different directions. DTIMS resolving power (R) is defined as the ratio between the drift time (t_d) and the full peak width at half-maximum ($w_{1/2}$) (i.e., $R = t_d/w_{1/2}$). Revercomb and Mason[63] first identified parameters affecting ion pulse width, as measured at the exit of the drift tube. In a first approximation, and assuming negligible contributions from ion-molecule reactions and coulombic repulsions, the pulse width at the detector (Δt) depends on the initial gate pulse width and the contribution of diffusional broadening as the pulse travels down the drift tube according to the following relationship:

$$(\Delta t)^2 = (\Delta t_0)^2 + \left(\frac{16kT \ln 2}{Ve} \right) \frac{t_d^2}{z} \tag{7.5}$$

where Δt_0 is the initial ion pulse width, k is Boltzmann's constant, T is the drift gas temperature, V is the potential difference across the drift tube, e is the elementary charge, and z is the number of charges on the ion. Assuming constant instrumental operating parameters, the most obvious way to increase the resolving power is to decrease the initial ion injection pulse width. From another perspective, an analyte peak width will never be narrower than the initial width of the packet of ions injected into the drift tube; however, as smaller ion packets are injected, there is a dramatic decrease in sensitivity. Multiplexing IMS offers an avenue for circumventing this problem by allowing the injection of narrow ion packets of a higher ionic density, as with ion trapping methods, or at a higher frequency as in PRS and arbitrary gating approaches. In both scenarios, sensitivity is maximized without sacrificing resolving power. With continued utilization and improvements of multiplexed IMS methods in conjunction with the efforts of various research groups tackling instrument improvements and miniaturization, DTIMS and DTIM-MS instruments will extend beyond tools for the advanced research scientist. It is our hope that their use will become routine in everyday places like hospitals, public transportation hubs, food and fuel production plants, and perhaps, one day, our homes.

REFERENCES

1. Clowers, B. H.; Siems, W. F.; Hill, H. H.; Massick, S. M., "Hadamard Transform Ion Mobility Spectrometry", *Anal. Chem.* 2006, *78*, 44–51.
2. Sloane, N. J. A.; Harwit, M. *Hadamard Transform Optics*; Academic: New York, 1979.
3. Zare, R. N.; Fernandez, F. W.; Kimmel, J. R., "Hadamard Transform Time-of-Flight Mass Spectrometry: More Signal, More of the Time", *Angew. Chem.* 2003, *42*, 30–35.
4. Busch, K. W.; Benton, L. D., "Multiplexed Methods in Atomic Spectroscopy", *Anal. Chem.* 1983, *55*, 445A–460A.
5. Chester, T. L.; Winefordner, J. D., "Evaluation of the Analytical Capabilities of Frequency Modulated Sources in Multielement Non-Dispersive Flame Atomic Fluorescence Spectrometry", *Spectrochimica Acta* 1976, *31B*, 21–29.
6. Iwata, T.; Muneshige, A.; Araki, T., "Analysis of Data Obtained From a Frequency-Multiplexed Phase-Modulation Fluorometer Using an Autoregressive Model", *Appl. Spectrosc.* 2007, *61*, 950–955.
7. Dong, C.; Wen-Qing, L.; Yu-Jun, Z.; Jian-Guo, L.; Qing-Nong, W.; Rui-Feng, K.; Min, W.; Yi-Ben, C.; Jiu-Ying, C., "Modulation Frequency Multiplexed Tunable Diode Laser Spectroscopy System for Simultaneous CO, CO_2 Measurements", *Chin. Phys. Lett.* 2006, *23*, 2446–2449.
8. Oh, D. B.; Paige, M. E.; Bomse, D. S., "Frequency Modulation Multiplexing for Simultaneous Detection of Multiple Gases By Use of Wavelength Modulation Spectroscopy With Diode Lasers", *Appl. Opt.* 1998, *37*, 2499–2501.
9. Dave, D. P.; Akkin, T.; Milner, T. E.; Rylander III, H. G., "Phase-Sensitive Frequency-Multiplexed Optical Low-Coherence Reflectometry", *Opt. Commun.* 2001, *193*, 39–43.
10. Zhang, J.; Yang, G.; Lee, Y. Z.; Chang, S.; Lu, J. P.; Zhou, O., "Multiplexing Radiography Using a Carbon Nanotube Based X-Ray Source", *Appl. Phys. Lett.* 2006, *89*, 064106/064101–064106/064103.
11. Al-Qaisi, M. K.; Akkin, T., "Polarization-Sensitive Optical Coherence Tomography Based on Polarization-Maintaining Fibers and Frequency Multiplexing", *Optics Express* 2008, *16*, 13032–13041.
12. Oh, W. Y.; Yun, S. H.; Vakoc, B. J.; Shishkoz, M.; Desjardins; A. E., P.; B. H.; de Boer, J. F.; Tearney, G. J.; Bouma, B. E., "High-Speed Polarization Sensitive Optical Frequency Domain Imaging with Frequency Multiplexing", *Optics Express* 2008, *16*, 1096–1103.
13. Takeda, M.; Gu, Q.; Kinoshita, M.; Takai, H.; Takahashi, Y., "Frequency-Multiplex Fourier-Transform Profilometry: A Single-Shot Three-Dimensional Shape Measurement of Objects with Large Height Discontinuities and/or Surface Isolations", *Appl. Opt.* 1997, *36*, 5347–5354.
14. Liu, T.; Fernando, G. F., "A Frequency Division Multiplexed Low-Finesse Fiber Optic Fabry-Perot Sensor System for Strain and Displacement Measurements", *Rev. Sci. Instrum.* 2000, *71*, 1275–1278.

15. Stern, A.; Javidi, B., "Three-Dimensional Image Sensing and Reconstruction With Time-Division Multiplexed Computational Integral Imaging", *Applied Optica* 2003, *42*, 7036–7042.

16. Stern, A.; Javidi, B., "Information Capacity Gain by Time-Division Multiplexing in Three-Dimensional Integral Imaging", *Opt. Lett.* 2005, *30*, 1135–1137.

17. Doriese, W. B.; Beall, J. A.; Deiker, S.; Duncan, W. D.; Ferreira, L.; Hilton, G. C.; Irwin, K. D.; Reintsema, C. D.; Ullom, J. N.; Vale, L. R.; Xu, Y., "Time-Division Multiplexing of High-Resolution X-Ray Microcalorimeters: Four Pixels and Beyond", *Appl. Phys. Lett.* 2004, *85*, 4762–4764.

18. Yong-Gang, Z.; Zhao-Bing, T.; Xiao-Jun, Z.; Yi, G.; Ai-Zhen, L.; Xiang-Rong, Z.; Yan-Lan, Z.; Sheng, L., "An Innovative Gas Sensor with On-Chip Reference Using Monolithic Twin Laser", *Chin. Phys. Lett.* 2007, *24*, 2839–2841.

19. Cooper, D. J. F.; Smith, P. W. E., "Simple High-Performance Method For Larger-Scale Time Division Multiplexing of Fibre Bragg Grating Sensors", *Meas. Sci. Technol.* 2003, *14*, 965–974.

20. Santini, R. E.; Milano, M. J.; Pardue, H. L., "Rapid Scanning Spectroscopy—Prelude to a New Era in Analytical Spectroscopy", *Anal. Chem.* 1973, *45*, 915A–927A.

21. Blaedel, W. J.; Boyer, S. L., "Programmable Monochromator for Accurate High-Speed Wavelength Isolation", *Anal. Chem.* 1973, *45*, 425–433.

22. Marshall, A. G.; Comisarow, M. B., "Fourier and Hadamard Methods in Spectroscopy", *Anal. Chem.* 1975, *47*, 491A–504A.

23. Fellgett, P., "Theory of Multiplex Interferometric Spectrometry", *J. Physique Radium* 1958, *19*, 187–191.

24. Gross, M. L.; Rempel, D. L., "Fourier Transform Mass Spectrometry", *Science* 1984, *226*, 261–268.

25. Knorr, F. J.; Eatherton, R. L.; Siems, W. F.; Hill, H. H., "Fourier Transform Ion Mobility Spectrometry", *Anal. Chem.* 1985, *57*, 402–406.

26. Lenk, R.; Lucken, E. A. C., "Nuclear Quadrupole Resonance by Fourieir Transfrom of the Free Induction Decay", *Pure Appl. Chem.* 1974, *40*, 199–299.

27. Gestblom, B.; Noreland, E., "Fourier Transform of Dielectric Time Domain Spectroscopy Data", *J. Phys. Chem.* 1976, *80*, 1631–1634.

28. McGurk, J. C., "Transient Emission, Off-Resonant Transient Absorption, and Fourier Transform Microwave Spectroscopy ", *J. Chem. Phys.* 1974, *61*, 3759–3767.

29. Farrar, T. C., "Pulsed and Fourier Transform NMR Spectroscopy", *Anal. Chem.* 1970, *42*, 109A–112A.

30. Glick, M. R.; Jones, B. T.; Smith, B. W.; Winefordner, J. D., "Molecular Absorption Measurements by FT/US-Vis Spectrometry", *Appl. Spectrosc.* 1989, *43*, 342–344.

31. Habib, M. A.; Bockris, J. O. M., "FT-IR Spectrometry for the Solid/Solution Interface", *J. Electroanal. Chem.* 1984, *180*, 287–306.

32. Marshall, A. G. *Fourier, Hadamard, and Hilbert Transforms in Chemistry*; Plenum: New York, 1982.

33. St. Louis, R. H.; Siems, W. F.; Hill, H. H., "Apodization Functions in Fourier-Transform Ion Mobility Spectrometry", *Anal. Chem.* 1992, *64*, 171–177.

34. Mariano, A. V.; Su, W.; Guharay, S. K., "Effect of Space Charge on Resolving Power and Ion Loss in Ion Mobility Spectrometry", *Anal. Chem.* 2009, *81*, 3385–3391.

35. Kimmel, J. R.; Fernandez, F. M.; Zare, R. M., "Effects of Modulation Defects on Hadamard Transform Time-of-Flight Mass Spectrometry (HT-TOFMS)", *J. Am. Soc. Mass Spectrom.* 2003, *14*, 278–286.

36. Hudgens, J. W.; Bergeron, D. E., "A Hadamard Transform Electron Ionization Time-of-Flight Mass Spectrometer", *Rev. Sci. Instrum.* 2008, *79*, 014102-014101–014102-014110.

37. Kwasnik, M.; Caramore, J.; Fernandez, F. M., "Digitally-Multiplexed Nanoelectrospray Ionization Atmospheric Pressure Drift Tube Ion Mobility Spectrometry", *Anal. Chem.* 2009, *81*, 1587–1594.

38. Szumlas, A. W.; Ray, S. J.; Hieftje, G. M., "Hadamard Transform Ion Mobility Spectrometry", *Anal. Chem.* 2006, *78*, 4474–4481.

39. Decker Jr., J. A., "Hadamard-Transform Spectrometry. New Analytical Technique", *Anal. Chem.* 1972, *44*, 127A–134A.

40. Kaiser, R., "Application of the Hadamard Transform to NMR Spectroscopy with Pseudonoise Excitation", *J. Magn. Reson.* 1974, *15*, 44–63.

41. Treado, P. J.; Morris, M. D., "Multichannel Hadamard Transform Raman Microscopy", *Appl. Spectrosc.* 1990, *44*, 1–4.

42. Chen, G.; Mei, E.; Gu, W.; Zeng, X.; Zeng, Y., "Instrument for Hadamrad Transform Three-Dimensional Fluorescence Microscope Image Analysis", *Analytical Chimica Acta* 1995, *300*, 261–267.

43. Kaneta, T.; Yamaguchi, Y.; Imasaka, T., "Hadamard Transform Capillary Electrophoresis", *Anal. Chem.* 1999, *71*, 5444–5446.

44. Lin, C. H.; Kaneta, T.; Chen, H. M.; Chen, W. X.; Chang, H. W.; Liu, J. T., "Applications of Hadamard Transform to Gas Chromatography/Mass Spectrometry and Liquid Chromatography/Mass Spectrometry", *Anal. Chem.* 2008, *80*, 5755–5759.

45. Brock, A.; Rodriguez, N.; Zare, R. N., "Hadamard Transform Time of Flight Mass Spectrometry", *Anal. Chem.* 1998, *70*, 3735–3741.

46. Brock, A.; Rodriguez, N.; Zare, R. N., "Characterization of a Hadamard Transform Time-of-Flight Mass Spectrometer", *Rev. Sci. Instrum.* 2000, *71*, 1306–1318.

47. Fernandez, F. M.; Vadillo, J. M.; Kimmel, J. R.; Wetterhall, M.; Markides, K.; Rodriguez, N.; Zare, R. N., "Hadamard Transform Time-of-Flight Mass Spectrometry: A High-Speed Detector for Capillary-Format Separations", *Anal. Chem.* 2002, *74*, 1611–1617.

48. Trapp, O.; Kimmel, J. R.; Yoon, O. K.; Zuleta, I. A.; Fernandez, F. M.; Zare, R. N., "Continuous Two-Channel Time-of-Flight Mass Spectrometric Detection of Electrosprayed Ions", *Angew. Chem.* 2004, *43*, 6541–6544.

49. Tarver, E. E., "External Second Gate, Fourier Transform Ion Mobility Spectrometry: Parametric Optimization for Detection of Weapons of Mass Destruction.", *Sensors* 2004, *4*, 1–13.

50. McLean, J. A.; Russell, D. H., "Multiplex Data Acquisition Based on Analyte Dispersion in Two Dimensions: More Signal More of the Time", *Int. J. Ion Mobility Spectrom.* 2005, *8*, 66–71.

51. Henderson, S. C.; Valentine, S. J.; Counterman, A. E.; Clemmer, D. E., "ESI/Ion Trap/Ion Mobility/Time-of-Flight Mass Spectrometry for Rapid and Sensitive Analysis of Biomolecular Mixtures", *Anal. Chem.* 1999, *71*, 291–301.

52. Myung, S.; Lee, Y. J.; Moon, M. H.; Taraszka, J. A.; Sowell, R.; Koeniger, S.; Hilderbrand, A. E.; Valentine, S. J.; Cherbas, L.; Cherbas, P.; Kaufmann, T. C.; Miller, D. F.; Mechref, Y.; Novotny, M. V.; Ewing, M. A.; Sporleder, C. R.; Clemmer, D. E., "Development of High-Sensitivity Ion Trap Ion Mobility Spectrometry Time-of-Flight Techniques: A High-Throughput Nano-LC-IMS-TOF Separation of Peptides Arising from a Drosophila Protein Extract", *Anal. Chem.* 2003, *75*, 5137–5145.

53. Creaser, C. S.; Benyezzar, M.; Griffiths, J. R.; Stygall, J. W., "A Tandem Ion Trap/Ion Mobility Spectrometer", *Anal. Chem.* 2000, *72*, 2724–2729.

54. Wyttenbach, T.; Kemper, P. R.; Bowers, M. T., "Design of a New Electrospray Ion Mobility Mass Spectrometer", *Int. J. Mass Spectrom.* 2001, *212*, 13–23.

55. Clowers, B. H.; Ibrahim, Y. M.; Prior, D. C.; Danielson, W. F.; Belov, M. E.; Smith, R. D., "Enhanced Ion Utilization Efficiency Using an Electrodynamic Ion Funnel Trap as an Injection Mechanism for Ion Mobility Spectrometry", *Anal. Chem.* 2008, *80*, 612–623.

56. Tang, K.; Shvartsburg, A. A.; Lee, H. N.; Prior, D. C.; Buschbach, M. A.; Li, F. M.; Tolmachev, A. V.; Anderson, G. A.; Smith, R. D., "High-Sensitivity Ion Mobility Spectrometry/Mass Spectrometry Using Electrodynamic Ion Funnel Interfaces", *Anal. Chem.* 2005, *77*, 3330–3339.

57. Belov, M. E.; Buschbach, M. A.; Prior, D. C.; Tang, K. Q.; Smith, R. D., "Multiplexed Ion Mobility Spectrometry-Orthogonal Time-of-Flight Mass Spectrometry", *Anal. Chem.* 2007, *79*, 2451–2462.

58. Clowers, B. H.; Belov, M. E.; Prior, D. C.; William, F. D.; Ibrahim, Y.; Smith, R. D., "Pseudorandom Sequence Modifications for Ion Mobility Orthogonal Time-of-Flight Mass Spectrometry", *Anal. Chem.* 2008, *80*, 2464–2473.

59. Belov, M. E.; Nikolaev, E. N.; Harkewicz, R.; Masselon, C. D.; Alving, K.; Smith, R. D., "Ion Discrimination During Ion Accumulation in a Quadrupole Interface External to a Fourier Transform Ion Cyclotron Resonance Mass Spectrometer", *Int. J. Mass Spectrom.* 2001, *208*, 205–225.

60. Belov, M. E.; Clowers, B. H.; Prior, D. C.; Danielson, W. F.; Liyu, A. V.; Petritis, B. O.; Smith, R. D., "Dynamically Multiplexed Ion Mobility Time-of-Flight Mass Spectrometry", *Anal. Chem.* 2008, *80*, 5873–5883.

61. Hanley, Q. S., "Masking, Photobleaching, and Spreading Effects in Hadamard Transform Imaging and Spectroscopy Systems", *Appl. Spectrosc.* 2001, *55*, 318–330.

62. Dahl, D. A.; McJunkin, T. R.; Scott, J. R., "Comparison of Ion Trajectories in Vacuum and Viscous Environments Using SIMION: Insights for Instrument Design", *Int. J. Mass Spectrom.* 2007, *266*, 156–165.

63. Revercomb, H. E.; Mason, E. A., "Theory of Plasma Chromatography Gaseous Electrophoresis—Review", *Anal. Chem.* 1975, *47*, 970–983.

8 IMS/MS Applied to Direct Ionization Using the Atmospheric Solids Analysis Probe Method

Charles N. McEwen, Hilary Major, Martin Green, Kevin Giles, and Sarah Trimpin

CONTENTS

8.1 INTRODUCTION

Numerous ambient direct ionization methods have been introduced for use with mass spectrometry over the last several years.[1–20] A major advantage of these methods is speed of analysis, which is achieved not only by the fast insertion and ionization of the sample, but by the elimination of most sample preparation and chromatographic separations. However, this presents a problem in materials analysis and for mixtures in general because of the complexity of the mass spectra that result from direct analysis of complex mixtures. The atmospheric solids analysis probe (ASAP)[2,21] mass spectrometry (MS) method offers some separation related to volatility by control of the heated gas used to effect vaporization, but this is not sufficient for many mixtures. Ion mobility spectrometry (IMS) offers rapid gas-phase separation of ions based on differences in charge state and collision cross section (CCS) (size/shape). Here we explore the utility of a commercial IMS/MS instrument with ASAP sample introduction for analysis of complex mixtures.

The ASAP method allows samples to be introduced into an atmospheric pressure (AP) ion source for rapid direct analysis using MS.[22] Sample vaporization is effected by using a heated nitrogen gas stream. The nitrogen gas and suitable power supplies to heat the gas stream are generally available on commercial mass spectrometers fitted with electrospray ionization (ESI) and/or atmospheric pressure ionization sources. Ionization is by corona discharge and thus is similar to the method initially described by Horning et al.[23,24] In fact, Horning's group introduced direct AP analysis in the 1970s, but sample introduction was from a heated platinum wire introduced into a nitrogen gas stream directed into the AP ionization chamber.[25] Heating the sample with a nitrogen gas stream has several advantages in addition to cost and simplicity. The nitrogen gas provides a dry and inert atmosphere for more reproducible results, and because the sample is heated from the exterior less thermal fragmentation is produced than heating from a hot metal surface. Samples can therefore be loaded onto a cool substrate and introduced into the gas stream at an elevated temperature, resulting in ballistic heating of the samples. This results in rapid vaporization of the sample components with the result that mass spectra can be obtained in a few seconds even from complex samples.

As noted above, temperature ramping capabilities offer only moderate separation capability according to compound volatility, but at the expense of speed of analysis, and the method is not adequate for complex mixtures. Ion mobility spectrometry/mass spectrometry (IMS/MS) offers two highly orthogonal dimensions of rapid separation allowing decongestion of complexity and is thus well suited for direct ionization methods as was shown by Weston et al.[26] in their study on the use of desorption electrospray ionization (DESI) combined with IMS/MS for the analysis of pharmaceutical formulations.

8.2 INSTRUMENTATION

All the data reported here were acquired on a hybrid quadrupole/ion mobility/orthogonal acceleration time-of-flight (oa-TOF) instrument (Synapt G2 HDMS, Waters Corp., Milford, MA). A schematic of the instrument is shown in Figure 8.1. Ions generated in the atmospheric pressure source enter the vacuum system and pass through an ion guide and quadrupole mass filter to the IMS section of the instrument, which comprises three ion guides.

The trap ion guide (Figure 8.1) accumulates ions while in the center ion guide mobility separation occurs from the previous packet of trap ions. Thus the sensitivity is not compromised by the time required to achieve mobility separation. The transfer ion guide delivers the mobility separated ions to the oa-TOF mass analyzer. Ion arrival time (or drift time) distributions are recorded by synchronizing the oa-TOF mass spectral acquisitions with the gated release of ions from the trap device. The mobility separator uses a repeating train of DC pulses (or travelling wave) to propel ions through the gas-filled cell in a mobility-dependent manner. An example of the mobility separation achievable for small molecules is shown in Figure 8.2 for the structural isomers 1,2- and 1,3-dinitrobenzene. Because these species are isomeric they cannot be separated by mass spectrometry alone.

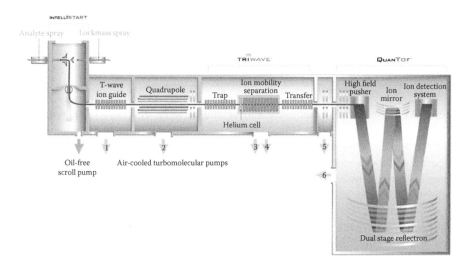

FIGURE 8.1 Schematic of the Synapt G2 HDMS mass spectrometer. (With permission of Waters Corporation.)

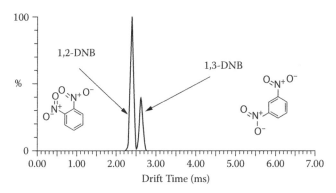

FIGURE 8.2 Ion mobility separation of 1,2- and 1,3-dinitrobenzene.

8.3 ANALYSIS OF OILS

Crude oils represent some of the most complex mixtures, with some samples having on the order of 20,000 to 50,000 components. One way of analyzing these complex mixtures is to use a mass spectrometer with very high resolution. FTMS (Fourier transform MS) instruments are able to resolve all ions except isomers in a crude oil mixture, at least in the mass range below mass-to-charge ratio (m/z) 1000. [27,28] Since resolution in FTMS decreases with increasing mass, the magnetic field strength needed to resolve higher mass components also increases. Thus for some crude oils it may take a 14 Tesla magnet to resolve all the ions at m/z 1000 and 18 Tesla at m/z 1500. Ultimately mass resolution alone is incapable of resolving the many isomeric species present in these mixtures. The current state of the art IMS/MS instrumentation does not have the resolving power in either the mass or IMS dimensions to completely separate such a mixture; nevertheless, the combination of IMS with MS can help to characterize complex mixtures such as crude oils. Recent work used IMS/MS and FTMS to look at crude oil classes.[29] IMS/MS was able to show that higher mass components took on a more compact structure.

Figure 8.3 is the mass spectrum of a crude oil obtained on the Synapt G2 using the ASAP method. The inset is the m/z region between 413 and 415 showing ca. 20,000 mass resolution (FWHM). The presence of peak shoulders demonstrates that this mixture contains many components that are not completely mass resolved. Figure 8.4 shows a plot of m/z against drift time. The raw data were peak detected using a multidimensional peak detection algorithm. The dots represent a single drift time (t_d), m/z, and intensity value for each species. A threshold has been applied to remove the weakest signals and simplify the data. The series of diagonal dots running from top right to bottom left represent ions that differ by one carbon atom, usually CH_2. This is best shown by the inset in Figure 8.4. Thus the IMS separates these homologous series. The extracted mobility drift times for peaks at around m/z 414 shown in Figure 8.3 (inset) are displayed in Figure 8.5 using a 5-mDa detection window. Each of the isobaric peaks has a slightly different extracted t_d value. Rapid fingerprinting of oil using IMS/MS to determine, for example, the source may be possible using a snapshot approach similar to that previously reported for polymers.[30] Such an approach provides a nearly instantaneous picture allowing small differences to be rapidly visualized.

Because the ionization mechanism for ASAP is a chemical ionization process there is the potential to generate analyte ions by either charge transfer, in the case of a dry nitrogen atmosphere in the source, or by proton transfer if there is residual

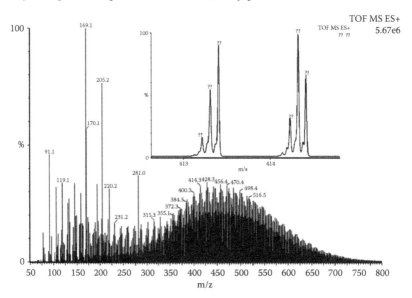

FIGURE 8.3 ASAP MS analysis of crude oil with an expansion of the region around the m/z 413–415 (inset) showing a resolution of 20,000 FWHM.

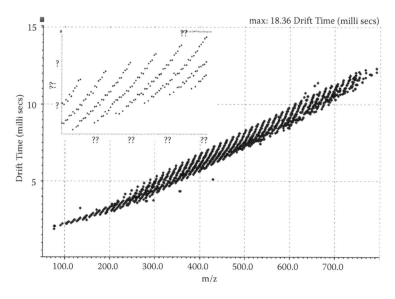

FIGURE 8.4   Plot of m/z against drift time for ASAP analysis of a crude oil sample (inset shows expanded region from m/z 380 to 460).

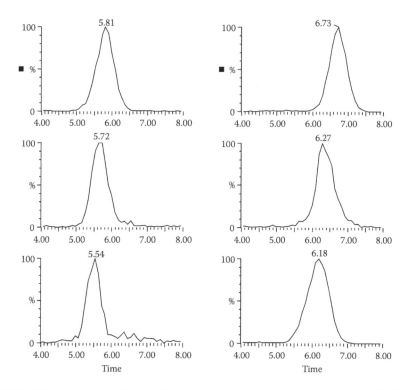

FIGURE 8.5   Extracted drift time chromatograms for ions detected around m/z 414 in the crude oil sample shown in Figure 8.3.

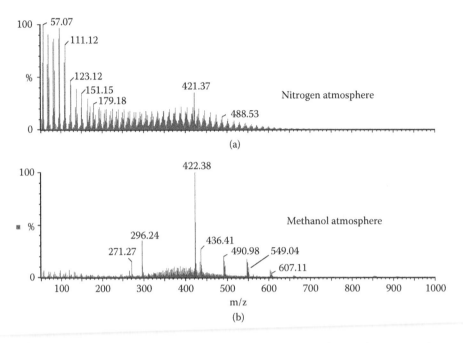

FIGURE 8.6 ASAP MS analysis of a lubricating oil formulation in a methanol atmosphere (b) showing ionization by proton transfer and in a dry nitrogen atmosphere (a) showing charge transfer ionization.

water vapor in the source or if a suitable modifier such as methanol or ammonium hydroxide solution is introduced. Figure 8.6 shows the ASAP analysis of a lubricating oil before and after the addition of methanol to the source region. The addition of the methanol clearly shows the enhancement of the amine antioxidants and ester additives, which have a higher proton affinity than the base mineral oil, resulting in the formation of the protonated molecules. The dry nitrogen atmosphere favors the formation of radical cations of the mineral oil by charge transfer. The peak at m/z 421.37 in the dry nitrogen atmosphere and m/z 422.38 in the methanol atmosphere are the charge transfer and proton transfer ions, respectively, from dinonyl diphenylamine, a commonly used antioxidant. The ability of ASAP to generate ions by both charge transfer and proton transfer enables the ionization of a wide range of compounds with varying polarity, and when coupled to IMS can help to characterize complex mixtures such as these.

8.4 SEPARATION SPACE

Fortunately, most samples are not as complex as crude oil. However, small isobaric molecules accessible by the ASAP method will also have relatively small differences in CCS. In order to explore the two-dimensional (2-D) (t_d vs. m/z) separation space of the Synapt G2 IMS/MS at relatively low mass, two sets of compounds with significantly different structures were run using the ASAP method. One of the samples was a distribution of polyphenylene molecules having –OCF$_3$ functional groups on each phenyl ring. This is a stiff, elongated polymer and thus is expected to have a large CCS versus molecular weight (MW). The other compound contains fullerenes with different numbers of perfluorodecyl pendant groups attached to C_{60} and represents a more compact series of structures. Figure 8.7a shows the m/z vs. t_d data of the positive ion IMS/MS separation of a mixture of the polyphenylene and the perfluoroalkyl substituted C_{60} compounds. The data area labeled 1 is from the perfluorodecyl C_{60} and the data labeled 2 is from the polyphenylene. Clearly, as expected, the more

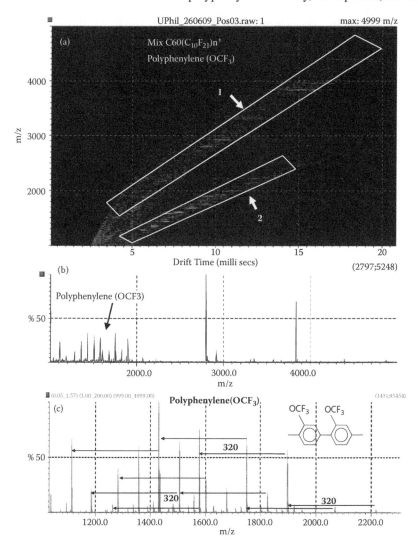

FIGURE 8.7 Positive ion plot of t_d versus m/z and mass spectra of a mixture of a polyphenylene (OCF_3) distribution and perfluorodecyl C_{60} compounds. (a) shows the t_d versus m/z representation, (b) is the summed mass spectrum of the mixture, and (c) is the extracted mass spectrum from region 2 in (a).

compact C_{60}-related compounds drift faster than the more elongated polyphenylene molecules of the same mass. Since these molecules are close to the extremes for size per mass increment, they provide a glimpse of the available separation space in the IMS dimension. In this sample, the π-bond interactions in polyphenylene (OCF_3) provide an energy barrier to bond rotation and produce a rather inflexible linear structure. On the other hand, the perfluoroalkyl C_{60} molecules have flexible pendant perfluoroalkyl groups that can be arranged on the C_{60} core to form numerous structural isomeric possibilities and thus cross sectional heterogeneity. Thus the perfluoroalkyl C_{60} might be expected to show a broader distribution of drift times for any one elemental composition than the polyphenylenes. This is observed in Figure 8.8 for equivalent drift times for the $C_{60}(C_{10}F_{21})_4$ and polyphenylene (OCF_3) having six $C_{14}H_6F_6O_2$ repeat units.

The summed positive ion mass spectrum obtained for the mixture is shown in Figure 8.7b. However, as seen in Figure 8.7c, the mass spectra of a distribution can be selectively extracted from the data. Thus the mass spectrum of polyphenylene (OCF_3) extracted from region 2 in Figure 8.7a is displayed in Figure 8.7c. Although this is a trivial example, it shows the ability to select and display only data of interest.

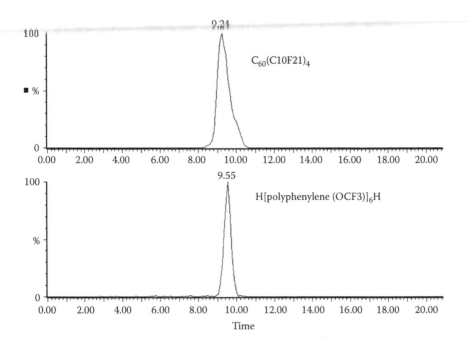

FIGURE 8.8 Extracted drift time chromatograms for ions detected at 9.24 and 9.55 ms and representing $C_{60}(C_{10}F_{21})_4$ (top) and $H(C_{14}H_6F_6O_2)_6H$ (bottom). The synthesis of the polyphenylene is not expected to produce isomeric structures and because of π-bond interactions is expected to also be "conformationally pure." The $C_{60}(C_{10}F_{21})_4$ is expected to have numerous structural isomers.

8.5   PERFLUORODECYL C$_{60}$

The negative ion IMS/MS driftscope representation of perfluorodecyl C$_{60}$ is shown in Figure 8.9a. There are obviously two distinct distributions labeled 1 (singly charged) and 2 (doubly charged). The mass spectrum of the singly charged distribution is

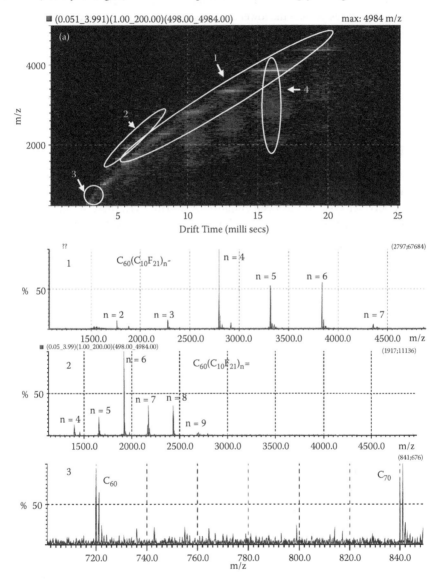

FIGURE 8.9   Negative ion mass spectrum of perfluorodecyl C$_{60}$. (a) is the t_d versus m/z plot showing four circled regions. The extracted mass spectrum of region 1 is shown in (1) for singly charged ions, region 2 shown in (2) is doubly charged ions, and region 3 shown in (3) is the mass spectrum of C$_{60}$ and C$_{70}$ extracted from low abundant ions.

readily extracted as shown in Figure 8.9(1). The doubly charged distribution is also readily observed (Figure 8.9[2]) and is the result of capture of two electrons, a rare event that was previously published.[31] Interestingly, the doubly charged ions show as many as nine perfluorodecyl groups on the core C_{60}, whereas only seven are observed in the singly charged distribution. Just as expected, the more highly charged ions drift faster than the lower charge state ions and are expected to produce an enhanced detector response. IMS/MS makes separating the two distributions simple (Figure 8.9 [1 and 2]). Figure 8.9(3) displays the mass spectrum of the data labeled 3 in the driftscope dataset and shows ions for C_{60} and C_{70}. Also of interest are the ions that have equal drift times but lower m/z values (circled area 4). This is the result of fragmentation of the n = 5 molecular ion occurring after the ion mobility separation.

8.6 POLYMERS

Polymer blends are another area of complexity that has been shown to benefit from IMS/MS. As noted earlier, a snapshot approach to observing small differences in complex polymer systems has been employed.[30,32] However, these studies involved multiply charged ESI mass spectra of polymers in which charged state manipulations aided the 2-D t_d versus m/z separation. Here, we generally deal with singly charged ions. Figure 8.10 shows the separation of a three-component blend of polystyrene with two low-molecular-weight polybutylene glycols having different end8 groups. IMS drift times are achieved that allow these polymers to be cleanly separated as shown by the total mass spectrum (Figure 8.10a) and the combined extracted mass spectrum (Figure 8.10b). Mass spectra can be extracted for each of the areas labeled in Figure 8.10a to obtain mass spectra of each polymer in the blend (Figure 8.11). This is a clear advantage of IMS/MS when using direct ionization to analyze complex mixtures. The unknown circled area labeled ? in Figure 8.10a is unexpected and involves species differing by CH_2 groups; this sequence would be difficult to isolate in the absence of IMS. Examination of the t_d versus m/z plot in Figure 8.10 compared with the composite mass spectrum in Figure 8.11 readily shows the power of IMS for giving a snapshot of a complex mixture.

8.7 LASERSPRAY IONIZATION

Recently, a new direct ambient method of producing multiply charged ions from volatile and nonvolatile compounds at atmospheric pressure using laser ablation of a matrix/analyte mixture (laserspray ionization) was introduced.[17,18] In this method, a nitrogen laser beam was fired through a glass microscope slide into the matrix/analyte mixture that was near the ion entrance aperture of the mass spectrometer. The mechanism for formation of multiply charged ions is believed to be laser initiated formation of charged clusters that are desolvated (evaporation of matrix but not charge) until the molten matrix/analyte droplet approaches the Rayleigh limit of charge where droplet fission occurs and eventually ionization by one of the mechanisms proposed for ESI.[18] Thus, sufficient heat to effect desolvation of a solid matrix is necessary to observe multiply charged ions. A system was set up with a Synapt G2

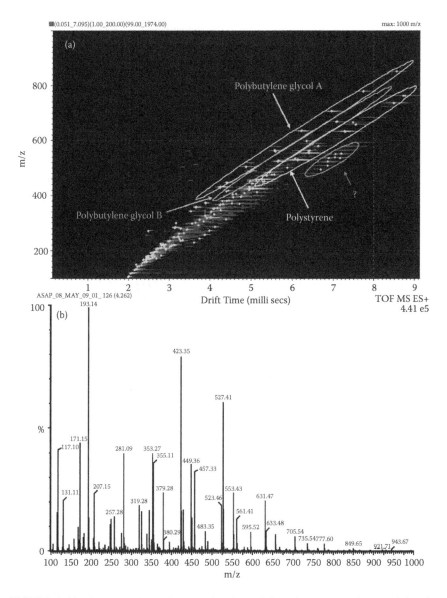

FIGURE 8.10 Polymer blend of low-molecular-weight polystyrene and two polybutylene glycols that differ in end groups. (a) is the t_d versus m/z representation showing distinct regions for the three polymers as well as an unknown region. (b) is the summed mass spectrum of the mixture.

mass spectrometer that allowed successful ionization of proteins and peptides (see Chapter 9). Using the standard dried droplet approach[33] commonly used in MALDI (matrix-assisted laser desorption/ionization), a mixture of 2,5-dihydroxybenzoic acid and insulin was added to the exterior of a melting point tube and dried. In an experiment similar to ASAP, except that instead of a hot gas, a nitrogen laser beam

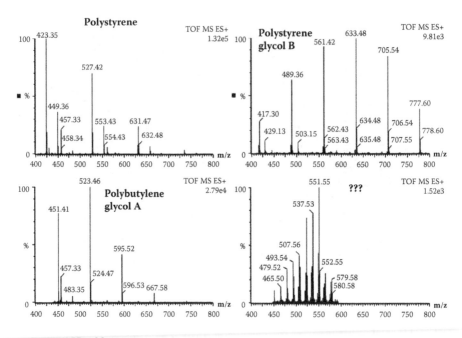

FIGURE 8.11 Mass spectra of individual components extracted from the highlighted regions of the mobility drift plot shown in Figure 8.10.

was fired at the matrix/analyte on the melting point tube. The mass spectrum from this experiment is shown in Figure 8.12a. The driftscope presentation of t_d vs. m/z is displayed in Figure 8.12b showing t_d separation of charge states +3 through +7. Note that charge states +3 to +7 drift with increasing shorter times with increasing charge as expected, but charge state +3 has an unexpectedly long drift time suggesting a more open structure.

8.8 CONCLUSION

The ASAP method is capable of producing high-quality data from complex mixtures, and with incorporation of the laserspray method extends the type of compounds that can be analyzed to higher mass and nonvolatile materials. The combination of orthogonal separation provided by mass spectrometry and IMS results in a 2-D data set with very high peak capacity. This approach provides a powerful tool for reducing spectral complexity and can be used to produce a characteristic "fingerprint" or "snapshot" for a particular complex mixture. IMS-MS can also provide insight into the structural relationships within a complex mixture, revealing the presence of homologous series of compounds and the presence of isomeric forms. Combining direct ionization with IMS and high-resolution MS provides an especially powerful method for studying complex systems in real time. As IMS/MS instruments are developed, these capabilities will continue to improve.

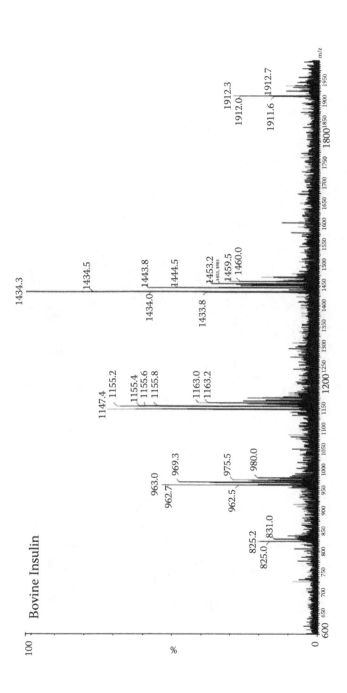

FIGURE 8.12 (a) Laserspray mass spectrum of bovine insulin obtained by laser ablation of a 2,5-dihydroxybenzoic acid/bovine insulin mixture prepared using the dried droplet MALDI sample preparation method.

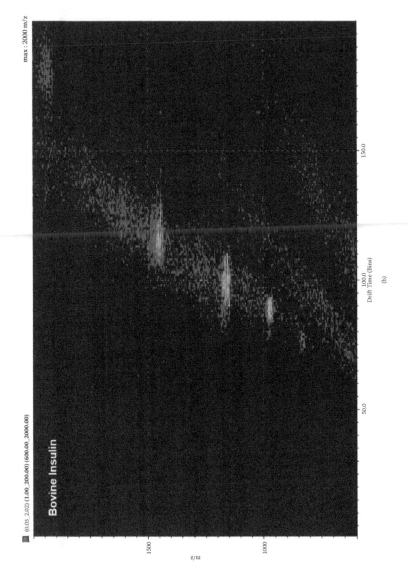

FIGURE 8.12 (b) The DriftScope presentation of the IMS/MS acquisition of the data.

ACKNOWLEDGMENTS

The authors gratefully acknowledge the Richard Houghton Fund at the University of the Sciences in Philadelphia (CNM) and the Wayne State University start-up funds as well as the NSF CAREER 0955975 funds (ST).

REFERENCES

1. Takats, Z.; Wiseman, J. M.; Gologan, B.; Cooks, R. G., Mass Spectrometry Sampling Under Ambient Conditions with Desorption Electrospray Ionization. *Science* 2004, *306*, 471–473.
2. McEwen, C. N.; McKay, R. G.; Larsen, B. S., Analysis of Solids, Liquids, and Biological Tissues Using Solids Probe Introduction at Atmospheric Pressure on Commercial LC/ MS Instruments. *Anal. Chem.* 2005, *77*, 7826–7831.
3. Whitson, S. E.; Erdodi, G.; Kennedy, J. P.; Lattimer, R. P.; Wesdemiotis, C., Direct Probe-Atmospheric Pressure Chemical Ionization Mass Spectrometry of Cross-Linked Copolymers and Copolymer Blends. *Anal. Chem.* 2008, *80*, (20), 7778–7785.
4. Haapala, M.; Pol, J.; Kauppial, T. J.; Kostiainen, R., Desorption Atmospheric Pressure Photoionization. *Anal. Chem.* 2007, *79*, 7867–7872.
5. Cody, R. B.; Laramée, J. A.; Durst, H. D., Versatile New Ion Source for the Analysis of Materials in Open Air Under Ambient Conditions. *Anal. Chem.* 2005, *77*, (8), 2297–2302.
6. Andrade, F. J.; Shelley, J. T.; Wetzel, W. C.; Webb, M. R.; Gamez, G.; Ray, S. J.; Hieftje, G. M., Atmospheric Pressure Chemical Ionization Source. 2. Desorption-Ionization for the Direct Analysis of Solid Compounds. *Anal. Chem.* 2008, *80*, 2654–2663.
7. Cotte-Rodriguez, I.; Takats, Z.; Talaty, N.; Chen, H.; Cooks, R. G., Desorption Electrospray Ionization on Surfaces: Sensitivity and Selectivity Enhancement by Reactive Desorption Electrospray Ionization. *Anal. Chem.* 2005, *77*, 6755–6764.
8. Balogh, M. P., Incipient Technologies and Thermal Desorption Techniques. *LC GC N. Am.* 2007, *25*, (12), 1184.
9. Balogh, M. P., Alternatives in the Face of Chemical Diversity. *LC GC N. Am.* 2007, *25*, (4), 368.
10. Harris, G. A.; Nyadong, L.; Fernandez, F. M., Recent Developments in Ambient Ionization Techniques for Analytical Mass Spectrometry. *Analyst* 2008, *133*, 1297–1301.
11. Popov, I. A.; Chen, H.; Kharybin, O. N.; Nikolaev, E. N.; Cooks, R. G., Detection of Explosives on Solid Surfaces by Thermal Desorption and Ambient Ion/molecule Reactions. *Chem. Commun.* 2005, 1953–1956.
12. Wu, J.; Hughes, C. S.; Picard, P.; Latarte, S.; Gaudreault, M.; L'evesque, J.-F.; Nicoll-Griffith, D. A.; Bateman, K. P., High-Throughput Cytochome P450 Inhibition Assays Using Laser Diode Thermal Desorption-Atmospheric Pressure Chemical Ionization Tandem Mass Spectrometry. *Anal. Chem.* 2007, *79*, 4657–4665.
13. Ratcliffe, L. V.; Rutten, F. J. M.; Barrett, D. A.; Whitmore, T.; Seymour, D.; Greenwood, C.; Aranda-Gonzalvo, Y.; Robinson, S.; McCoustrat, M., Surface Analysis Under Ambient Conditions using Plasma-Assisted Desorption/Ionization Mass Spectrometry. 2007, *79*, 6094–6101.
14. Na, N.; Zhao, M. X.; Zhang, S. C.; Yang, C. D.; Zhang, X. R., Development of a Dielectric Barrier Discharge Ion Source for Ambient Mass Spectrometry. *J. Am. Soc. Mass Spectrom.* 2007, *18*, 1859.
15. Van Berkel, G. J.; Pasilisi, S. P.; Ovchinnikovia, O., Establishment and Emerging Atmospheric Pressure Surface Sampling/Ionization Techniques for Mass Spectrometry. *J. Mass Spectrom.* 2008, *43*, 1161–1180.

16. Dixon, R. B.; Sampsona, J. S.; Muddiman, D. C., Generation of Multiply Charged Peptides and Proteins by Radio Frequency Acoustic Desorption and Ionization for Mass Spectrometric Detection. *J. Am. Soc. Mass Spectrom.* 2009, *20*, 597–600.

17. Trimpin, S.; Inutan, E. D.; Herath, T. N.; McEwen, C. N., Matrix-Assisted Laser Desorption/Ionization Mass Spectrometry Method for Selectively Producing Either Singly or Multiply Charged Molecular Ions. *Anal. Chem.* 2010, *82*, 11–15.

18. Trimpin, S.; Inutan E. D.; Herath, T. N.; McEwen, C. N. Laserspray ionization, a new atmospheric pressure MALDI method for producing highly charged gas-phase ions of peptides and proteins directly from solid solutions. *Mol. Cell Proteomics* 2010, *9*, 362–367.

19. Sampson, J. S.; Hawkridge, A. M.; Muddiman, D. C., Generation and Detection of Multiply-Charged Peptides and Proteins by Matrix-Assisted Laser Desorption Electrospray Ionization (MALDESI) Fourier Transform Ion Cyclotron Resonance Mass Spectrometry. *J. Am. Soc. Mass Spectrom.* 2006, *17*, 1712–1716.

20. Haung, M. Z.; Hsu, H. J.; Lee, J. Y.; Jeng, J.; Shiea, J., Direct Protein Detection from Biological Media through Electrospray-Assisted Laser Desorption Ionization/Mass Spectrometry. *J. Proteome Res.* 2006, *5*, 1107–1116.

21. McEwen, C.; Gutteridge, S., Analysis of the Inhibition of the Ergosteral Pathway in Fungi Using the Atomospheric Solids Analysis Probe (ASAP) Method. *J. Am. Soc. Mass Spectrom.* 2007, *17*, 1274–1278.

22. Lloyd, J. A.; Harron, A. F.; McEwen, C., Combination Atmospheric Pressure Solids Analysis Probe and Desorption Electrospray Ionization Mass Spectrometry Ion Source. *Anal. Chem.* 2009, *81*, 9158–9162.

23. Horning, E. C.; Horning, M. G.; Carroll, D. I.; Dzidic, I.; Stillwell, R. N., New Picogram Detection System Based on a Mass Spectrometer with an External Ionization Source at Atmospheric Pressure. *Anal. Chem.* 1973, *45*, 936–943.

24. Caroll, D. I.; Dzidic, I.; Haegele, K. D.; Stillwell, R. N.; Horning, E. C., Atmospheric Pressure Ionization Mass Spectrometry Corona-Discharge Ion Source for Use in Liquid Chromatography-Mass Spectrometry Computer Analytical System. *Anal. Chem.* 1975, *47*, 2369–2373.

25. Horning, E. C.; Carroll, D. I.; Dzidic, I.; Haegele, K. D.; Lin, S.-N., Development and Use of Analytical Systems Based on Mass Spectrometry. *Clin. Chem.* 1977, *23*, 13–21.

26. Weston, D. J.; Bateman, R.; Wilson, I. D.; Wood, T. R.; Creaser, C. S., Direct Analysis of Pharmaceutical Drug Formulations Using Ion Mobility Spectrometry/Quadrupole-Time-of-Flight Mass Spectrometry Combined with Desorption Electrospray Ionization. *Anal. Chem.* 2005, *77*, 7572–7580.

27. Purcell, J. M.; Rodgers, R. P.; Hendrickson, C. L.; Marshall, A. G.; McEwen, C. N.; Larsen, B. S., Petroleum Molecular Speciation by ASAP, APPI, and APCI Fourier Transform Ion Cyclotron Resonance Mass Spectrometry. *Proceedings of the 56th ASMS Conference on Mass Spectrometry and Allied Topics* 2008, Denver, CO, (June 1–5), WPG 145.

28. Marshall, A. G.; Rodgers, R. P., Petroleomics: The Next Grand Challenge for Chemical Analysis. *Acc. Chem. Res.* 2004, *37*, 53–59.

29. Fernandez-Lima, F. A.; Becker, C.; McKenna, A. M.; Rodgers, R. P.; Marshall, A. G.; Russill, D. H., Petroleum Crude Oil Characterization by IMS-MS and FTICR MS. *Anal. Chem.* 2009, *81*, 9941–9947.

30. Trimpin, S.; Clemmer, D. E., Ion Mobility Spectrometry/Mass Spectrometry Snapshots for Assessing the Molecular Compositions of Complex Polymeric Systems. *Anal. Chem.* 2008, *80*, 9073–9083.

31. McEwen, C. N.; Fagan, P. J.; Krusic, P. J., Mass Spectrometry of Perfluoroalkylated Buckminsterfullerene: A Case for Sequential Gas Phase Double Electron Capture. *Int. J. Mass Spectrom.* 1995, *146–147*, 297–304.

32. Trimpin, S.; Plasencia, M.; Isailovic, D.; Clemmer, D. E., Resolving Oligomers from Fully Grown Polymers with IMS-MS. *Anal. Chem.* 2007, *79*, 7965–7974.
33. Karas, M.; Hillenkamp, F., Laser Desorption Ionization of Proteins with Molecular Masses Exceeding 10,000 Daltons. *Anal. Chem.* 1988, *60*, 2299–2301.

9 Total Solvent-Free Analysis, Charge Remote Fragmentation, and Structures of Highly Charged Laserspray Ions Using IMS-MS

Ellen D. Inutan, Emmanuelle Claude, and Sarah Trimpin

CONTENTS

9.1 INTRODUCTION

9.1.1 MOTIVATION

Insoluble or even low solubility materials frequently present nearly insurmountable analytical challenges. Cell extraction is one such example. After cell opening, several extractions are carried out using different conditions and solvents. A residual "pellet" is obtained containing the insoluble fraction of the cell extract. This "pellet" is discarded because of the lack of appropriate analytical technology for analyses. [1,2] Other examples, such as phospholipids and phosphopeptides, can be solubilized and successfully extracted, but have an extreme tendency to adhere to column materials and are lost during attempts at purification (Scheme 9.1).[3–8] Perhaps the most well-known examples are membrane proteins that are poorly soluble or entirely insoluble if extracted from their native environment. Less than 1% of known protein structures are membrane proteins[9] even though it is estimated that 20–30% of the human genome encodes for them.[10,11] This, along with, for example, the production and deposition of insoluble plaques formed by β-amyloid peptides from amyloid-precursor protein (APP) limit our understanding of Alzheimer's disease as is the case with many age-related diseases.[12] Future success is anticipated in discovering novel means to characterize insoluble materials on a molecular level in order to relate molecular structure to biological, chemical, or physical processes.[9,13–21] There is also a pressing need for rapid analysis of structurally, chemically, and dynamically complex systems, especially those of unadulterated tissue sections.

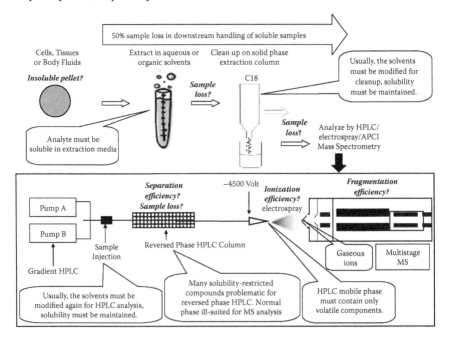

SCHEME 9.1   Representation of a liquid-based separation scheme showing numerous areas for sample loss.

9.1.2 Current Limitations

Analysis of complex mixtures using direct methods such as infusion electrospray ionization (ESI) or matrix-assisted laser desorption/ionization (MALDI) produce congested mass spectra that are difficult to interpret, especially if multiply charged ions are involved. Multidimensional condensed-phase separation methods such as liquid chromatography (LC) and size exclusion chromatography (SEC) interfaced with mass spectrometry (MS) are powerful for analyzing complex mixtures, but require solubility in appropriate solvents (Scheme 9.1) and increase the fractions produced, thus increasing the analysis time.[21–24] These liquid-phase separation methods are not applicable to tissue imaging analysis. Ion mobility spectrometry (IMS) MS, similar to condensed-phase separation methods, is a powerful solvent-free next-generation technology for reducing mass spectral complexity in a continuous fashion and extending the dynamic range while also providing structural information using cross section analysis.

9.1.3 Inroads into the Insolubleome

Total solvent-free analysis (TSA) by MS[17–19] is carried out using the dry solvent-free sample preparation method in which matrix and analyte are ground together[13,15] (or matrix is sprinkled on top of/below the analyte)[19,25] irrespective of solubility, and the homogeneous (or layered) matrix/analyte sample is ionized using laser ablation and the ions separated according to size (shape) and charge by IMS[26–28] in the absence of any solvent followed by MS detection. TSA addresses numerous challenges related to solvent such as the inability to solubilize the sample, solvent-induced ionization suppression, limited dynamic range, sample loss frequently observed in downstream handling, as well as sample alterations such as those of ex vivo chemical oxidation of, for example, methionine residues.[29–32] However, the TSA approach also simplifies sample preparation, separation, and acquisition, and less-acquainted operators can be rapidly familiarized.[19,33,34] The search for appropriate solvents is eliminated and desorption/ionization and separation conditions for the molecule of interest are minimized. The operator-to-operator and sample-to-sample reproducibility is also improved and sufficient separation of as little as isomeric compositions can be achieved in some cases.[26,27,35,36]

This book chapter focuses on the analyses of lipids, peptides, and proteins including those directly from mouse brain tissue. Examples described here were obtained using a homebuilt multidimensional IMS³-MS instrument developed by Clemmer and coworkers (instrument scheme and experimental details provided in Chapter 10)[37–39] or on commercial Waters Corporation SYNAPT[40] and SYNAPT G2 (see Chapter 8, Figure 8.1) IMS-MS[41] instruments with MS/MS capabilities.

9.2 RAPID SOLVENT-FREE SEPARATION BY ESI-IMS-MS

The introduction of IMS-MS as an analytically powerful tool by Clemmer and others[42–45] is the foundation for TSA providing a solvent-free separation method. Even for soluble materials, liquid-based separations can be demanding in terms of time, sufficient knowledge, and training to achieve satisfactory results. As amply noted in other chapters, solvent-free separation is possible because IMS separates ions drifting in an applied electric field through a buffer gas by charge and cross section (size and shape), and MS by an ion's mass-to-charge (m/z) ratio. A two-dimensional (2D)-plot of drift time distribution (t_d) $vs.$ m/z provides a powerful pictorial representation of complex mixtures. IMS-MS is also able to provide *structural information* based on drift times (cross sections). Nearly instantaneous separation of the components of complex mixtures is achieved, and reproducible three-dimensional (3D) images containing information related to molecular size, mass, and abundance are generated. Perhaps the most powerful analytical aspect of IMS is its prospect to separate isomeric components of materials by exploiting the dimension of shape. Various complex materials such as human plasma,[26] lipids,[27,28] and synthetic materials (see Chapter 10)[35,36] can be separated in a continuous fashion. Some of these results are described in greater detail below, highlighting the many applications and benefits of IMS-MS for reducing complexity without prior solvent-based separation approaches or time-consuming fragmentation analysis, indicating applicability for analyses in which condensed-phase separation is not applicable as is the case in tissue imaging by IMS-MS.

9.2.1   IMS-MS of Lipids

Initial work using the homebuilt ESI-IMS-MS instrument previously described[37-39] impressively demonstrated the usefulness of the solvent-free IMS separation technique for lipids (Figure 9.1). For example, sphingomyelins (Scheme 9.2) are baseline separated in the drift time dimension with $\Delta \sim 1.2$ ms for ethylene chain length differences (Figure 9.1c.2).[27] Partial separation is achieved for single unsaturation differences with values of $\Delta \sim 0.3$ ms (Figure 9.1c.3). These results reveal, as expected, that the most compact sphingomyelin structures are those with shorter chain length and increasing unsaturation. The type of cation has little influence on the shape of the sphingomyelin ion as indicated by the nearly identical drift times of ions differing only in the attached cation. Defined mixtures of the chemically different lipids, *N*-palmitoyl glycine, *N*-arachidonoyl ethanolamide (anandamide), and phosphatidyl choline, demonstrated the existence of a variety of charge states and sizes (inverted micelles) and shows that IMS-MS assists in improving the dynamic range for the detection of lower abundant and/or ionization retarded *N*-palmitoyl glycine.[27]

An example of inverted micelles is shown for phosphatidyl inositol (Scheme 9.2).[27] The 2D plot of t_d $vs.$ m/z (Figure 9.2) shows aggregation up to $[PI_{16}+6cat]^{6+}$. Charge states run diagonally through the image. The extracted embedded mass spectral information is shown in Figure 9.3 for each charge state. The right-hand panel shows the importance of incorporating the IMS dimension to separate isobaric composition caused by m/z ($[PI_2+2cat]^{2+}$ $vs.$ $[PI+cat]^+$ (Figure 9.3.II.A and B) and charge convolution (faster drifting $[PI+cat]^+$ $vs.$ slower drifting $[PI+cat]^+$, Figure 9.3.II.D and E). Aggregation is confirmed by selection and activation studies of phosphatidyl inositol in which $[PI_5+3cat]^{3+}$ dissociated into mainly $[PI_4+2cat]^{2+}$, $[PI+cat]^+$, and some smaller aggregates. The highest order aggregates are observed for phospholipids, and smaller and structurally simpler lipids such as *N*-acyl amino acids and mono-

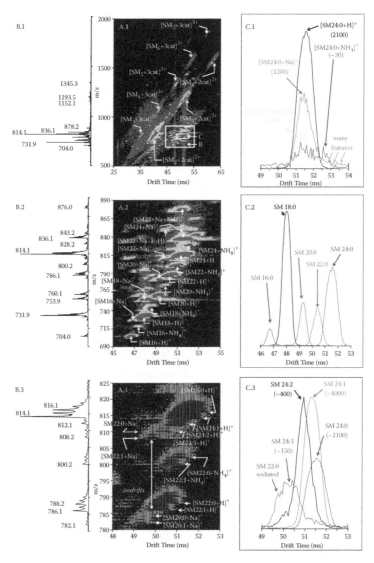

FIGURE 9.1 Mass spectra extracted along the diagonal of the t_d vs. m/z images (center) and drift time distributions of sphingomyelins (SM). (A.1) is the full range showing charge-state aggregates. (A.2) is an inset of singly charged ions showing separation of about 5 ms between $[SM16+H]^+$ and $[SM24+H]^+$. Different cations populate the image; the features assigned $[SM22+Na+K-H]^+$ and $[SM24+Na+K-H]^+$ could also be oxidized $[SM22ox+2Na-H]^+$ and $[SM24ox+2Na-H]^+$. (A.3) is the inset of singly charged ions of $[SM24+H]^+$ and reveals that the lower value t_d features, thus faster drift times relate to an increasing degree of unsaturation. (B.1–3) are the mass spectra of diagonal slices of $[SM+cat]^+$ in the mass range shown. (C.1–3) are the t_d distributions with the numbers in brackets providing the ion abundances. (C.1) shows the effect of different cations on SM24:0. (C.2) shows the effect of different chain lengths of protonated SM. (C.3) shows different degrees of unsaturations of protonated SM. For additional details see ref. 27. (From Trimpin, S.; Tan, B.; Bohrer, B. C.; O'Dell, D. K.; Merenbloom, S. I.; Pazos, M. X.; Clemmer, D. E.; Walker, J. M. *Int. J. Mass Spectrom.* 2009, *287*, 58–69. With permission.)

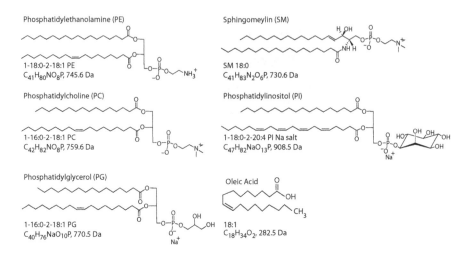

SCHEME 9.2   Structures of *N*-acyl amino acids and related lipid compounds.

glycerides also show aggregation. The inverted micelles might be formed by salt bridges of head groups while the fatty acid chain provides the hydrophobic shell.

The following relative ion mobilities were observed for phospholipids (Scheme 9.2) separated according to their polar head groups (Figure 9.4): phosphatidyl ethanolamine < phosphatidyl choline < phosphatidyl glycerol < sphingomyelin < phosphatidyl inositol.[27] The phospholipids with the same core separate according to the size requirements of the head group, thus the phosphatidyl ethanol amine is smaller and travels faster than the phosphatidyl glycerol with a t_d difference of about 1.2 ms. Phospholipids with the same choline head group separate according to the core as is the case for phosphatidyl choline and sphingomyelin with a drift time separation of about 2.3 ms. The sphingomyelin with the ceramide core has a significantly more open conformation as compared to the glycerol core of phosphatidyl choline.

A number of low-molecular-weight *N*-acyl amino acid lipids of biological significance were identified based on values of t_d and *m/z* values provided by IMS-MS. Both data acquisition and visualization of the separated isomeric lipids is rapidly achieved.[27] For example, *N*-arachidonoyl isoleucine, and *N*-arachidonoyl leucine exhibit t_d fingerprints that are sufficiently distinct to delineate isomeric composition. It is often difficult to obtain meaningful MS/MS fragment ions from these low abundant lipids for structural characterization. The mass window for parent ion selection for most MS instruments is insufficient to cleanly select isobaric compounds, and the fragmentation pattern of isomers is commonly very similar.

More recently we investigated the ability to separate lipid compositions using the SYNAPT G2 instrument (see Chapter 8, Figure 8.1). A typical example is shown in Figure 9.5 for sphingomyelin. As described above (Figure 9.1), inverted micelles are observed. The clear drift time separation of, for example, the isobaric families (Figure 9.5a, inset) in the form of $[M + cat]^+$ (~110 to 140 Bins) and $[M_2 + 2 cat]^{2+}$ (~70 to 80 Bins) allows for efficient charge state deconvolution. Further, a partial separation of chain length differences is achieved as can be seen in the extracted drift time

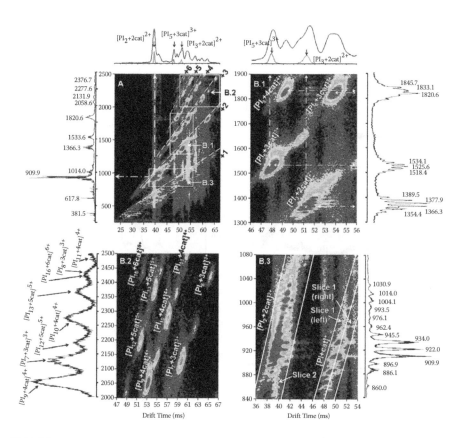

FIGURE 9.2 Center contains t_d *vs. m/z* images: (A) Full-range image showing different charge states of what can be described as inverted micelles. (B.1) to (B.3) are insets displaying different charge states. To the right and left of each image are the summed mass spectra and on top are drift time distributions of phosphatidylinositols (PI). For more details see ref. 27. (From Trimpin, S.; Tan, B.; Bohrer, B. C.; O'Dell, D. K.; Merenbloom, S. I.; Pazos, M. X.; Clemmer, D. E.; Walker, J. M. *Int. J. Mass Spectrom.* 2009, *287*, 58–69. With permission.)

distributions of the protonated sphingomyelin (Figure 9.5b). As expected, the separation is smaller than with the homebuilt instrument (Figure 9.1C.2). Though different sphingomyelin samples were investigated with two different instruments, the same trend and similar pictorials are observed (Figure 9.1A.2 and Figure 9.5a).

The commonality between any IMS-MS instrument is that the data acquisition is rapid and usually unproblematic, but the time-limiting factor is frequently data extraction and interpretation. IMS-MS provides the ability to detect relatively minor differences almost instantaneously as discussed in Chapter 10 for polymer systems using the "snapshot" approach.[35,36] Differences in lipid and protein mixtures are also captured nearly instantaneously using IMS-MS instrumentation even when isomeric components are present and can be displayed visually as a snapshot (Figures 9.1 to 9.5), similar to 2D-gels, providing distinctive visual patterns that through pattern recognition

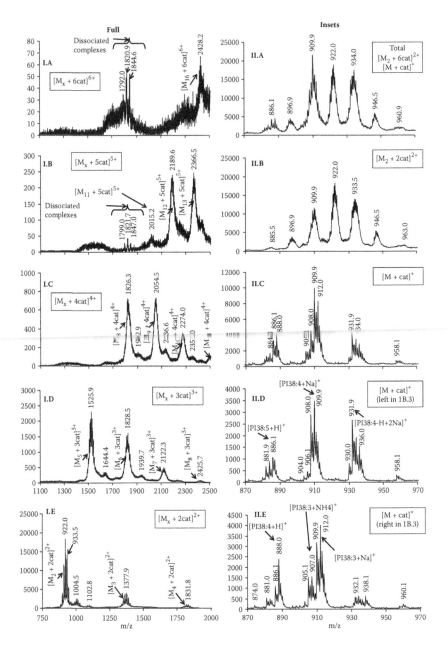

FIGURE 9.3 Mass spectra of phosphatidylinositol (PI) micelle charge state families. The mass spectra are of slices integrated along a diagonal in the $t_d(m/z)$ distribution ranging from the +6 series (slice 6, [I.A]) to the +1 series (slice 6, [II.C]). Dissociated complexes (I.A, I.B) indicate fragmentation of higher charge-state complexes. M_x indicates aggregation for each respective charge state. For additional details see ref. 27. (From Trimpin, S.; Tan, B.; Bohrer, B. C.; O'Dell, D. K.; Merenbloom, S. I.; Pazos, M. X.; Clemmer, D. E.; Walker, J. M. *Int. J. Mass Spectrom.* 2009, *287*, 58–69. With permission.)

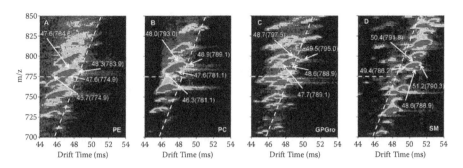

FIGURE 9.4 Limited range t_d *vs.* *m/z* image insets displaying the singly charged ions of phospholipids (A) PE, (B) PC, (C) PG, and (D) SM. The $t_d(m/z)$ values for features within the *m/z* range 775–795 are shown for the various features. The white lines are included to guide the eye for changes within (A) through (D), respectively. For more details see ref. 27. (From Trimpin, S.; Tan, B.; Bohrer, B. C.; O'Dell, D. K.; Merenbloom, S. I.; Pazos, M. X.; Clemmer, D. E.; Walker, J. M. *Int. J. Mass Spectrom.* 2009, *287*, 58–69. With permission.)

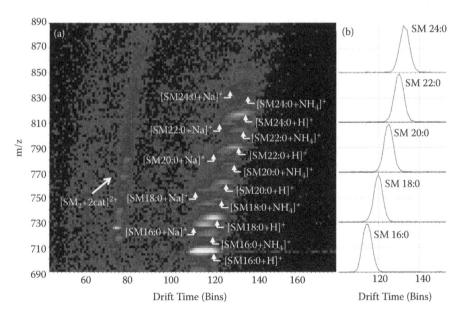

FIGURE 9.5 Data acquired for sphingomyelin using the SYNAPT G2. (a) compares with Figure 9.1A.2 and (b) compares with Figure 9.1C.2 both obtained on a homebuilt instrument. Drift time separation of lipids differing by degree of unsaturation is achieved but with less resolution than with the 3-m homebuilt instrument. Inverted micelles are observed to the left of the t_d *vs.* *m/z* image in (a).

software or the human eye can be used to rapidly evaluate complex materials by a comparative means. It is anticipated that in the future, a "snapshot" approach should be extremely powerful in, for example, biomarker discovery using the laserspray ionization method, which provides the necessary ability for rapid acquisition *and* data analyses of a large number of samples with high sensitivity.[25,46,47]

9.2.2   FRAGMENTATION ANALYSIS OF LIPIDS USING MULTIDIMENSIONAL IMS-MS

Multidimensional ESI-IMS^3-MS (homebuilt instrument) in substitution of LC MS^2 provides high-efficiency separation, reduces the time per analysis, and achieves not only unique m/z fragment ion fingerprints similar to MS/MS but also t_d fingerprints. [26,27] In a multidimensional IMS^n experiment, as opposed to a multidimensional MS^n experiment, ions are t_d selected (and not m/z selected) for fragment ion analysis. Therefore, multidimensional IMS^3 experiments introduce an additional dimension to fragment ion analysis by separating the parent ion in the t_d dimension according to size, shape, and charge and also the fragment ions. For example, using the ESI-IMS^3-MS method, the *sn*-1 and *sn*-2 composition of the t_d selected parent ions of phosphatidyl glycerol were determined.

9.2.3   CONFORMATIONAL ANALYSIS OF LIPIDS USING IMS-MS

Only a few MS methods (e.g., proton/deuterium exchange)[48] are available to perform conformational analysis of ions; IMS-MS in combination with molecular modeling of experimentally derived collision cross-sections is one of the most precise. [42,49,50] As an example, two stable conformations, a compact structure and an elongated structure, were observed for *N*-acyl amino acids.[27] Unexpectedly, when additional energy was supplied to these small lipids, the elongated structures were converted to drastically more compact structures. Thus, in addition to its ability as a separation tool, IMS is also capable of examining conformations and conformational changes of small molecules. Collision cross-section analysis coupled to molecular dynamics modeling will provide deeper insight in the future.

9.3   TOTAL SOLVENT-FREE ANALYSIS: SOLVENT-FREE MALDI AND IMS-MS

9.3.1   CHARGE REMOTE FRAGMENTATION OF FATTY ACIDS BY TSA

Solvent-free separation is especially powerful in partnership with solvent-free ionization. [17–19] Here we discuss a few milestones critically influencing the production of ions of the entire sample to be separated and mass detected. We demonstrated that solvent-free MALDI[13,15] and matrix-ionized laser/desorption (MILD) methods[30] overcome many restrictions encountered using the traditional solvent-based ESI and MALDI methodology. [16,51–54] In MILD, conditions provide the ability to convert fatty acids to doubly lithiated, singly positively charged ions with the general formula [M-H+2Li]$^+$.

Obviously, only components of mixtures that are converted to ions can be separated in the IMS dimension and detected by MS. Simultaneous ionization of compounds, such as basic and acidic lipids, is critical and is generally unsuccessful using ESI. The synergy of TSA (homebuilt instrument) using MILD conditions for ionization provided us with the ability to ionize and gas-phase separate fatty acids in the positive mode by facile adduction of lithium cation(s).[55] Figure 9.6 shows a typical example obtained on a MALDI-IMS-MS instrument (SYNAPT) for M being oleic

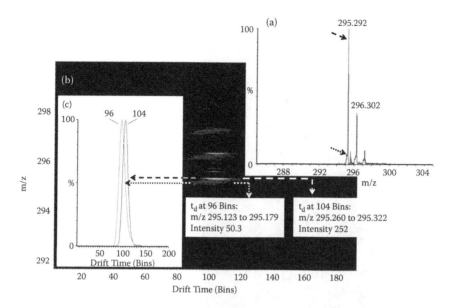

FIGURE 9.6 The mass spectrum (a) and the t_d *vs.* *m/z* inset (b) of the region around the [M + 2Li –H]⁺ ion of oleic acid produced using the solvent-free MILD method.[30] Inset (c) shows the extracted drift times for the isobaric components shown by arrows in (a). The extracted information can be displayed as a nested dataset of {drift time *vs.* mass-to-charge *vs.* ion intensity}; for example {104;295.292;252} for the abundant ion of [M-H+2Li]⁺.

acid, a monounsaturated fatty acid (Scheme 9.2). The ions of interest are selected in a quadrupole analyzer before entering the trapping region of the Tri-Wave. The clean selection of [M-H+2Li]⁺ with *m/z* 295.292 along with its two higher mass isotopes is shown in Figure 9.6a. The incorporation of the IMS dimension (Figure 9.6b) shows the presence of two different species, at drift times of 96 and 104. The drift time distributions along with intensity and *m/z* information for both isobaric ions are displayed in Figure 9.6c.

These IMS gas-separated ions are subjected to subsequent fragmentation in the transfer region, thus after the ions exit the drift dimension, of the Tri-Wave cell and the results can be seen in the 2D plot of t_d *vs.* *m/z* in Figure 9.7. The ions with the lower mobility, thus more open gas phase structure, fragments to lower molecular weight ions. The fragmentation is distinctively different from what is observed for the higher mobility ions.

When extracting the respective mass spectra for the higher (Figure 9.7C.1) and lower mobility ions (Figure 9.7C.3) observed in the 2D IMS-MS plot (Figure 9.6B), two different precursor ions are obtained. The most abundant ion at *m/z* 295.292 corresponds to the expected [M-H+2Li]⁺. The isobaric compound that only produced higher-molecular-weight fragment ions (Figure 9.7C.1 and C.2) is of low abundance in the precursor selected IMS-MS spectrum (Figure 9.6a) as a tiny shoulder of the expected and high abundant [M-H+2Li]⁺. The lower mobility ions, in addition to the more open structure, show charge remote fragmentation, as indicated in Figure 9.7C3

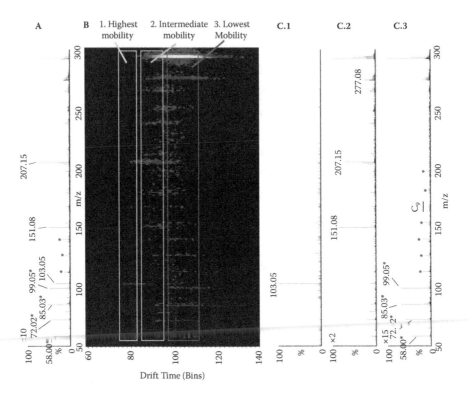

FIGURE 9.7   Fragmentation of the [M + 2Li −H]⁺ nominal mass ions of oleic acid (m/z 295) ionized using solvent-free MILD conditions. (A) shows the total fragment ion mass spectrum of m/z 295. (B) shows that the m/z 295 ion has three distinct components (isobars) with slow (B.1), intermediate (B.2), and fast (B.3) drift times. (C) shows the extracted fragment ion spectrum of the slow (C.1), intermediate (C.2), and fast (C.3) drift time components. Only (C.3) shows charge remote fragmentation.

by asterisks. The clean precursor selection afforded by the quadrupole is a beneficial asset using IMS-MS instrumentation.

The identical TSA approach using the SYNAPT G2 (higher mobility at 95 *vs.* low mobility at 125) instead of the first generation SYNAPT instrument (higher mobility at 89 *vs.* low mobility at 101) provided a drift resolution enhancement of 18 Bins for the doubly lithiated palmitic acid over the isobaric ion. Palmitic acid is a saturated fatty acid and one C2 unit shorter as compared to oleic acid (Scheme 9.2).

9.3.2   Tissue Imaging by TSA

Mouse brain tissue sections, covered with MALDI matrix (e.g., 2,5-dihydroxybenzoic acid, α-cyano-4-hydroxycinnamic acid) using a device we call a *SurfaceBox* and described elsewhere[19] were imaged by TSA using the MALDI-IMS-MS (SYNAPT) instrument. The *SurfaceBox* achieves one-step solvent-free automatic matrix deposition through vigorous movements of beads pressing the matrix material through a metal or nylon mesh into an air gap and onto the tissue section. Using a 3-μm mesh produces homogeneous coverage of <5 μm crystals in 5 minutes permitting fast, uniform coverage of analyte and possible high-spatial resolution surface analysis. The 2D-plot (Figure 9.8.1A) of t_d vs. m/z dimension provides a display of gas-phase separated ions. A wealth of ions is observed in the total mass spectrum (this is the mass spectral output as if IMS dimension were not applied) in the region from m/z 700 to 900 indicative of lipids (Figure 9.8.1B). Only limited information can be extracted

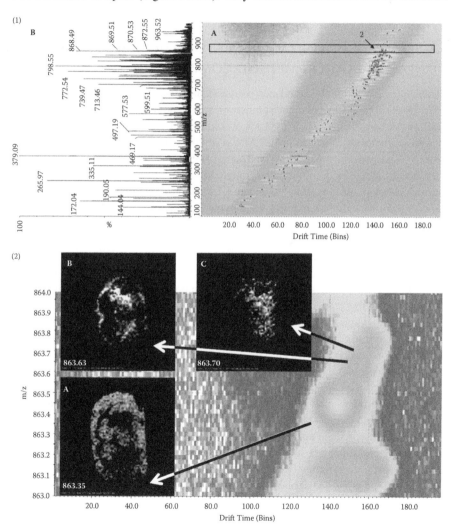

FIGURE 9.8 Total solvent-free (TSA) imaging of mouse brain tissue using CHCA as matrix: (1A) drift time (t_d) vs. m/z 2D representation of data. (1B) Total mass spectrum of all ions detected. (2) Insets of the 2D data and ion images of m/z values of (A) 863.35, (B) 863.63, and (C) 863.7. For additional details, see ref. 19. (From Trimpin, S.; Herath, T.N.; Inutan, E.D.; Wager-Miller, J.; Kowalski, P.; Claude, E., Walker, J.M.; Mackie, K. *Anal. Chem.* 2010, *82*, 359–367. With permission.)

with respect to low abundant ions due to the complexity of the sample. An inset of the 2D-plot is provided in Figure 9.8(2). This view permits the visualization of low abundant ions. These ions have slightly different t_d distributions and therefore must be isobaric (and not isotopic) in origin. Ion images (Figure 9.8(2)A–C) were obtained that clearly delineate isobaric compositions of low abundant ions in a mass window of m/z 863.0 to 864.0. The location of each ion in the image and considerations of the decimal values of the m/z indicate that ions at m/z 863.35 and 863.63 are phospholipids and at m/z 863.70 a peptide.

9.4   LASERSPRAY IONIZATION: A NEW ATMOSPHERIC PRESSURE LASER ABLATION METHOD COMPATIBLE WITH TSA

Electrospray ionization (ESI) and matrix-assisted laser desorption/ionization (MALDI) differ by the sample environment (solution *vs.* solid) and the observable charge state(s) (multiply *vs.* singly charged). The yields of fragment ions are enhanced using multiply charged parent ions observed in ESI-MS, a key benefit in structure characterization. Multiply charged ions also allow analysis of high-molecular-weight compounds on mass spectrometers with a limited m/z range. MALDI MS, in contrast, is ideal for the analysis of heterogeneous samples, because it is easier to interpret mass spectra of singly charged ions and often requires less sample. On the other hand, ESI is easily interfaced with liquid separation methods, whereas MALDI is most often used as a direct ionization method. For complex samples, MALDI-MS benefits from post-ionization separation methods such as high-resolution MS or IMS-MS.

A new method has recently been described that uses laser ablation of common MALDI matrices at atmospheric pressure (AP) to produce either MALDI-like singly or ESI-like multiply charged ions by operator choice.[46,47] We called this new approach laserspray ionization (LSI). It operates on the principle that the analyte/matrix sample is ablated by the use of a laser operating at AP and ions are subsequently formed from highly charged matrix/analyte clusters during a desolvation process.[47] The choice of charge-state selection is a valuable attribute in the analysis of complex mixtures.[46] Highly charged ions are produced for the first time by laser ablation of larger molecules such as proteins (and synthetic polymers) with analysis on high-performance but mass-range-limited atmospheric pressure ionization (API) mass spectrometers. The ability of LSI to produce multiply charged ions enhances fragmentation as was demonstrated by obtaining nearly complete protein sequence coverage of ubiquitin using electron transfer dissociation (ETD) technology on an LTQ (Thermo Fisher Scientific) mass spectrometer.[47] The applicability for mouse brain tissue has been shown.[25]

9.4.1   LSI-IMS-MS

The following considerations needed to be addressed for initial source modification and LSI-IMS-MS applications. The mechanism for formation of highly charged ions by LSI is proposed to be initial formation of charged clusters or liquid droplets of matrix/analyte by laser ablation followed by matrix evaporation similar to the ion formation mechanism in ESI. Increasing the temperature of the AP to vacuum ion transfer capillary on an Ion Max source of a Thermo Fisher Scientific Orbitrap Exactive increased the observed ion abundances of multiply charged ions using the matrix 2,5-dihydroxybenzoic acid (2,5-DHB).[47] To introduce LSI on the SYNAPT G2, we therefore fabricated a simple desolvation device[41] consisting of a 1/16-inch long by 1/8-inch outer diameter (OD) by 1/16-inch inner diameter (ID) copper tube heated using a nichrome wire coil. This device was shown to be effective in desolvating clusters produced by laser ablation of a 2,5-DHB matrix to provide multiply charged ions of incorporated proteins by LSI on the SYNAPT G2 mass spectrometer. The copper tubing was heated to ca. 250 °C. A typical example is shown in Figure 9.9 for a mixture of bovine insulin and β-amyloid (42-1). The convolution of different charge states of protein mixtures can be seen in Figure 9.9C.1. Separation of the individual charge states of proteins by IMS is demonstrated (Figure 9.9A.2 and B.2), as well as separation of both proteins in the IMS dimension (Figure 9.9C.2).

An exciting new development is the discovery of a new LSI matrix material, 2,5-dihydroxyacetophenone (2,5-DHAP), that can be used with laser ablation at AP to produce multiply charged ions of proteins and does not require the high temperature desolvation region necessary with 2,5-DHB.[56] Figure 9.10 shows an example of a mixture of isomeric β-amyloid peptide (1-42) and β-amyloid peptide (42-1) analyzed by LSI-IMS-MS. Extracting the drift time distributions for the individual charge states shows that baseline separation is achieved for each charge state with the exception of +3 (Figure 9.10B, inset). Not unexpectedly, the β-amyloid peptide (1-42) reveals a much more compact structure then the reversed peptide (42-1). The unifying theme of amyloid formation is the structural transition from an initial globular form, as is observed here, to a highly ordered regular form.[57,58] The successful separation of two isomers within this two-component mixture was confirmed by analyzing the individual isomer samples. Solvent-free gas-phase separation cleanly separates and unequivocally delineates both isomeric mixture components. Because the number of charges and the size of amyloid (1-42) and (42-1) are identical, the successful gas-phase separation of this isomeric mixture using IMS indicates that the shapes of the desorbed/ionized macromolecules are retained. This is the first evidence that LSI is a similarly "soft" ionization method as compared to ESI though operating from the solid state and directly from a surface.

LSI is an excellent fit with our goal of chemical analysis related to both complexity and insolubility and has now been interfaced with an IMS-MS (SYNAPT G2) instrument and the solvent-free gas-phase separation of LSI-generated protein ions demonstrated. With the new LSI matrix, anyone can obtain LSI mass spectra on any Waters instrument (and likely any instrument from any other company) with a simple laser setup.

9.4.2 Structures of Ions Produced by LSI-IMS-MS

The *charge state distribution* and *abundances* of multiply charged ions by the LSI method are similar to those obtained by ESI. One important question that can be answered using IMS-MS is how similar is the energy regime for formation of ions by LSI and ESI? Certainly considerable differences exist in producing ions by the two methods even if the charge state distributions are similar. In LSI, the solvent is

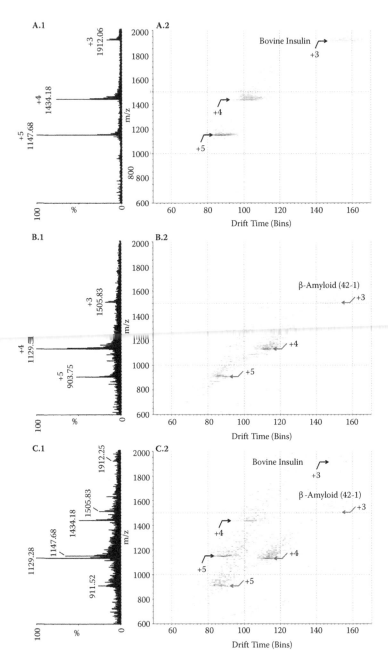

FIGURE 9.9 Water solutions of β-amyloid (42-1), bovine insulin, and a mixture of the two small proteins were prepared and 50 pmol applied to the glass slide using the dried droplet method and 2,5-DHB as matrix. Heat (ca. 225 °C) was applied to the copper transfer tube with the ion source set at 150 °C. Mass spectra of (A.1) β-amyloid (42-1), (B.1) bovine insulin, and (C.1) of the mixture. The DriftScope t_d vs. m/z representations are displayed in (A.2) to (C.2) showing clean separation of the two peptides in (C.2).

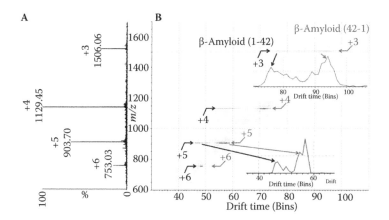

FIGURE 9.10 LSI mass spectrum of a 10 pmol μL^{-1} solution of an equimolar mixture of β-amyloid (1-42) and (42-1) isomers using the matrix 2,5-DHAP. The mass spectrum shown in (A) does not distinguish that two compounds are present, but in (B) the t$_d$ vs. *m/z* DriftScope snapshot clearly shows the two components. The inset in (B) shows the solvent-free gas phase separation in the drift time dimension of the isomeric mixture of charge states +3 and +5. For more information refer to Trimpin S. 2010, 58th ASMS Conference on Mass Spectrometry and Allied Topics, Salt Lake City, Utah. (With permission.)

a solid matrix and ions are observed only if sufficient heat is applied to evaporate or sublime the matrix within the timeframe required for ion formation to occur prior to reaching the ion optics. In LSI, ions are produced in the AP to vacuum ion transfer region and no voltage is applied in the ion source, but a rather energetic laser ablation process is required. ESI also needs desolvation, but of a more volatile liquid solvent; requires a high voltage; and the ions are produced at atmospheric pressure before entering the AP to vacuum ion transfer region. With these considerable differences, it might be expected that the multiply charged ions would be formed with different energies, and thus the ion structures would be different.

IMS-MS is well suited for examining molecular structures to determine the influence of solvent and acquisition conditions (voltage, temperature).[48,49,59] LSI in combination with IMS-MS (SYNAPT G2) can be used to separate charge states of multiply charged ions of β-amyloid (1-42) and examine each drift time distribution. Results are compared to the same samples ionized by ESI on the same instrument. In addition to its ability as a separation tool, IMS is also capable of examining conformations by cross-section analysis coupled to molecular dynamics modeling. Here, we make use of a comparative approach between LSI and ESI protein ions.

The new LSI matrix was selected that produces ions at the maximum ion source temperature of the SYNAPT G2 Z-spray source (150 °C), thus allowing the ion source temperature and source voltage (20 V) to be the same with ESI and LSI. As much as possible, all conditions were maintained exactly identical between ESI and LSI. The results of this experiment are shown in Figure 9.11. The ion abundances and charge state distributions obtained by the two different ionization methods are shown in the left panel. LSI produces lower abundant ions and slightly lower charge states as compared to the ESI method. Higher charge states in MS relate to more open and

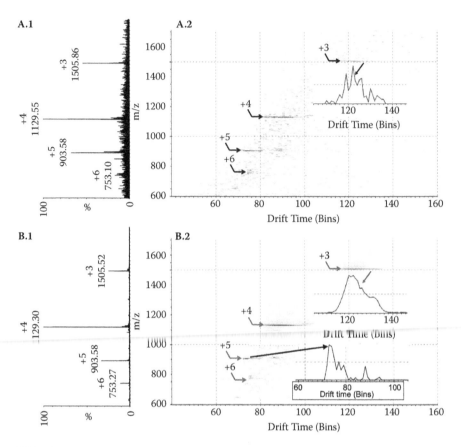

FIGURE 9.11   A 5 pmol μL-1 solution of beta-amyloid (1-42) prepared in 50:50 ACN:H2O:TFA acquired in (A) by LSI and (B) by ESI. For LSI, 2,5-DHAP was used as matrix and the data acquired using the copper transfer tube device but without applying heat other than that produced ESI was obtained at a flow rate of 5 μL/min–1 and a source at 150 °C. To ensure beneficial and comparative conditions for both ionization methods, LSI and ESI acquisitions were obtained at 20 V cone voltage. The mass spectral charge states in (A.1) and (B.1) are similar and the drift times in (A.2) and (B.2) nearly identical demonstrating very similar or identical gas phase ion structures by the two ionization methods. For more details to this aspect, see Trimpin S. 2010, 58th ASMS Conference on Mass Spectrometry and Allied Topics, Salt Lake City, Utah. (With permission.)

elongated structures. The right panel compares the t_d vs. m/z dataset as well as Inset drift time distributions for charge states +3 and +5. The drift time distributions of the LSI ions are slightly sharper than those of the ESI ions indicating fewer conformations. Sharper drift times are analytically useful for improved IMS separations. Most importantly, the β-amyloid (1-42) ions appear to have essentially identical drift times, and thus structures, using Tri-Wave Technology (SYNAPT G2). This is direct evidence that both ionization methods are energetically very similar. Temperature effects on ion structure likely occur after the ions are released into the gas phase and not in the droplet. Thus, in ESI and LSI, the naked ions experience 150 °C.

9.4.3 LSI FOR TISSUE IMAGING APPLICATIONS

Tissue imaging (see Chapters 12 and 13) has the potential to be immensely useful in areas as diverse as detecting cancer boundaries, determining drug uptake locations, and in mapping signaling molecules in brain tissue. Imaging by MS is well established, especially using secondary ion mass spectrometry (SIMS), but SIMS is only marginally useful with intact biological tissue. MALDI MS, on the other hand, has been employed for tissue imaging with success, especially for high-abundant components such as membrane lipids, drug metabolites, and proteins. However, there are a number of disadvantages in using vacuum-based MS for tissue imaging, especially in relation to unadulterated tissue. AP-MALDI tissue imaging circumvents many of the disadvantages of vacuum MALDI but is limited because of sensitivity issues at high spatial resolution. We presented initial work on a new AP imaging method with high spatial resolution.[25] Here, we expand on aspects for the potential use to analyze high-mass compounds through multiple charging.

The xyz-stage of the Waters SYNAPT G2 nanospray ion source is used as a solid support for a sample placed on a moveable glass microscope slide and positioned in front of the mass spectrometer orifice. Source geometry and conditions were optimized on lipids and peptides from purchased samples. This arrangement enables application of transmission geometry AP laser ablation of tissue covered with a matrix material to image lipids. Optical microscopy is employed to determine the size of the ablated areas under various focusing conditions. Preliminary tissue imaging results show that spatial ablations of ~15 μm^2 and spatial volume ablations < 300 μm^3 are achieved.[60] Matrix coverage, the kind of matrix used, solvent-based versus solvent-free preparation, and whether the matrix was applied on top or below the matrix all influence the analytical outcome.

Previously reported results with LSI only observed multiply charged ions for solvent-based MALDI sample preparation methods,[46,47] but the new LSI matrix produces multiply charged ions using TSA at AP without using any heat as is shown for the non-β amyloid component of Alzheimer's disease (NAC) in Figure 9.12.[31] The discovery of this new LSI matrix has wide-ranging implications for undisturbed (solvent-free sample preparation) while possibly permitting the production of highly charged LSI ions directly from tissue.

9.5 CONCLUSION

Solvent-free separation using IMS-MS when combined with solvent-free sample preparation and appropriate ionization methods provides TSA that are independent of analyte solubility and is applicable to complex mixtures. A new ionization method, LSI, which operates at AP with laser ablation of samples prepared in a MALDI matrix, has been effectively interfaced with a SYNAPT G2 IMS-MS instrument. Early results show separation of protein mixtures and that protein ion structures from LSI and ESI are similar. Further, a new matrix allows LSI multiply charged ion formation to be extended to solvent-free matrix preparations. Thus TSA by LSI-IMS-MS is a new approach to tissue imaging at high spatial resolution on high-end mass spectrometers such as the SYNAPT G2 and Orbitrap instruments. LSI has

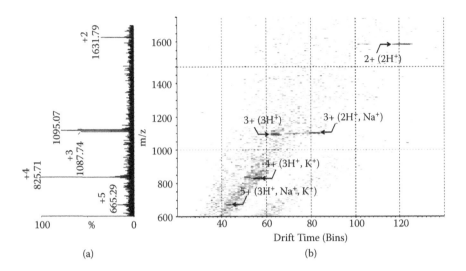

FIGURE 9.12 Mass spectrum of the non-β amyloid component of Alzheimer's disease (NAC) peptide prepared by first dissolving NAC into water and drying followed by solvent-free sample preparation using 2,5-DHAP as matrix. The dried mixture was vibrated with metal beads for 5 minutes at 20 kHz using a Tissuelyser. The mass spectra were acquired using the copper transfer tube device but without applying heat other than that produced by placing the source heater of the SYNAPT G2 at 150°C for LSI studies. (A) Shows the mass spectrum and (b) the t_d vs. m/z DriftScope representation demonstrating the efficient production of large multiply charged peptide ions directly from a surface at atmospheric pressure. The higher charge states show cation addition as well as proton addition, similar to observations in vacuum MALDI (ref. 31). Lower charge states show two distinct shapes.

also been shown to be compatible with electron transfer dissociation (ETD) further enhancing its analytical utility[47] expanding sequence analysis and identification of peptides and proteins directly from tissue.[60] Combinations of these high-end technologies such as LSI-IMS-MS-ETD in conjunction with TSA at AP for structural elucidation of primary sequences, post-translational modifications, and conformations will provide a wealth of information and empower many new applications in the future. This technology will be especially powerful for protein characterization studies directly from tissues.

ACKNOWLEDGMENTS

The author gratefully acknowledges all the vital contributions to this chapter during her education by Prof. Klaus Müllen and Dr. Hans-Joachim Räder (Max-Planck-Institute for Polymer Research, Mainz, Germany), Profs. Max L. Deinzer (State University, Oregon), Peter S. Spencer (Oregon Health & Science University), David E. Clemmer (Indiana University), and J. Michael Walker (Indiana University 1950–2008). Collaborators, Profs. Charles N. McEwen (The University of the Sciences in Philadelphia) and Ken Mackie (Indiana University) are gratefully acknowledged for providing instrument access and samples. Financial support is gratefully acknowledged from National Science Foundation CAREER 0955975, Wayne State University (startup funds), American Society for Mass Spectrometry Research Award (financially supported by Waters Company), and DuPont Company Young Investigator Award.

REFERENCES

1. Meitzler, J. L.; Gray, J. J.; Hendrickson, T. L. Truncation of the caspase-related subunit (Gpi8p) of *Saccharomyces cerevisiae* GPI transamidase: Dimerization revealed. *Arch. Biochem. Biophys.* 2007, *462*, 83–93.
2. Tan, B.; O'Dell, D. K.; Yu, Y. W.; Monn, M. F.; Hughes, H. V.; Burstein, S.; Walker, J. M. Identification of endogenous acyl amino acids based on a targeted lipidomics approach, *J. Lipid Res.* 2010, *51*, 112–119.
3. Cavanagh, J.; Benson, L. M.; Thompson, R.; Naylor, S. In-line desalting mass spectrometry for the study of noncovalent biological complexes. *Anal. Chem.* 2003, *75*, 3281–3286.
4. Raska, C. S.; Parker, C. E.; Dominski, Z.; Marzluff, W. F.; Glish, G. L.; Pope, R. M.; Borchers, C. H. Direct MALDI-MS/MS of phosphopeptides affinity-bound to immobilized metal ion affinity chromatography beads, *Anal. Chem.* 2002, *74*, 3429–3433.
5. Zhang, G.; Fan, H.; Xu, C.; Bao, H.; Yang, P. On-line preconcentration of in-gel digest by ion-exchange chromatography for protein identification using high-performance liquid chromatography-electrospray ionization tandem mass spectrometry. *Anal. Biochem.* 2003, *313*, 327–330.
6. Wang, J.; Jiang, X.; Sturm, R. M.; Li, L. Combining tissue extraction and off-line capillary electrophoresis matrix-assisted laser desorption/ionization Fourier transform mass spectrometry for neuropeptide analysis in individual neuronal organs using 2,5-dihydroxybenzoic acid as a multi-functional agent. *J. Chromatogr. A* 2009, *1216*, 8283–8288.
7. Sun, B.; Ranish, J. A.; Utleg, A. G.; White, J. T.; Yan, X.; Lin, B.; Hood, L. Shotgun glycopeptide capture approach coupled with mass spectrometry for comprehensive glycoproteomics. *Mol. Cell. Proteomics* 2007, *6*, 141–149.
8. He, H.; Conrad, C. A; Nilsson, C. L.; Ji, Y.; Schaub, T. M; Marshall, A. G.; Emmett, M. R. Method for lipidomic analysis: p53 expression modulation of sulfatide, ganglioside, and phospholipid composition of U87 MG glioblastoma cells. *Anal. Chem.* 2007, *79*, 8423–8430.
9. Wu, C. C.; Yates, J. R. The application of mass spectrometry to membrane proteomics. *Nat. Biotechnol.* 2003, *21*, 262.
10. Wallin, E.; Von Heijne, G. Genome-wide analysis of integral membrane proteins from eubacterial, archaean, and eukaryotic organisms. *Protein Sci.* 1998, *7*, 1029.
11. Stevens, T. J.; Arkin, I. T. Do more complex organisms have a greater proportion of membrane proteins in their genomes? *Proteins* 2007, *39*, 417.
12. Butterfield, D. A.; Kanski, J. Brain protein oxidation in age-related neurodegenerative disorders that are associated with aggregated proteins. *Mech. Ageing Dev.* 2001, *122*, 945–962.
13. Trimpin, S.; Deinzer, M. L. Solvent-free mass spectrometry for hydrophobic peptide sequence analysis and protein conformation studies. *Biotechniques* 2005, *39*, 799–805.
14. Pan, Y.; Stocks, B. B.; Brown, L.; Konermann, L. Structural characterization of an integral membrane protein in its natural lipid environment by oxidative methionine labeling and mass spectrometry. *Anal. Chem.* 2009, *81*, 28–35.

15. Trimpin, S. Solvent-free matrix-assisted laser desorption ionization. In *Encyclopedia of mass spectrometry: Molecular ionization*, Vol. 6. Elsevier, Maryland Heights, MO, 2006, 683–688.

16. Nordhoff, E.; Lehrach, H.; Gobom, J. Exploring the limits and losses in MALDI sample preparation of attomole amounts of peptide mixtures. *Int. J. Mass Spectrom.* 2007, *268*, 139–146.

17. Trimpin, S.; Brizzard, B. Analysis of insoluble proteins. *Biotechniques* 2009, *46*, 409–419.

18. Trimpin, S. A perspective on MALDI alternatives—Total solvent-free analysis and electron transfer dissociation of highly charged ions by laserspray ionization. *J. Mass Spectrom.* 2010, *45*, 471–485.

19. Trimpin, S.; Herath, T.N.; Inutan, E.D.; Wager-Miller, J.; Kowalski, P.; Claude, E.; Walker, J.M.; Mackie, K. Automated solvent-free matrix deposition for tissue imaging by mass spectrometry. *Anal. Chem.* 2010, *82*, 359–367.

20. Trimpin S. Alternatives in IMS-MS—Total solvent-free analysis and structures of highly charged laserspray ions. 58th ASMS Conference on Mass Spectrometry and Allied Topics May 23–27, 2010, Salt Lake City, Utah.

21. Li, L.; Mustafi, D.; Fu, Q.; Tereshko, V.; Chen, D. L.; Tice, J. D.; Ismagilov, R. F. Nanoliter microfluidic hybrid method for simultaneous screening and optimization validated with crystallization of membrane proteins. *Proc. Natl. Acad. Sci. U.S.A.* 2006, *103*, 19243–19248.

22. McComb, M.; Perlman, D.H.; Huang, H.; Costello, C.E. Evaluation of an on-target sample preparation system for matrix-assisted laser desorption/ionization time-of-flight mass spectrometry in conjunction with normal-flow peptide high-performance liquid chromatography for peptide mass fingerprint analyses. *Rapid Commun. Mass Spectrom.* 2006, *21*, 44–58.

23. Lohaus, C.; Nolte, A.; Blueggel, M.; Scheer, C.; Klose, J.; Gobom, J.; Schueler, A.; Wiebringhaus, T.; Meyer, H. E.; Marcus, K. Multidimensional chromatography: A powerful tool for the analysis of membrane proteins in mouse brain. *J. Proteome Res.* 2007, *6*, 105–113.

24. Trimpin, S.; Mixon, A. E.; Stapels, M.; Kim, M.-Y.; Spencer, P. S.; Deinzer, M. L. Identification of endogenous phosphorylation sites of bovine medium and low molecular weight neurofilament proteins by tandem mass spectrometry. *Biochemistry* 2004, *43*, 2091–2105.

25. Trimpin, S.; Herath, T. N.; Inutan, E. D.; Cernat, S. A.; Wager-Miller, J.; Mackie, K.; Walker, J. M. Field-free transmission geometry atmospheric pressure matrix-assisted laser desorption/ionization for rapid analysis of unadulterated tissue samples. *Rapid Commun. Mass Spectrom.* 2009, *23*, 3023–3027.

26. Liu, X.; Valentine, S. J.; Plasencia, M. D.; Trimpin, S.; Naylor, S.; Clemmer, D. E. Mapping the human plasma proteome by SCX-LC-IMS-MS. *J. Am. Soc. Mass Spectrom.* 2007, *18*, 1249–1264.

27. Trimpin, S.; Tan, B.; Bohrer, B. C.; O'Dell, D. K.; Merenbloom, S. I.; Pazos, M. X.; Clemmer, D. E.; Walker, J. M. Profiling of phospholipids and related lipid structures using multidimensional ion mobility spectrometry-mass spectrometry. *Int. J. Mass Spectrom.* 2009, *287*, 58–69.

28. Trimpin, S.; Tan, B.; Bohrer, B. C.; Merenbloom, S. I.; Clemmer, D. E.; Walker, J. E. Lipidome profiling with ion mobility spectrometry-mass spectrometry. 56th ASMS Conference on Mass Spectrometry and Allied Topics June 1–5, 2008, Denver, Colorado, Session: Lipid Analysis by Mass Spectrometry, 2:50 pm.

29. Trimpin, S.; Grimsdale, A. C.; Räder, H. J.; Müllen, K. Characterization of an insoluble poly (9,9-diphenyl-2,7-fluorene) by solvent-free sample preparation for MALDI-TOF mass spectrometry. *Anal. Chem.* 2002, *74*, 3777–3782.

30. Trimpin, S.; Clemmer, D. E.; McEwen, C. N. Charge-remote fragmentation of lithi-ated fatty acids on a TOF-TOF instrument using matrix-ionization. *J. Am. Soc. Mass Spectrom.* 2007, *18*, 1967–1972.

31. Trimpin, S.; Deinzer, M. L. Solvent-free MALDI-MS for the analysis of β-amyloid peptides via the Mini-Ball Mill Approach: Qualitative and quantitative improvements. *J. Am. Soc. Mass Spectrom.* 2007, *18*, 1533–1543.

32. Trimpin, S.; Deinzer, M. L. Solvent-free MALDI-MS for the analysis of a membrane protein via the Mini-Ball Mill Approach: A case study of bacteriorhodopsin. *Anal. Chem.* 2007, *79*, 71–78.

33. Trimpin, S.; Keune, S.; Räder, H. J.; Müllen, K. Solvent-free MALDI-MS: Developmental improvements in the reliability and the potential of MALDI analysis of synthetic poly-mers and giant organic molecules. *J. Am. Soc. Mass Spectrom.* 2006, *17*, 661–671.

34. Trimpin, S.; Wijerathne, K.; McEwen, C. N. Rapid methods of polymer and poly-mer additives identification: Multi-sample solvent-free MALDI, pyrolysis at AP, and atmospheric solids analysis probe mass spectrometry. *Anal. Chim. Acta* 2009, *65*, 20–25.

35. Trimpin, S.; Clemmer, D. E. Ion mobility spectrometry/mass spectrometry snapshots for assessing the molecular compositions of complex polymeric systems. *Anal. Chem.* 2008, *80*, 9073–9083.

36. Larsen, B. S.; Wijerathne, K.; Bohrer, B.; Inutan, E. D.; Karunaweera, S.; Clemmer, D. E.; Trimpin, S. Ion mobility spectrometry-mass spectrometry of star-branched poly(ethylene glycols). 57th ASMS Conference on Mass Spectrometry and Allied Topics May 31–June 4, 2009, Philadelphia, Pennsylvania, WOF 4:10 pm.

37. Merenbloom, S. I.; Koeniger, S. L.; Valentine, S. J.; Plasencia, M. D.; Clemmer, D. E. IMS-IMS and IMS-IMS-IMS/MS for separating peptide and protein fragment ions. *Anal. Chem.* 2006, *78*, 2802–2809.

38. Koeniger, S. L.; Merenbloom, S. I.; Valentine, S. J.; Jarrold, M. F.; Udseth, H.; Smith, R.; Clemmer, D. E. An IMS-IMS analogue of MS-MS. *Anal. Chem.* 2006, *78*, 4161–4174.

39. K. Merenbloom, S. I.; Bohrer, B. C.; Koeniger, S. L.; Clemmer, D. E. Assessing the peak capacity of IMS-IMS separations of tryptic peptide ions in He at 300. *Anal. Chem.* 2007, *79*, 515–522.

40. Pringle, S. D.; Giles, K.; Wildgoose, J. L.; Williams, J. P.; Slade, S. E.; Thalassinos, K.; Bateman, R. H.; Bowers, M. T.; Scrivens, J. H. An investigation of the mobility separa-tion of some peptide and protein ions using a new hybrid quadrupole/travelling wave IMS/oa-ToF instrument. *Intern. J. Mass Spectrom.* 2007, *261*, 1–12.

41. Inutan, E. D.; Trimpin, S. Laserspray ionization (LSI) ion mobility spectrometry (IMS) mass spectrometry (MS). *J. Am. Soc. Mass Spectrom.* 2010, *21*, 1260–1264.

42. Bohrer, B. C.; Merenbloom, S. I.; Koeniger, S. L.; Hilderbrand, A. E.; Clemmer, D. E. Biomolecule analysis by ion mobility spectrometry. *Ann. Rev. Anal. Chem.* 2008, *1*, 293–327.

43. Kanu, A. B.; Hill, H. H. Ion mobility spectrometry detection for gas chromatography. *J. Chromatogr. A* 2008, *1177*, 12–27.

44. McLean, J. A.; Ruotolo, B. T.; Gillig, K. J.; Russell, D. H. Ion mobility-mass spectrom-etry: A new paradigm for proteomics. *Int. J. Mass Spectrom.* 2005, *240*, 301–315.

45. Kaur-Atwal, G.; O'Connor, G.; Aksenov, A. A.; Bocos-Bintintan, V.; Paul Thomas, C. L.; Creaser, C. S. Chemical standards for ion mobility spectrometry: A review. *Int. J. Ion Mobility Spectrom.* 2009, *12*, 1–14.

46. Trimpin, S.; Inutan, E. D.; Herath, T. N.; McEwen, C. N. Matrix-assisted laser des-orption/ionization mass spectrometry method for selectively producing either singly or multiply charged molecular ions. *Anal Chem.* 2010, *82*, 11–15.

47. Trimpin, S.; Inutan E. D.; Herath, T. N.; McEwen, C. N. Laserspray ionization, a new atmospheric pressure MALDI method for producing highly charged gas-phase ions of peptides and proteins directly from solid solutions. *Mol. Cell. Proteomics* 2010, *9*, 362–367.
48. Maier, C. S.; Deinzer, M. L. Protein conformations, interactions, and H/D exchange. *Meth. Enzymology* 2005, *402*, 312–360.
49. Baumketner, A.; Bernstein, S. L.; Wyttenbach, T.; Bitan, G.; Teplow, D. B.; Bowers, M. T.; Shea, J. Amyloid β-protein monomer structure: A computational and experimental study. *Protein Sci.* 2006, *15*, 420–428.
50. Segev, E.; Wyttenbach, T.; Bowers, M. T.; Gerber, R. B. Conformational evolution of ubiquitin ions in electrospray mass spectrometry: Molecular dynamics simulations at gradually increasing temperatures. *Phys. Chem. Chem. Phys.* 2008, *10*, 3077–3082.
51. Bouschen, W.; Spengler, B. Artifacts of MALDI ample preparation investigated by high-resolution scanning microprobe matrix-assisted laser desorption/ionization (SMALDI) imaging mass spectrometry. *Int. J. Mass Spectrom.* 2007, *266*, 129–137.
52. Cohen, S. L. Ozone in ambient air as a source of adventitious oxidation. A mass spectrometric study. *Anal. Chem.* 2006, *78*, 4352–4362.
53. Froelich, J. M.; Reid, G. E. The origin and control of ex vivo oxidative peptide modifications prior to mass spectrometry analysis. *Proteomics* 2008, *8*, 1334–1345.
54. Speicher, K. D.; Kolbas, O.; Harper, S.; Speicher, D. W. Systematic analysis of peptide recoveries from in-gel digestions for protein identifications in proteome studies. *J. Biomol. Tech.* 2000, *11*, 74–86.
55. Trimpin, S. Total solvent free analysis and tissue imaging by ion mobility spectrometry-mass spectrometry. 18th International Mass Spectrometry Conference, Bremen, Germany, September 3, 2009, 11:30–11:50.
56. McEwen, C. N.; Larsen, B. S.; Trimpin, S. Laserspray ionization on a commercial AP-MALDI mass spectrometer ion source: Selecting singly or multiply charged ions. *Anal. Chem.* 2010, *82*, 4998–5001.
57. Mager, P. P. Molecular simulation of the primary and secondary structures of the Aβ (1-42)-peptide of Alzheimer's disease. *Med. Res. Rev.* 1998, *18*, 403–430.
58. Tompa, P. Structural disorder in amyloid fibrils: Its implication in dynamic interactions of proteins. *FEBS J.* 2009, *276*, 5406–5415.
59. Li, J.; Taraszka, J. A.; Counterman, A. E.; Clemmer, D. E. Influence of solvent composition and capillary temperature on the conformations of electrosprayed ions: Unfolding of compact ubiquitin conformers from pseudonative and denatured solutions. *Int. J. Mass. Spectrom.* 1999, *185/186/187*, 37–47.
60. Inutan, E. D.; Richard, A. L.; Wager-Miller, J.; Mackie, K.; McEwen, C. N.; Trimpin, S. Laserspray—A new method for protein analysis directly from tissue at atmospheric pressure with ultra high mass resolution and electron transfer dissociation. *Mol. cell. Proteomics*, under revision.

Applications

10 Snapshot, Conformation, and Bulk Fragmentation

Characterization of Polymeric Architectures Using ESI-IMS-MS

Sarah Trimpin, David E. Clemmer,
and Barbara S. Larsen

CONTENTS

10.1 INTRODUCTION

10.1.1 MOTIVATION

Polymers,[1] which are ubiquitous in modern society,[2] are chemically complex mixtures made from molecular building blocks assembled to provide two- (2-D) and three-dimensional (3-D) structures with desirable properties and functions.[3–7] More complexity is promised for future polymer generations through new synthetic methods in which the 3-D shape has functional importance similar to that observed with proteins.[8] Additionally, the increased use of polymers in association with drugs and drug delivery,[9–11] medical devices,[12] foods,[13] cosmetics,[11] perfumes,[14] and nanoparticles[3,11] is generating regulatory scrutiny that necessitates detailed manufacturing control and knowledge of composition. Partly for these reasons a workshop held in Arlington, Virginia, on the Future of Polymer Science and was attended by representatives from NIST, NSF, DOE, NIH, and NASA, who concluded, "More research is needed to improve polymer characterization techniques, tailor make polymers, and to control processing of polymers".[15]

No single analytical technique can provide detailed characterization of complex mixtures of closely related molecules as is frequently encountered with polymeric systems. Mass spectrometry (MS) is one of the most promising because of its high sensitivity and mass resolution, and especially because of its ability to characterize individual oligomeric molecules of a polymer distribution. Many polymers and polymeric blends are too complex to be detected and analyzed directly by MS techniques, and condensed-separation methods are employed but suffer the drawbacks discussed earlier.

The determination of the molecular heterogeneity of synthetic polymers (Scheme 10.1) remains an analytical challenge for a number of reasons. Besides having a molecular weight distribution, polymers have functional (end groups) and topological (blocks, random stars, etc.) variations that strongly influence their utility. In addition to these compositional, functional, and topological alterations, further complexity arises from ter- and higher order polymers, topological factors associated with isomeric compositions, as well as from blends. An example in which topology in solution is of interest is the thermosensitive behavior of dendrimers at a temperature near that of the human body, which can result in a conformational change of the molecule. Such properties might be attractive for their use as intelligent nanocapsules for drug delivery and catalysis. Other topological characteristics such as the tacticity of homopolymers are important for the structure formation in the solid state

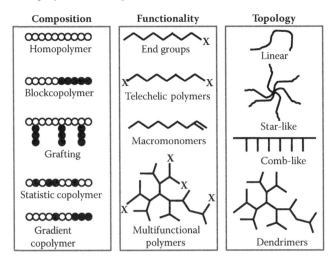

SCHEME 10.1   Types of polymeric heterogeneity influencing architecture and potential use.

(e.g., crystalline or amorphous). The formation of various microphase structures can be defined simply by changing the order and the length of blocks in a block copolymer. The reason for such behavior can be seen in a thermodynamic incompatibility of chemically different block structures. Therefore, the arrangement of blocks, the various block length, and block length distributions need to be exactly determined.

10.1.2 TRADITIONAL ANALYTICAL INSTRUMENTATION AND METHODS FOR POLYMER CHARACTERIZATION

Numerous analytical techniques have been developed or applied in polymer analyses, which led to a better understanding of polymer complexity even for insoluble polymers,[16–18] but with new knowledge came new questions.[19–21] Techniques such as size exclusion chromatography (SEC),[22] applied to bulk molecular weight distribution analysis, and contemporary "critical" chromatography,[23] separating polymers according to their end group structures and tacticity, are essential for complete polymer characterization, but neither provide molecular structural information. By measuring mass-to-charge (m/z) values, electrospray ionization (ESI) and matrix-assisted laser desorption/ionization (MALDI) MS, with superb detection sensitivities, are used to determine molecular compositions.[24–28] However, MS analysis fails to differentiate isomeric structures frequently encountered with polymeric systems (e.g., block copolymers, linear vs. starlike or branched, degree of branching, etc.) because of identical m/z values. For these reasons, the coupling of separation methods with MS for polymer analysis is especially powerful.

Condensed-phase chromatographic methods[22,23] are necessary for complete characterization of complex polymer mixtures (e.g., blends of copolymers). However, these separation methods can be problematic producing, for example, an increased number of fractions that need to be analyzed.[22] Eluting molecules at the fraction boundaries are observed multiple times, increasing the amount of data acquired and stored while reducing the sample concentration without gaining additional information. Besides the increased data and interpretation time, these boundaries limit accurate quantitative measurements. The development of "critical" separation[29] applied to many polymer heterogeneities (chirality, tacticity, block length of copolymer blocks, etc.) overcomes some condensed-phase limitations, but the determination of "critical separation conditions" remains a challenge that requires expertise and considerable effort.[29] Thus the development of "simple to use," rapid, and sensitive separation techniques to determine polymer heterogeneities is of great practical importance. ESI-IMS-MS was therefore explored to establish new methodology in hopes of answering increasingly comprehensive questions related to complex polymeric systems.[30–32]

10.1.3 IMS-MS INSTRUMENTATION METHODS FOR POLYMER CHARACTERIZATION

Synthetic polymers have been studied using IMS-MS by a number of different groups.[33–37] Here we discuss new aspects and applications of gas-phase separations using "Conformational Analysis," "Snapshots," and "Bulk Fragmentation"[30–32] obtained on homebuilt 2-m (Scheme 10.2) and 3-m drift tube instruments developed by Clemmer and coworkers.[38–40] IMS separates ions drifting in an applied electric field through a buffer gas by charge and cross section (size and shape), and MS by an ion's m/z ratio. Though homebuilt instrumentation can be extremely powerful and has led the way in demonstrating the power of IMS-MS, the instruments are limited to a few academic laboratories. Here, we also describe the applicability to solve industrial polymer challenges using commercial instrumentation (SYNAPT G2 [see Chapter 8, Figure 8.1], Waters Company, Manchester, UK).

10.2 CHARACTERIZATION OF POLYMERIC ARCHITECTURES BY MULTIDIMENSIONAL ESI-IMS-MS

One of the homebuilt multidimensional instruments incorporates two consecutive 1-m-long drift tubes (Scheme 10.2). In addition to the three drift regions the instrument incorporates three ion funnels, two ion gates, and two ion activation regions. The instrument can be operated in several different modes,[30–32,38–40] the simplest of which is the nested IMS-MS experiment. Datasets are acquired by operating all drift regions and funnels in transmission mode. Selection of ions of specified mobilities can be achieved at an ion gate and activated in the consecutive activation region.[30,31,38] If sufficient voltage is supplied in an activation experiment, fragmentation can be introduced. Selections of specific ions can be also obtained to determine collision cross sections.[30,31,38]

10.2.1 Sensitive and Rapid Analysis of Homopolymers

ESI of polyethylene glycol (PEG) provides multiply charged ions for each oligomer mass in the total mass spectrum (mass spectral result as if no IMS dimension is included) making interpretation difficult due to the ensuing mass spectral complexity (Figure 10.1, left- and right-hand panels). As can be seen in Figure 10.1, IMS-MS plot of drift time (t_d) vs. m/z vs. intensity provides an additional and powerful

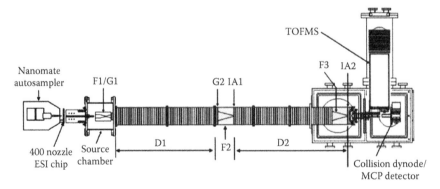

SCHEME 10.2 Schematic diagram of the 2-m ESI-IMS-IMS-MS instrument showing the ion funnels (F), the ion gates (G) and the two drift regions D1 and D2. The ESI source can be fitted with a nanomate for automatic acquisition. The mass analyzer is an orthogonal reflectron time-of-flight (TOF). For more details on the instrument refer to Trimpin, S.; Plasencia, M. D.; Isailovic, D.; and Clemmer, D. E. *Anal. Chem.* 2007, *79*, 7965–7974. (With permission.)

separation of individual polymer chains. From this dataset, the deconvolution of the isobaric multiply charged ESI ions observed in the total mass spectrum is achieved by extracting mass spectral data to obtain for each charge state baseline separated polymer distributions (Figure 10.2). This is a remarkable accomplishment, otherwise only obtainable with high-resolution mass spectrometry with considerable difficulty in spectral interpretation.[26] The charge state is "inverted" relative to what is observed for proteins and can be explained by solvation of the cation by the polymer backbone. In this synthesized sample, we observe in low abundance another polymer distribution at notable faster drift times and higher m/zs, thus smaller cross sections, which we assign to smaller oligomers. Because ions of different cross sections are time resolved in the IMS drift region, the dynamic range is enhanced in the MS dimension, similar to chromatographic methods interfaced to MS.

The usefulness for industrial challenges is shown in Figure 10.3 in which two different PEG polymers were readily characterized using IMS-MS. The traditional

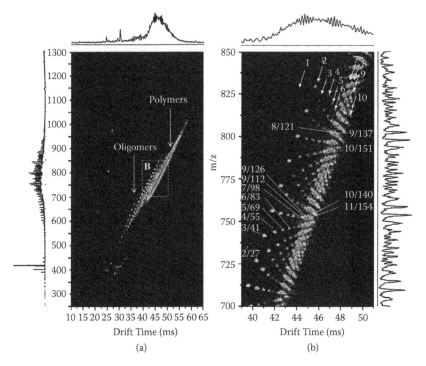

FIGURE 10.1 (a) Plot of t_d *vs.* m/z of PEG 6550 doped with cesium acetate acquired on the homebuilt IMS-MS instrument shown in **Scheme 10.1** using ESI. (b) Inset of panel (a) to show more details of the polymer distribution. The integrated mass spectrum from panel (a) is shown to the left of panel (a) and the integrated mass spectrum for panel (b) to the right of (b). Above each panel are the integrated drift time distributions. The charge state resolved components, as outlined in **Figure 10.2**, are labeled by charge state families ranging from low-intensity +2 oligomers to the more intense +10 and +11 polymers (white arrows). For more information see Trimpin, S.; Plasencia, M. D.; Isailovic, D.; and Clemmer, D. E. *Anal. Chem.* 2007, *79*, 7965–7974. (With permission.)

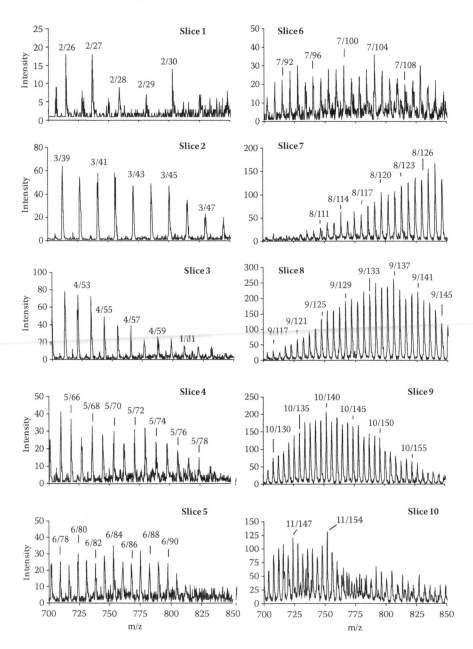

FIGURE 10.2 Mass spectra of several charge state families from PEG 6550 doped with cesium acetate. The mass spectrum obtained by integration along a diagonal in the $t_d(m/z)$ distribution of **Figure 10.1** is used to generate slices in the m/z dimension for charge state families ranging from the +2 series (slice 1) to the +11 series (slice 10). Prominent features in the charge state resolved mass spectra are labeled by charge state/PEG residue number. For more details refer to Trimpin, S.; Plasencia, M. D.; Isailovic, D.; and Clemmer, D. E. *Anal. Chem.* 2007, *79*, 7965–7974. (With permission.)

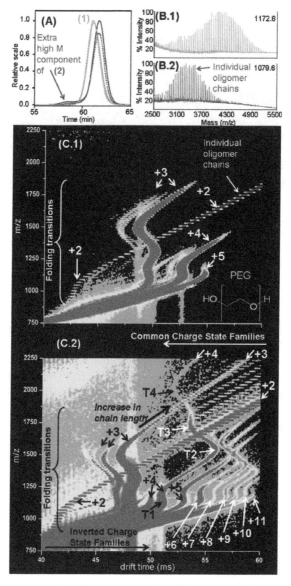

FIGURE 10.3 Analyses of two similar PEG samples, by (A) size exclusion chromatography (SEC) showing (1) PEG ~4250 (green) and (2) PEG ~3400 Da (blue), which vary only slightly in molecular weight and the lower MW PEG having a high-mass tail. The MALDI mass spectra of these polymers doped with cesium chloride (b) identifies the MW differences but not the high-mass tail on PEG 3400 (B.2). (C) 3-D IMS-MS images (color coding: red most to light blue least abundant) of the samples doped with cesium chloride as charge provider using ESI on the homebuilt 3-m drift tube IMS-MS instrument. Multiple charge state families are denoted for charge state +2 through +5 (sample 1) and +2 through +11 (sample 2), extracted from the mass spectra that are obtained by integration along a diagonal in the $[t_d(m/z)]$ distribution. The snapshot images not only show the polymers to be different but (C.2) shows the PEG 3400 has a high-mass tail. For discussion of the folding transition refer to Figure 10.8.

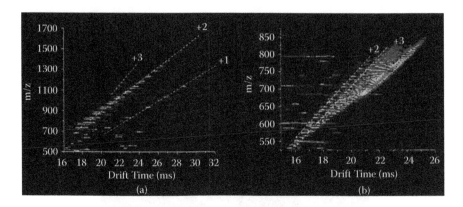

FIGURE 10.4 Snapshots of t_d *vs.* *m/z* (a) PMMA ~6 kDa and (b) PEG ~6 kDa both doped with CsCl using 60 V activation energy obtained with the 2-m drift tube instrument (Scheme 10.2). Higher charge states usually drift faster as observed in (a), but for PEG charge state inversion occurs and the consecutively higher charge states drift slower. The charge state separation of both polymers provides baseline separated ion signals in the extracted mass spectra as described in Trimpin, S.; and Clemmer, D. E. *Anal. Chem.* 2008, *80*, 9073–9083. (With permission.)

MALDI technique for polymer characterization fails to detect the high-mass tail of PEG sample 2. SEC verifies the presence of the high mass component of sample 2. The IMS-MS plot shows charge-state families that are highly specific of polymer size, type, and cation used and are thus unique to each polymer composition as one can see for poly(methyl methacrylate) (PMMA) ~6 kDa (Figure 10.4a) and PEG ~6 kDa, both doped with CsCl (Figure 10.4b).

A clear advantage of IMS-MS is the speed with which separation is achieved and the ability to automate the process. Substituting ESI-IMS-MS for LC-MS/MS reduces the time per analysis (order of milliseconds separation time). In the order of 100 samples were IMS separated and mass detected within a 12-hour time period.[31] By comparison, condensed-phase separations frequently take 1 hour per sample.[22]

10.2.2 BULK FRAGMENTATION ANALYSIS OF HOMOPOLYMERS AND BLENDS

MS/MS fragmentation of polymers is often most useful with complex molecular architecture, but any single selected mass may have multiple isomers, thus providing numerous low abundance fragments. The favorable loss of the metal cationizing agent with no charge left on the fragment ions further decreases the sensitivity.[41] The ion yields are therefore low even for multiply charged polymer ions compared to MS/MS of biological polymers such as peptides and proteins.

One way to get around the low abundances of polymer fragment ions is through bulk fragmentation, which is possible with the homebuilt IMS-MS instrument. Bulk fragmentation analysis is based on fragmenting ions without selection of drift times or *m/z* values. The improvement of bulk fragmentation for structure analysis of synthetic polymers is that the fragment ion yields are greatly enhanced because all oligomers are subjected to fragmentation simultaneously, thus common fragment ions are additive. The principle of bulk fragment ion analysis is shown in Figure 10.5 for PEG ~2 kDa using increasing energy for fragmentation (0–160 V) in the first activation region of the 2-m drift tube instrument (Scheme 10.2). Unique to this approach is that bulk fragmentation operates without the use of any m/z or t_d selection so that polymeric blends can be analyzed all at the same time (Figure 10.6a).

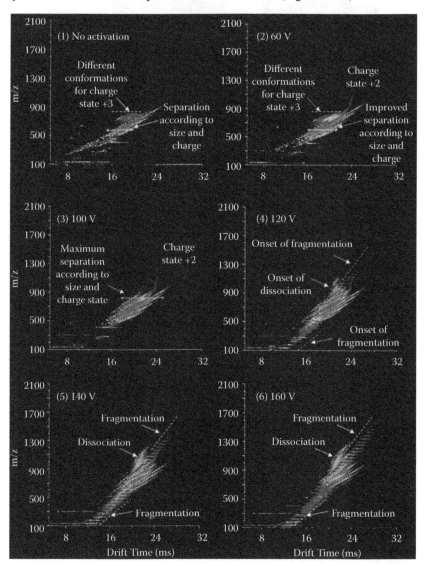

FIGURE 10.5 Visualization of changes in the t_d *vs.* m/z image of PEG ~2 kDa ionized with Cs+ upon increasing voltages [(1) 0 to (6) 160 V] applied in the middle of the drift tube (here, 2-m IMS instrument). Activation with loss of one or more charges occurs up to 100 V, whereas fragmentation is obtained greater than 140 V. For additional information see Trimpin, S.; and Clemmer, D. E. *Anal. Chem.* 2008, *80*, 9073–9083. (With permission.)

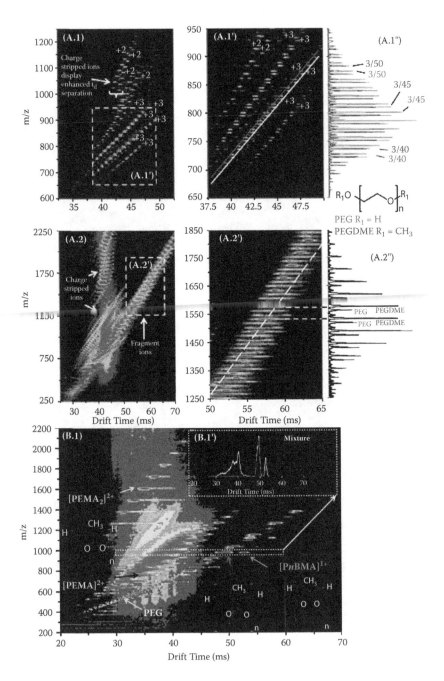

FIGURE 10.6 IMS-MS snapshots of blends of polydisperse macromolecules using the homebuilt 3-m drift tube instrument. (A) Binary blend of PEG and PEGDME (2000 Da) doped with cesium chloride and differing only by methyl groups located on the end of each

oligomer. (A.1) Full 60 V activated snapshot and inset (A.1′) of the most prominent features. Slices of one doublet feature, with charge state +3 extracted, showing the two mass spectra overlaid to the right (A.1″) labeled by charge state/number of repeat units (PEG red, PEGDME blue). (A.2) Full 200 V activation snapshot and inset (A.2′) of the region of fragmentation. A.2″; extracted mass spectrum of one slice. (B) Blend of PEMA and the two isomers, PtBMA and PnBMA (2000 Da), only differing by the side chains doped with sodium chloride along with a low-abundance PEG contamination. (B.1) Full IMS-MS snapshot image. Inset (B.1′) is the td distribution of the blend compared with the pure PtBMA sample showing separation of the two isomeric polymers. For further details see Trimpin, S.; and Clemmer, D. E. *Anal. Chem.* 2008, *80*, 9073–9083. (With permission.)

10.2.3 SNAPSHOT ANALYSIS OF BLENDS AND COPOLYMERS

MS methods measure only ion abundance and m/z ratios. However, with IMS-MS, t_d vs. m/z fingerprints can delineate isomeric composition of, for example, poly-*n*-butylmethacrylate (P*n*BMA) and poly-*t*-butylmethacrylate (P*t*BMA) in a four-component blend as shown in Figure 10.6b and verified against the drift time of the pure isomeric samples.[31] As expected, the more compact P*t*BMA isomeric structure travels faster through the drift gas than P*n*BMA, the more elongated side-chain. The ability to baseline separate ions by differences as little as the shape of isomers is a significant step forward in characterizing polymeric architectures.

The speed of analysis in IMS-MS is of less value if interpretation is a time-consuming process. Fortunately, IMS-MS is ideal for computer-generated 2-D (3-D with ion intensity) plots of t_d vs. m/z, which, similar to 2-D gels in protein analysis, allow rapid visualization of differences between samples. The "snapshot" approach is exemplified in two typical examples including homploymers with an "invisible" high-mass tail in a sole MS-based analysis but is readily observed with IMS-MS (Figure 10.3) and a copolymer versus its homopolymer blend analogue. Visualization, by eye or by pattern recognition software, can be extremely rapid complementing the speed of data acquisition by IMS-MS. Thus minor changes in composition such as the low- (Figure 10.1) and high-mass tail of a PEG polymer (Figure 10.3) are easily distinguished using the snapshot approach and the targeted areas of interest can then be analyzed.

Once a system is fully characterized, however, the snapshot approach is even more powerful in observing and understanding changes to a system with a great savings of interpretation time. Of course, this is only possible with sufficient reproducibility of the sample preparation and instrument conditions. The one-time data analysis of the remarkable distinctive snapshot pictorial of a JEFFamine copolymer (Figure 10.7) provides a good example of the power of this approach. The pictorial representation of a blend of polymers made from the same monomer groups (PEG, ~2000 Da, and polypropylene glycol [PPG], ~2200 Da) allows ready differentiation from a random copolymer of ethylene oxide (EO, which can polymerize to PEG) and propylene oxide (PO, which can polymerize to PPG) (JEFFamine, ~2000 Da), all ionized using ESI with Cs⁺ adduction.[31] Not only is the method able to distinguish each component, but the various charge states produced by ESI separate in the IMS-MS dimension so that complexity that prevented analysis by MS alone is removed. The unique 3-D image formed by the copolymer (Figure 10.7a.2 and inset

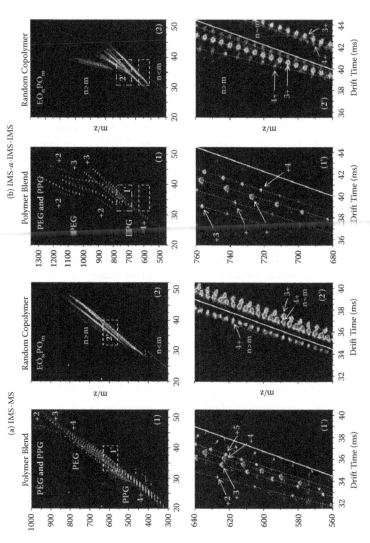

FIGURE 10.7 Image analysis (a) IMS-MS and (b) IMS-a-IMS-MS of (1) a blend of PEG (~2000 Da) and PPG (~2200 Da) with inset (1') as well as (2) random copolymer of EO and PO (JEFFamine, ~2000 Da) doped with cesium chloride with inset (2'). Inset regions (1', 2') are indicated with dashed squares in (1, 2); note that the inset regions were taken from areas with the most prominent features and thus differ between (a) and (b). A white line is included to guide the eye for changes within insets of (a) and of (b), respectively. Charge state families for both samples of different polymeric materials are denoted as extracted from the mass spectra. For additional information refer to Trimpin, S.; and Clemmer, D. E. *Anal. Chem.* 2008, *80*, 9073–9083. (With permission.)

10.7a.2′) makes this dataset visually striking and exhibits a pattern representative of a random copolymer. By employing lower activation energy at the gate as described for polymer fragmentation, it is possible to cause charge stripping and produce a new pictorial representation (Figure 10.7b) that aids in verification of composition. For example, in contradiction to the blend that shows the distinctive charge stripping of PEG (Figure 10.7B.1 and inset 10.7b.1′), the random copolymer holds the cation more tightly (Figure 10.7b.2 and inset 10.7b.2′). Thus by activation it becomes possible to distinguish the EO/PO blend from the random copolymer.

10.2.4 CONFORMATIONAL ANALYSIS OF POLYMERIC STRUCTURES

IMS-MS (Scheme 10.2) operated under low field conditions permits the determination of collision cross sections of polymeric ions that, when applied to molecular modeling approaches, can be used to gain insight into structure (Scheme 10.3). The number of charges observed on any oligomer increases with increasing size of the polymer backbone and is directly related to the density of cation solvating functional groups. Coulombic repulsion of the cations in multiply charged PEG oligomers causes the polymer backbone to elongate as can be seen in the structures labeled +2 (a pretzel-like structure), +3, and +9 (a beads-on-a-string structure) obtained from collision cross sections. This elongation causes dramatic structural changes enabling the solvent-free separation in the gas-phase of ionized oligomers by charge and cross section.

Unique to IMS and demonstrating its ability to interrogate ionic structures, folding transitions are observed upon increasing polymerization degree for the PEG ions. [31] These folding transitions (Ts) occur at a specific polymer m/z (~900–1200) for each charge state, +2 to +11 (Figure 10.8a), showing that these multiply charged ions begin to fold, producing increasingly globular structures in which the molecular size (as measured by t_d) does not increase and may actually decrease.[31] These transitions are seen in Figure 10.8 as distinctive sigmoidal contours. The structural differences in shape shown in Figure 10.8c are obtained from preliminary modeling based on the experimentally determined cross-section analysis. Interestingly, PEG with 170

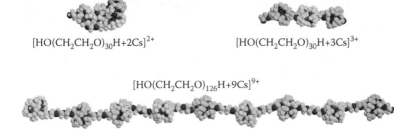

$[HO(CH_2CH_2O)_{30}H+2Cs]^{2+}$ $[HO(CH_2CH_2O)_{30}H+3Cs]^{3+}$

$[HO(CH_2CH_2O)_{126}H+9Cs]^{9+}$

SCHEME 10.3 Linear structure modeled for PEG from experimentally derived cross sections from ions having 30 repeat units ionized with two and three Cs+ cations and PEG with 126 repeat units and ionized by adduction of nine Cs+ cations. For more details refer to Trimpin, S.; Plasencia, M. D.; Isailovic, D.; and Clemmer, D. E. *Anal. Chem.* 2007, *79*, 7965–7974. (With permission.)

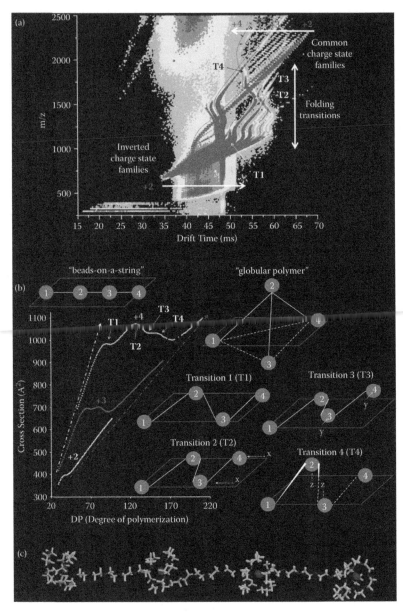

FIGURE 10.8 Folding transitions of PEG ~3400 with Cs⁺ as charge carrier. (a) t_d *vs. m/z* snapshot. (b) Cross sections were extracted from (a) for each charge state and plotted against the degree of polymerization. Various folding transitions are observed for each charge state. With increasing numbers of charges the number of folding transitions increases. To the right are possible folding pathways that may describe the drastic structural changes from "beads-on-a-string" to the "globular polymer" for the polymer with 4 Cs⁺ cations attached. (c) The PEG oligomer with 55 repeat units and 4 Cs⁺ ions with an experimental cross-section of 714.5 Å² is best described as "beads-on-a-string." For further description of the folding transition see Trimpin, S.; and Clemmer, D. E. *Anal. Chem.* 2008, *80*, 9073–9083. (With permission.)

monomer units (charge state +4) is found to be significantly smaller (~1000 A^2) than PEG with 120 monomer units (~1080 A^2) and very similar in size to PEG with 80 monomer units even though the oligomers differ in molecular weights by more than a factor of two. When a critical mass is reached for any charge state, the coulombic forces are insufficient to overcome hydrogen bonding and van der Waals forces, producing 2-D plots of t_d vs. m/z that are highly specific of polymer size. These results correlate well with those from Ude et al.[36] for the compilation of all the Z(m,z) data obtained for PEG samples doped with ammonium acetate.

10.3   CHARACTERIZATION OF INDUSTRIAL POLYMERIC ARCHITECTURES USING COMMERCIAL IMS-MS INSTRUMENTATION

The homebuilt IMS-MS instrument was initially used to define high-resolution drift time separation characteristics of star-branched 4-arm poly(ethylene glycol) (PEG) obtained from the DuPont Company (Figure 10.9a).[32] Results are compared to those obtained on the commercial IMS-MS SYNAPT G2 instrument (Figure 10.9b). The combination of adequate IMS drift time resolution, high m/z resolution, high sensitivity, and the ease of operation and speed of data manipulation makes the new SYNAPT G2 (see Chapter 8, Figure 8.1) an especially useful analytical tool. Both IMS-MS instruments were able to separate by shape differences as small as a single ethylene oxide unit. These studies indicate that the multiple charging imparted by ESI is crucial for separation of arm-length differences. This can be rationalized by what we learned from the collision–cross section analysis described above. That is, crucial to the analytical success is the ability to introduce charge repulsion into the polymeric molecules being separated by using an ionization method that produces multiply charged ions. Consequently the doubly and not singly charged ions provided sufficiently different shape to be separated. In the molecular weight range (~800 to ~2000 Da) studied, the branched 4-arm polymers did not show charge inversion as expected because they are inherently more globular than multiply charged linear PEG. The linear PEG component of the mixture did show charge inversion where consecutively higher charge state ions travel slower through the drift region.

   The isomers of these branched polymers offer an opportunity to observe drift time resolution differences in the homebuilt 3-m drift tube instrument (similar to Scheme 10.2) and the SYNAPT G2 instrument. Neither instrument shows any differences in the drift time distributions for the singly charged ions of the 4-arm polymer; however, for the doubly charged ions, isomeric structures are apparent with both instruments. Figure 10.9 shows a comparison of the same 4-arm PEG polymer doped with lithium acetate acquired on both instruments. Figure 10.9a (inset) is the drift time plot of the m/z 492 ion. In comparison, Figure 10.9b (inset) shows the drift time results obtained for this polymer on the SYNAPT G2. Clearly the homebuilt instrument has superior drift time resolution, cleanly separating at least three isomeric structures in the doubly charged ions as well as separating the doubly (25–28 ms) from the singly (28–29 ms) charged ions. With the commercial instrument, the doubly charged ions are also drift-time separated from the singly (bins 40–50) charged

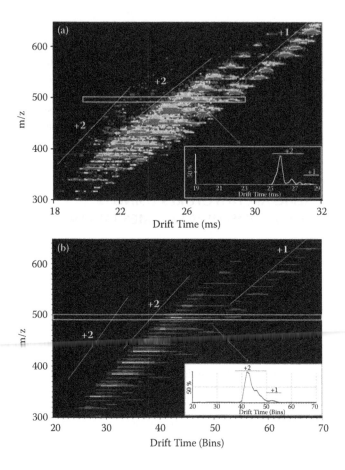

FIGURE 10.9   IMS-MS t_d *vs.* *m/z* snapshot of the acquisition of a 4-arm PEG 800 using ESI with ionization by Li+. (a) Data obtained from the homebuilt 3-m IMS-MS instrument with the inset showing the drift time distribution for *m/z* 492. (b) Data obtained using the SYNAPT G2 with the inset showing the drift time distribution at *m/z* 492. For additional information see Larsen, B. S.; Wijerathne, K.; Bohrer, B.; Inutan, E. D.; Karunaweera, S.; Clemmer, D. E.; Trimpin, S. Ion mobility spectrometry-mass spectrometry of star-branched poly(ethylene glycols). 57th ASMS Conference on Mass Spectrometry and Allied Topics May 31–June 4, 2009, Philadelphia, Pennsylvania, and Trimpin, S.; Larsen, B. S. Ion Mobility Spectrometry-Mass Spectrometry for Star-branched Poly(Ethylene Glycols). Symposium on IMS-MS for Polymer Analysis. Pittcon, February 28–March 5, 2010, Orlando, Florida, 1450–2. (With permission.)

ions (bins 50–55). It is obvious with the commercial instrument that the doubly charged ions have a broadened drift time distribution and at least two inflections are observed demonstrating isomeric separation. In neither instrument did the singly charged ions show isomeric structure.

An advantage of the commercial IMS-MS instrument is the resolution obtained in the MS dimension. An example is given for a protein obscured by PEG, something occasionally observed due to contamination during workup or an incomplete reaction of a pegylated protein. Figure 10.10B shows the t_d vs. m/z snapshot of a mixture of 4-arm PEG 800 doped with sodium chloride and the protein ubiquitin, and Figure 10.10A shows the mass spectrum of the mixture in which the ubiquitin component is not apparent. However, by interrogating the features within Figure 10.10B′ in the t_d vs. m/z plot, the presence of ubiquitin becomes readily apparent (Figure 10.10C1

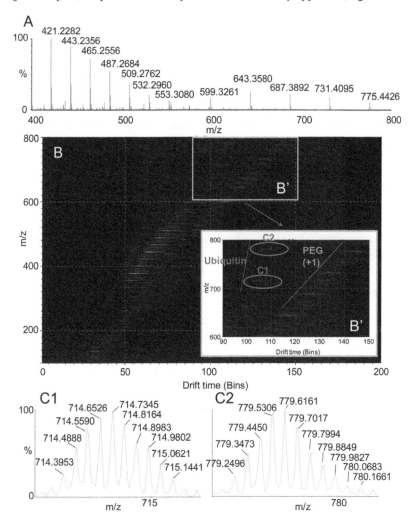

FIGURE 10.10 IMS-MS acquisition of a mixture of 4-arm PEG 800 and the protein ubiquitin ionized using ESI with Na+ charge carrier. (A) Mass spectrum of the mixture without IMS separation demonstrating the difficulty of observing the ubiquitin component using a solely MS approach. (B) Snapshot of t_d vs. m/z of the mixture showing a low abundant feature within inset B. (C1) +12 and (C2) +11 charge states and respective isotopic distributions in the mass spectral Inset extracted from the region shown in B′ rapidly determine the identity of ubiquitin. For further details see Trimpin, S.; Larsen, B. S. Ion Mobility Spectrometry-Mass Spectrometry for Star-branched Poly(Ethylene Glycols). Symposium on IMS-MS for Polymer Analysis. Pittcon, February 28–March 5, 2010, Orlando, Florida, Tuesday afternoon 2:40 1450–2. (With permission.)

and 10.10C2). The ^{13}C isotope peaks for these charge states are almost baseline resolved (Figure 10.10C), allowing the charge state and thus the molecular weight to be readily determined even without using the "W" mode, which provides even higher mass resolution.

10.4   SCOPE OF POTENTIAL USE

Developments that lead to the rapid and reliable analysis of polymeric architectures will have wide impact in quality control, regulatory, and health related areas. The IMS-MS snapshot method is especially useful in determining small changes in complex systems provided sample preparation/ionization and instrument conditions are identical between analyses. Interfacing various condensed-phase separation methods to an IMS-MS instrument will enhance mass-selective separations and detection of complexity, even allowing the differentiation of isomers and wide polydisperse polymers,[42,43] and providing improved dynamic range to the MS as well as the opportunity for architectural analysis.[45] Additionally, the complexity and challenges arising from poorly soluble mixtures may likely be addressed using total solvent-free analysis by MS (Chapter 9). Further, laserspray ionization, a new ionization approach for producing multiply charged ions (Chapter 8 and 9) from the solid state and directly from a surface promises more future advances in polymer analyses,[46] especially in combination with IMS-MS.

ACKNOWLEDGMENTS

The author gratefully acknowledges the indispensable contributions to this chapter during her education by Prof. Klaus Müllen and Dr. Hans-Joachim Räder (Max-Planck-Institute for Polymer Research, Mainz, Germany), Prof. David E. Clemmer (Indiana University) and Prof. J. Michael Walker (Indiana University 1950–2008). Collaborators, Prof. Charles N. McEwen (The University of the Sciences in Philadelphia) is gratefully acknowledged for providing instrument access and samples for supporting measurements. Financial support is also gratefully acknowledged from National Science Foundation CAREER 0955975, Wayne State University (start-up funds), American Society for Mass Spectrometry Research Award (financially supported by Waters Company), and DuPont Company Young Investigator Award.

REFERENCES

1. Staudinger, H. Ber. Über polymerization. *Dtsch. Chem. Ges. B: Abhandlungen* 1920, *53*, 1073–1085.
2. Ober, C. K. Polymer science—Shape persistence of synthetic polymers. *Science* 2000, *288*, 448–449.
3. Frechet, J. M. J. Functional polymers and dendrimers—Reactivity, molecular architecture, and interfacial energy. *Science* 1994, *263*, 1710–1715.
4. Stupp, S. I.; LeBonheur, V. V.; Walker, K.; Li, L. S.; Huggins, K. E.; Keser, M.; Amstutz, A. Supramolecular materials: Self-organized nanostructures. *Science* 1997, *276*, 384–389.

5. Friend, R. H.; Gymer, R. W.; Holmes, A. B.; Burroughes, J. H.; Marks, R. N.; Taliani C.; Bradley, D. D. C.; Dos Santos, D. A.; Bredas, J. L.; Logdlund, M.; Salaneck, W. R. Light emitting-diodes based on conjugated polymers. *Nature* 1999, *397*, 121–128.

6. Forrest, S. R. The path to ubiquitous and low-cost organic electronic appliances on plastic. *Nature 428*, 911–918.

7. Hoger, S. Shape-persistent macrocycles: From molecules to materials, *Chem. Europ. J.* 2004, *10*, 132–1329.

8. Hawker, C. J.; Wooley, K. L. The convergence of synthetic organic and polymer chemistries. *Science* 2005, *309*, 1200–1205.

9. Discher, D. E.; Eisenberg, A. Polymer vesicles, *Science* 2002, *297*, 967–973.

10. Allen, T. M.; Cullis, P. R. Drug delivery systems: Entering the mainstream. *Science* 2004, *303*, 1818–1822.

11. Nel, A.; Xia, T.; Madler, L.; Li, N. Toxic potential of materials at the nanolevel. *Science* 2006, *311*, 622–627.

12. Langer, R.; and Tirrell, D. A. Designing materials for biology and medicine, *Nature* 2004, *428*, 487–492.

13 Kaiser, J. Endocrine disrupters—Controversy continues after panel rules on bisphenol A. *Science* 2007, *317*, 884–885.

14. Fieber, W.; Herrmann, A.; Ouali, L.; Velazco, M. I.; Kreutzer, G.; Klok, H.-A.; Ternat, C.; Plummer, C. J. G.; Månson, J.-A. E.; Sommer, H. NMR diffusion and relaxation studies of the encapsulation of fragrances by amphiphilic multiarm star block copolymers. *Macromolecules* 2007, *40*, 537–5378.

15. People.ccmr.cornell.edu/~cober/nsfpolymerworkshop/index.html

16. Trimpin, S.; Grimsdale, A. C.; Räder, H. J.; Müllen, K. Characterization of an insoluble poly(9,9-diphenyl-2,7-fluorene) by solvent-free sample preparation for MALDI-TOF mass spectrometry. *Anal. Chem.* 2002, *74*, 3777–3782.

17. Trimpin, S.; Keune, S.; Räder, H. J.; Müllen, K. Solvent-free MALDI-MS: Developmental improvements in the reliability and the potential of MALDI analysis of synthetic polymers and giant organic molecules. *J. Am. Soc. Mass Spectrom.* 2006, *17*, 661–671.

18. Trimpin, S.; Weidner, S.M.; Falkenhagen, J.; McEwen, C.N. Fractionation and solvent-free MALDI-MS analysis of polymers using liquid adsorption chromatography at critical conditions in combination with a novel multi-sample on-target homogenization/transfer sample preparation method. *Anal. Chem.* 2007, *79*, 7565–7570.

19. Hanton, S. D. Mass spectrometry of polymers and polymer surfaces, *Chem. Rev.* 2001, *101*, 527–569.

20. Montaudo, G.; Samperi, F.; Montaudo, M. S. Characterization of synthetic polymers by MALDI-MS. *Prog. Polym. Sci.* 2006, *31*, 277–357.

21. Weidner, S. M.; and Trimpin, S. Mass spectrometry of synthetic polymers. *Anal. Chem.* 2008, *80*, 4349–4361.

22. Trimpin, S.; Eichhorn, P.; Räder, H. J.; Müllen, K.; Knepper, T.P. Recalcitrance of poly(vinylpyrrolidone): Evidence through matrix-assisted laser desorption/ionization time-of-flight mass spectrometry. *J. Chrom. A* 2001, *938*, 67–77.

23. Gorshkov, A. V.; Much, H.; Becker, H.; Pasch, H.; Evreinov, V. V.; Entelis, S. G. Chromatographic investigations of macromolecules in the "critical range" of liquid chromatography. I. Functionality type and composition distribution in polyethylene oxide and polypropylene oxide copolymers. *J. Chrom.* 1990, *523*, 91–102

24. Brown, R. S.; Weil, D. A.; Wilkins, C. L. Laser desorption Fourier-transform mass-spectrometry for the characterization of polymers. *Macromolecules* 1986, *19*, 1255–1260.

25. Wong, S. F.; Meng, C. K.; Fenn, J. B. Multiple charging in electrospray ionization of poly(ethylene glycols). *J. Phy. Chem.* 1988, *92*, 546–550.

26. O'Connor, P. B.; McLafferty, F. W. Oligomer characterization of 4-23 kDa polymers by electrospray Fourier transform mass spectrometry. *J. Am. Chem. Soc.* 1995, *117*, 12826–12831.

27. McEwen, C. N.; Simonsick, W. J.; Larsen, B. S.; Ute, K.; Hatada, K. The fundamentals of applying electrospray ionization mass spectrometry to low mass poly(methyl methacrylate) polymers. *J. Am. Soc. Mass Spectrom.* 1995, *6*, 906–911.

28. Falkenhagen, J.; Weidner, S. M. Detection limits of matrix-assisted laser desorption/ionisation mass spectrometry coupled to chromatography—A new application of solvent-free sample preparation. *Rapid Commun. Mass Spectrom.* 2005, *19*, 3724–3730.

29. Weidner, S.; Falkenhagen, J.; Krueger, R. P.; Just, U. Principle of two-dimensional characterization of copolymer. *Anal. Chem.* 2007, *79*, 4814–4819.

30. Trimpin, S.; Plasencia, M. D.; Isailovic, D.; Clemmer, D. E. Resolving oligomers from fully grown polymers with IMS-MS. *Anal. Chem.* 2007, *79*, 7965–7974.

31. Trimpin, S.; Clemmer, D. E. Ion mobility spectrometry/mass spectrometry snapshots for assessing the molecular compositions of complex polymeric systems. *Anal. Chem.* 2008, *80*, 9073–9083.

32. Larsen, B. S.; Wijerathne, K.; Bohrer, B.; Inutan, E. D.; Karunaweera, S.; Clemmer, D. E.; Trimpin, S. Ion mobility spectrometry-mass spectrometry of star-branched poly(ethylene glycols). 57th ASMS Conference on Mass Spectrometry and Allied Topics May 31–June 4, 2009, Philadelphia, Pennsylvania, WOF 4:10 pm.

33. Wyttenbach, T.; von Helden, G.; Bowers, M. T. Conformations of alkali ion cationized polyethers in the gas phase: Polyethylene glycol and bis[(benzo-15-crown-5)-15-ylmethyl] pimelate. *Int. J. Mass Spectrom.* 1997, *165/166*, 377–390.

34. Gidden, J.; Wyttenbach, T.; Jackson, A. T.; Scrivens, J. H.; Bowers, M. T. Gas-phase conformations of synthetic polymers: Poly(ethylene glycol), poly(propylene glycol), and poly(tetramethylene glycol). *J. Am. Chem. Soc.* 2000, *122*, 4692–4699.

35. Robinson, E. W.; Sellon, R. E.; Williams, E. R. Peak deconvolution in high-field asymmetric waveform ion mobility spectrometry (FAIMS) to characterize macromolecular conformations. *Int. J. Mass Spectrom.* 2007, *259*, 87–95.

36. Ude, S.; de la Mora, J. F.; Thomson, B. A. Charge-induced unfolding of multiply charged polyethylene glycol ions. *J. Am. Chem. Soc.* 2004, *126*, 12184–12190.

37. Bagal, D.; Zhang, H.; Schnier, P. D. Gas-phase proton-transfer chemistry coupled with TOF mass spectrometry and ion mobility-MS for the facile analysis of poly(ethylene glycols) and PEGylated polypeptide conjugates. *Anal. Chem.* 2008, *80*, 2408–2418.

38. Merenbloom, S. I.; Koeniger, S. L.; Valentine, S. J.; Plasencia, M. D.; Clemmer, D. E. IMS-IMS and IMS-IMS-IMS/MS for separating peptide and protein fragment ions. *Anal. Chem.* 2006, *78*, 2802–2809.

39. Koeniger, S. L.; Merenbloom, S. I.; Valentine, S. J.; Jarrold, M. F.; Udseth, H.; Smith, R.; Clemmer, D. E. An IMS-IMS analogue of MS-MS. *Anal. Chem.* 2006, *78*, 4161–4174.

40. Merenbloom, S. I.; Bohrer, B. C.; Koeniger, S. L., Clemmer, D. E. Assessing the peak capacity of IMS-IMS separations of tryptic peptide ions in He at 300 K. *Anal. Chem.* 2007, *79*, 515–522.

41. Laine, O.; Trimpin, S.; Räder, H. J.; Müllen, K. Europ. changes in post-source decay fragmentation behavior of poly(methylmethacrylate) polymers with increasing molecular weight studied by matrix-assisted laser desorption/ionization time-of-flight mass spectrometry. *J. Mass Spectrom.* 2003, *9*, 195–201.

42. Martin, K.; Spickermann, J.; Räder, H. J.; Müllen, K. Why does matrix-assisted laser desorption/ionization time-of-flight mass spectrometry give incorrect results for broad polymer distributions? *Rapid Commun. Mass Spectrom.* 1996, *10*, 1471–1474.

43. Montaudo, G.; Scamporrino, E.; Vitalini, D.; Mineo, P. Novel procedure for molecular weight averages measurement of polydisperse polymers directly from matrix-assisted laser desorption/ionization time-of-flight mass spectra. *Rapid Commun. Mass Spectrom.* 1996, *10*, 1551.
44. Trimpin, S.; Larsen, B. S. Ion Mobility Spectrometry-Mass Spectrometry for Star-branched Poly(Ethylene Glycols). Symposium on IMS-MS for Polymer Analysis. Pittcon, February 28–March 5, 2010, Orlando, Florida, Tuesday afternoon 2:40 (1450–2).
45. Trimpin, S.; Hoskins, J.; Wijerathne, K.; Inutan, E. D. Grayson, S. M. Library of Polymer Architectures Examined by Ion Mobility Spectrometry-Mass Spectrometry. 58th ASMS Conference on Mass Spectrometry and Allied Topics, May 23–27, 2010, Salt Lake City, Utah, MOC 2:50 pm.
46. Trimpin, S.; Inutan, E. D.; Herath, T. N.; McEwen, C. N. Matrix-assisted laser desorption/ionization mass spectrometry method for selectively producing either singly or multiply charged molecular ions. *Anal Chem.* 2010, *82*, 11–15.

11 Metabolomics by Ion Mobility–Mass Spectrometry

Kimberly Kaplan and Herbert H. Hill, Jr.

CONTENTS

11.1 ANALYTICAL METHODS FOR MEASURING METABOLOMES

One of the major applications of ion mobility and ion mobility–mass spectrometry (IMS and IM-MS) is in the separation, detection, and identification of metabolites in complex biological matrices in an analytical process known as metabolomics. Standard analytical methods, such as gas chromatography (GC), liquid chromatography (LC), capillary electrophoresis (CE), LC-MS, and nuclear magnetic resonance (NMR), have limitations of speed, specificity, and/or sensitivity when applied to metabolic analysis and metabolomics. Emerging analytical methods of ion mobility and ion mobility–mass spectrometry offer complementary approaches for separating, detecting, and quantifying metabolites in complex biological mixtures.

Metabolomics is a developing field in bioanalytical chemistry with a primary focus on early disease detection and prevention. An individual's metabolome is uniquely dynamic and based on genetics, environment, and nutrition.[1] Metabolites are small molecules (<1000 Da) that range in extremes of polarity, volatility, solubility, and concentration. They represent the end product of various biological regulatory processes,[2] and their chemical diversity creates a significant challenge for their qualitative and quantitative analysis. Currently five different strategies exist for metabolomic analysis:

1. Target metabolomics: Analysis restricted to specific metabolites (e.g., determining glucose levels in blood).
2. Metabolite profiling: Analysis focused on a group of metabolites such as carbohydrates, amino acids, or those associated with a specific pathway.
3. Metabolomics: Comprehensive analysis of the whole metabolome under a given set of conditions.
4. Metabolic fingerprinting: Classification of samples on the basis of either their biological relevance or origin. Metabolic fingerprinting is commonly used in clinical and pharmaceutical analysis to trace the fate of a drug or metabolite.
5. Metabonomics: Measure of the fingerprint of biochemical perturbations caused by disease, drugs, and toxins.[3]

The first two approaches, target metabolomics and metabolite profiling, focus on the detection and quatification of biomarkers. Biomarkers are specific metabolites used to distinguish a normal biological state from a pathogenic process.[4] Analytical methods have been developed to monitor selected biomarkers by well-calibrated methods of analyses.[3] If a patient is already sick, target or profile metabolomics are capable of fast diagnostics since patient symptoms can narrow the analysis to the most affected metabolite(s).

Metabolomics is the comprehensive analysis of an individual's metabolome that provides insight to prevent illness or disease by monitoring metabolic shifts in the overall metabolome prior to disease or infection.[3] The ability to measure a wide range of diverse chemical compounds rather than a handful of specific targeted compounds is the primary analytical challenge of metabolomics. One of the major analytical limitations in the field of metabolomics is the inability of current commercially available instruments to measure all metabolites simultaneously.[5] Conventional methods commonly applied to blood serum and plasma metabolomics include gas chromatography–mass spectrometry (GC-MS),[4] liquid chromatography–mass spectrometry (LC-MS),[6] proton nuclear magnetic resonance (^{1}H-NMR),[7,8] Fourier transform infrared spectroscopy (FT-IR),[9] and capillary electrophoresis–mass spectrometry (CE-MS).[10] A summary of the advantages and disadvantages for the different conventional methods of analysis can be found in Table 11.1.

In addition to the disadvantages and limitations listed in Table 11.1, sample preparation for conventional instruments can be labor intensive and time-consuming. Common metabolite extraction methods for serum include liquid-liquid extraction,[11] protein precipitation by the addition of an organic solvent such as methanol or acetonitrile,[12] solid-phase extraction (SPE),[13] and acid protein precipitation.[12] Most extraction procedures reported in the literature miss metabolites and result in the representation of only a portion of the metabolome (i.e., metabolite profiles).[3] The ideal instrument must be able to measure polar and nonpolar compounds, have a large dynamic range, provide rapid measurements, exhibit universal response, and to be compatible with a simple but efficient extraction method.

TABLE 11.1
Summary of Advantages and Disadvantages from Conventional Instrumentation Methods Used in Metabolomics

Analytical Method	Advantage	Disadvantage
NMR	• Rapid analysis • High resolution • No derivatization needed • Nondestructive	• Low sensitivity • Convoluted spectra • More than one peak per component • Libraries of limited use due to complex matrix
GC-MS	• Sensitive • Robust • Large linear range • Large commercial and public library	• Slow • Often requires derivatization • Many analytes thermally unstable or too large for analysis
LC-MS	• No derivatization required (usually) • Many modes of separation available • Large sample capacity	• Slow • Limited commercial libraries
CE-MS	• High separation power • Small sample requirements • Rapid analysis • Can analyze neutrals, anions, and cations in a single run • No derivatization (usually)	• Limited commercial libraries • Poor retention time reproducibility
FTIR	• Rapid analysis • Complete fingerprint of sample chemical composition • No derivatization needed	• Extremely convoluted spectra • More than one peak per component • Metabolite identification nearly impossible • Requires sample drying

Source: Reprinted from Shulaev, V. Metabolomics Technology and Bioinformatics. *Briefings in Bioinformatics.* 2006, 7, 128–139. Copyright (2006) with permission from Oxford University Press.

Ion mobility and ion mobiltiy–mass spectrometry offer a number of advantages for the field of metabolomics. These include the following: (1) noise reduction with a concomitant increase in signal-to-noise ratio; (2) rapid separation of ions based on size/charge ratio; (3) ion filter to help keep the mass spectrometer clean from contamination with complex samples; (4) good reproducibility; (5) qualitative analysis using mobility-mass correlation; (6) ease of quantification with isotope dilution; and (7) excellent compatabilty with both chromatography and mass spectrometry for two to three dimensions of separation. This chapter provides examples of these value-added data when ion mobility is used for metabolic analysis.

11.2 ION MOBILITY SPECTROMETRY BACKGROUND

Ion mobility spectrometry is a rapid gas-phase separation technique that has commonly been used to separate small molecules such as drugs and explosives.[14] A drift tube ion mobility spectrometer (DTIMS) consists of alternating conducting rings and insulators where the conducting rings are connected in series with resistors to create a uniform electric field (E). DTIMS usually has two sections, a reaction region and a drift region, that are separated by an ion gate. When electrospray ionization or a ^{63}Ni radioactive foil is used as the ionization source, ions enter the reaction region where they are desolvated or undergo chemical ionization. Stable product ions are then pulsed by an ion gate into the drift region of the spectrometer. A counter current gas flow is introduced into the DTIMS to keep neutral contaminants out of the drift region. Desolvated ions in the drift region of the spectrometer are accelerated in the electric field and decelerated by collisions with the buffer gas. Repetition of this acceleration-deceleration process over the length of the drift tube (L) results in a constant ion velocity (v_d), which is proportional to the electric field (E) such that the ratio of ion velocity to electric field is a constant called mobility (K).

$$K = \frac{v_d}{E} = \frac{L^2}{t_d V} \tag{11.1}$$

where V is the voltage drop over the drift region and t_d is the time the ion takes to migrate through the drift tube. The reduced mobility constant (K_0) incorporates the pressure (P) and temperature of the drift region (T) as shown in Equation (11.2).

$$K_0 = \frac{L^2}{t_d V} \frac{273}{T} \frac{P}{760} \tag{11.2}$$

For accurate determination of mobility, the drift time (t_d) of the ion must be corrected for half of the gate pulse width (t_g) and the time the ion spends in the mass spectrometer ($t_{m/z}$). For instance, for a time-of-flight mass spectrometer (TOF) with a pinhole interface and efficient focusing lens, ions are detected in microseconds; therefore, the time the ion spends in the TOF is negligible.

$$t_d = t_{\text{measured}} - 0.5t_g - t_{m/z} \tag{11.3}$$

Once the corrected drift time is calculated, the collision cross section (Ω) for that ion can be determined.

$$\Omega = \frac{(18\pi)^{\frac{1}{2}}}{16} \frac{ze}{(k_b T)^{\frac{1}{2}}} \left[\frac{1}{m_I} + \frac{1}{m_B} \right]^{\frac{1}{2}} \frac{t_d E}{L} \frac{760}{P} \frac{T}{273} \frac{1}{N_0} \tag{11.4}$$

where ze is the ion's charge; k_b is Boltzmann's constant; m_I and m_B are the masses of the ion and buffer gas, respectively; and N_0 is the number density at standard temperature and pressure. Ion mobility separations are dependent on the collision cross section of the ion.

Other commercially available ion mobility spectrometers used in metabolomics include traveling wave (T-wave) IMS[15,16] and high field asymmetric ion mobility spectrometry (FAIMS).[16–18] The difference between drift time IMS and T-wave

IMS is the mechanism of the electric field used to separate the ions. A T-wave IMS differs from drift time IMS through the use of low-voltage waves to push ions down the drift tube instead of the constant electric field used in a drift time IMS. With the T-wave, the separation has not been sufficiently characterized to determine collision cross section directly from drift times. T-wave IMS has to be calibrated against a drift time IMS.[15]

FAIMS is significantly different than drift-time and T-wave IMS. A FAIMS device is a continuous flow-through ion filter where a radio frequency (RF) voltage forces ions to oscillate between two electrodes as they are carried along by a stream of gas. [19] FAIMS has been referred to as an "ion mobility filter instead of an ion mobility analyzer."[16] Both FAIMS and T-wave IMS systems are relatively low-resolution ion mobility separation devices. Thus the primary focus of this chapter is on the DTIMS, which can have sufficient resolution to separate ions in complex mixtures based on ion mobility alone.

11.3 VOLATILE METABOLITES

Gas chromatography coupled with ion mobility spectrometry provides a rapid non-invasive method for the determination of volatile metabolites. One instrument that is commercially available and has been used for metabolomics incorporates multicapillary columns (MCCs) for the preseparation of metabolite mixtures prior to ion mobility spectrometry. The MCCs consist of ~1000 glass capillaries held at a constant temperature (30°C). Coupling the GC retention times with the ion mobility spectra provides two-dimensional spectra.[20]

Figure 11.1a illustrates a two-dimensional GC-IMS spectrum in which the chromatogram is shown on the x-axis and is obtained within 120 seconds (s) and the ion mobility spectrum is shown on the y-axis and is obtained within 24 milliseconds (ms). Thus many ion mobility spectra are obtained for each gas chromatogram. The sample shown in Figure 11.1a is of the breath of a human who was infected with *angina lateralis*, a bacterial infection that causes inflammation of the throat. As shown in Figure 11.1a, this infection produced a major metabolite that occurred at a chromatographic retention time of 56 s. Several ion mobility spectra of this compound are shown in Figure 11.1b. As can be seen from the mobility spectra that were obtained at a specific retention time (56 s in this case), the ion mobility spectra are relatively clean and uncluttered with peaks from other metabolites. Chromatographic preseparation improves both the resolving power of the IMS as well as the quantification. The most intense peak in Figure 11.1b was obtained 24 hours after treatment of the patient with an antibiotic and the other spectra were taken at 48 and 72 hours after treatment. Figure 11.1c shows the intensity of the ion mobility peak as a function of time after treatment and demonstrates the exponential decline in concentration of the biomarker after treatment. Unfortunately, the biomarker could not be identified from the GC-IMS spectra and illustrates the need for the addition of mass spectrometry to aid in the identification of unknown compounds.

For these experiments, a radioactive ionization source (^{63}Ni) was used to create reactant ions that chemically ionize the analyte to form unique response ions. ^{63}Ni

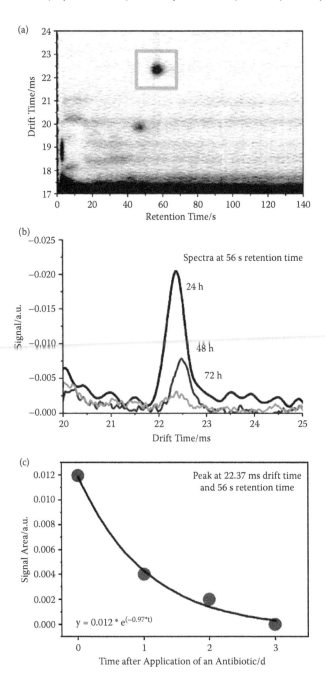

FIGURE 11.1 ^{63}Ni-MCC-IMS topographic plot of a breath analysis of a patient diseased with *angina lateralis* before medication (a); drift time spectra of the relevant peak during medication (b) and signal area of the peak during medication (c). (Reprinted from Vautz, W. & Baumbach, J. I. *Int. J. Ion Mobil. Spec.* 2008, 11, 35–41. Copyright 2008, with permission from Springer-Verlag.)

is a β-emitter, a source that emits electrons that can ionize the drift gas that reacts typically with water or water clusters to cause the formation of a reactant ion.[21] The reactant ion concentration stays constant if nothing else is present; however, as an analyte is introduced, the reactant ion will decrease and the analyte ion will increase proportional to the concentration as shown in Figure 11.2. Quantitative analysis in IMS depends on the ionization reproducibility, transmission of ions through the tube, and resolving power.

Approximately 200 volatile metabolites have been identified in breath and can be related to diseases such as diabetes, heart disease, bacterial infections, and lung cancer.[22] In another study using GC-IMS, human breath (10 mL) from 36 patients suffering from lung cancer and a control group of 56 healthy subjects were compared. Differentiation of healthy subjects from lung cancer patients was possible by examining biomarkers using discriminant analysis. Discriminant analysis is a statistical tool that uses a predictive model of group membership where the set of observations for group memberships are known.[23] The results are shown in Figure 11.3 and had an error rate of less than 5%. Accurate measurement of relevant biomarkers allowed a quantitative measure of the lung cancer progression.

In addition to human breath, the qualitative and quantitative analysis of volatiles from bacteria (*Pseudomonas aeruginosa*) and fungi (*Candida albicans*) have been demonstrated with a multicapillary column ion mobility spectrometer (MCC-IMS). [20] GC-IMS demonstrates a rapid, noninvasive analytical tool for medical applications such as early diagnosis, therapy, and medication control.[20] However, one of the limits of GC-IMS is the inability to identify unknown compounds. By coupling ion mobility to mass spectrometry, ions can be mass identified.

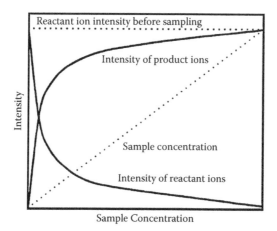

FIGURE 11.2 Intensity of reactant and product ions in dependence on sample concentration for a charge transfer reaction. (Reprinted from Stach, J. & Baumbach, J. I. *Int. J. Ion Mobil. Spec.* 2002, 5, 1–21. Copyright 2002, with permission from International Society for Ion Mobility Spectrometry.)

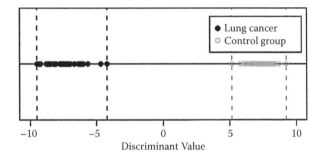

FIGURE 11.3 Discriminant values calculated from peak patterns detected in 2D GC-IMS topographic plots of breath analyses from 36 lung cancer patients and a healthy control group of 54 subjects. (Reprinted from Bader, S.; Urfer, W.; & Baumbach, J. I. *J. Chemometrics.* 2006, 20, 128–135. Copyright 2006, with permission from Wiley InterScience.)

11.4   NONVOLATILE METABOLITES

Many metabolites are polar, ionic, and nonvolatile. Thus comprehensive metabolic analysis is not well suited for gas chromatographic techniques. Metabolite mixtures in complex biological samples such as lymph and blood, however, can be rapidly characterized by electrospray ionization with ion mobility coupled to mass spectrometry. As with GC-IMS, coupling ion mobility spectrometers with mass spectrometers produces two dimensions of information. There are many types of ion mobility–mass spectrometers (IM-MS) and most of these will be discussed in other chapters of this book. The focus of this chapter, however, is on atmospheric pressure IMS with quadrupole or time-of-flight mass spectrometry. Other types of IM-MS have also been recently been reviewed.[24] Advantages of coupling atmospheric pressure IMS with a mass spectrometer include:

- Atmospheric pressure IMS can achieve high-resolution separations in milliseconds prior to a mass spectrometer where mass/charge separation occurs in microseconds.[25,26]
- The ion mobility spectrometer (IMS) provides rapid separation of isomers.[27–29]
- Unlike chromatography, IMS provides qualitative information that is universal, reproducible, and dependent on the collision cross section of the ion and not dependent on interactions with a dynamic and unstable chromatographic stationary phase.[30]
- IM-MS provides two-dimensional (2D) datasets and metabolites in the same class fall on mobility-mass correlation lines (MMCL).[31,32]
- IM-MS requires no sample derivatization and can detect polar to nonpolar compounds.[32]
- IM-MS can be coupled with both liquid and gas chromatography to provide three dimensions of separation.[33,34]

- Separation selectivity can be achieved using different IMS drift gases and drift gas modifiers.[35]
- Fragmentation of a mobility selected ion can lead to MS/MS analysis providing additional qualitative information.[24]

These advantages of IM-MS are particularly applicable to metabolomics due to the complex nature of the sample and the large number of isomers and isobars that are present in a metabolome. For example, the human blood metabolome is thought to have about 2500 metabolites with molecular weights less than 800. Thus on average there are about three isomers or isobars per nominal mass. The following sections provide examples of IM-MS as applied to various types of metabolic analysis.

11.5 TARGET METABOLOMIC ANALYSIS USING IM-MS

For forensic analysis, target metabolomics is focused on the quantitative and qualitative analysis of specific metabolites such as drugs in blood, urine, tissue, or hair. In a recent study using IM-MS, metabolites in hair samples were analyzed.[36] Four of the nine hair samples were positive for caffeine and nicotine and one sample tested positive for methamphetamine in the mass spectrum. However, when mobility data was considered the caffeine and nicotine were found to be true positives while the methamphetamine was determined to be a false positive. The mobility values for the unknown with the mass of methamphetamine did not match that of the methamphetamine standard. This is an excellent example of utility of isomeric/isobaric separation in mobility space.

Another application of target metabolomics illustrative of the importance of isomeric/isobaric separation is in the field of pharmacokinetics. Pharmacokinetic studies focus on the absorption and distribution of drugs and their metabolites in biological systems.[37–39] A limiting factor in pharmacokinetic research is a requirement of high-throughput screening of isomeric metabolites. IM-MS is a rapid technique where isomeric separations occur within minutes. For example, IM-MS has been used to separate opiate metabolic isomers in less than 1 minute.[38]

Selectivity of isomeric separation can be affected by the use of different drift gases.[39] Nineteen compounds, including cocaine and metabolites, amphetamines, benzodiazepines, and small peptides, were separated in one of four drift gases: helium, nitrogen, argon, or carbon dioxide. One example of drift gas selectivity is shown in Figure 11.4 where two isomers, diazepam and loradiazepam, were separated in helium and carbon dioxide but were not resolved in nitrogen or argon.

For metabolomics studies using IM-MS, electrospray ionization (ESI) is commonly used and the quantitative analysis will be different than the ^{63}Ni. With ESI, internal standards, standard addition, isotope dilution, or external calibration curves are common practices for relative quantitative analysis. An example of ESI-IM-MS quantitative analysis was demonstrated for 20 phenylthiohydantoin (PTH)-derivatized amino acids, the final products in the Edman sequencing process of peptides and proteins. Detection limits for these amino acid derivatives ranged from 1.04 to 3.52 ng (less than 17 pmol).[40] Quantitative analysis has not yet been established with ESI-IM-MS for applications to metabolomics.

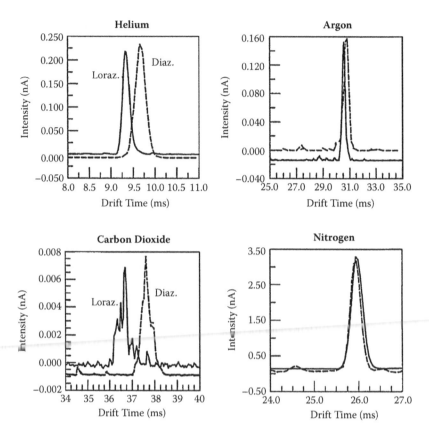

FIGURE 11.4 Comparison of the ion mobility spectrometry of two isomeric benzodiazepines (loradiazepam and diazepam) in the four different drift gases: helium, argon, nitrogen, and carbon dioxide. Ions were mass identified and the mobility spectra were obtained with selective ion monitoring. These isobaric ions separated in helium and carbon dioxide but not in argon or nitrogen. (Reprinted from Matz, L.; Hill, H. H. Jr.; Beegle, L.; & Kanik, I. *J. Am. Soc. Mass Spectrom.* 2002, 13, 300–307. Copyright 2002, with permission from Elsevier Science Inc.)

11.6 METABOLITE PROFILING BY IM-MS USING MOBILITY-MASS CORRELATION

For specific classes of metabolites, mobility is often correlated with mass. Early work on the correlation of mobility with mass was conducted by Karasek and co-workers in the 1970s.[41–43] They found that a correlation exists between mobility and mass for compounds in a homologous class, which they referred to as a trend line or mass-mobility correlation. Because metabolite profiling focuses on the determination of specific classes of metabolites such as carbohydrates, lipids, etc., MMCs have been applied for metabolite identification where different classes of metabolites form different MMCs, such as, lipids, oligonucleotides, and peptides shown in

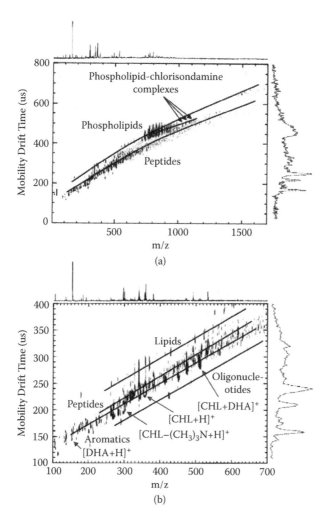

FIGURE 11.5 (a) MALDI-IM 2D plot of brain tissue with 1 nmol of chlorisondamine added using DHA matrix. (b) Close-up of MALDI-IM 2D plot from (a) below 700 Da. (Reprinted from Jackson, S; Wang, H.-Y.; Woods, A. S.; Ugarov, M.; Egan, T.; & Shultz, A. *J. Am. Soc. Mass Spectrom.* 2005, 16, 133–138. Copyright 2005, with permission from Elsevier Inc.)

Figure 11.5.[44] One of the major components in brain matter are lipids, and changes in phospholipid concentrations in tissue are associated with Farber disease, Alzheimer disease, and Gaucher disease. Therefore, by profiling the rat brain lipid metabolism using IM-MS, early detection could be beneficial for prevention or intervention of these diseases.[31] The focus of this work was measuring lipids in rat brain tissue using a low-pressure matrix-assisted laser desorption/ionization (MALDI)-low pressure IM-MS and determining noncovalent complexes between lipids in the tissue and chlorisondamine, a nicotinic antagonist.

Based on the understanding that homologous compounds from the same class will form unique MMC relationships, tentative identification of unknown metabolites in complex biological samples can be determined. One application of this approach was used for the tentative identification of metabolites found in human blood based on MMCs found in Figure 11.6.

The metabolites found in the human blood study were identified using standards or the following online databases: "METLIN-A metabolite mass spectral database," "LMSD: LIPID MAPS structure database," and "Golm Metabolome Database," and the classes of metabolites detected included amino acids, carbohydrates, endogenous amines, purines and pyrimidines, organic acids, sterols, estrogens, prostaglandins, phosphocholines, mono- and di-acylglycerophosphoethanolamines, mono- and di-acylglycerols, sphingolipids, isoprenoids, and various metabolic intermediates. [45] In this simple extraction of a droplet of blood, a wide range of compounds was detected using IM-MS, demonstrating the capabilities of IM-MS for the application to metabolomics.

11.7   METABOLOMICS BY IM-MS

The goal of metabolomics is to identify and quantify each metabolite in a metabolome rapidly and reproducibly. Target and profile metabolomics are often implemented because the overall goal of total metabolomics is not yet feasible. IM-MS is perhaps the analytical method with the best potential to achieve the goal of total metabolomics. It has been demonstrated to separate a vast range of metabolites in *Escherichia coli*,[36] fed and unfed rat lymph samples,[46] and human blood.[45] In the *E. coli* study, GC-MS and IM-MS were compared and it was found that IM-MS had higher overall metabolite detection and utilized a simple extraction method without derivatization (Figure 11.7).

IM-MS has also shown separation in a range of volatile to nonvolatile metabolites. More than 500 features were detected in the ~µM concentration range and were tentatively assigned as *E. coli* metabolites. For another example, 50 µL of human blood was added to a hot methanol and acetic acid mixture. Approximately ~1100 metabolite peaks with ion counts ≥3 and ~850 metabolite peaks with ion counts ≥5 with the simultaneous separation of over 200 isomeric metabolites with ion counts ≥10 were detected (Figure 11.8).[45]

11.8   METABONOMICS BY IM-MS

Because all metabolites cannot routinely be identified and quantified in a complex metabolome it is often satisfactory to investigate patterns of the metabolome to determine changes due to external stress on the biosystem. Data from metabolome analysis are complex and large. Thus multivariant analyses are often used to provide meaningful data. There are two types of multivariant analysis approaches used to statistically analyze metabolic data: supervised and unsupervised methods. As shown above in the volatile breath analysis by IMS, discriminant analysis was used to determine healthy patients from patients suffering from lung cancer. Discriminant analysis is a supervised method, meaning the classification of the sample must be

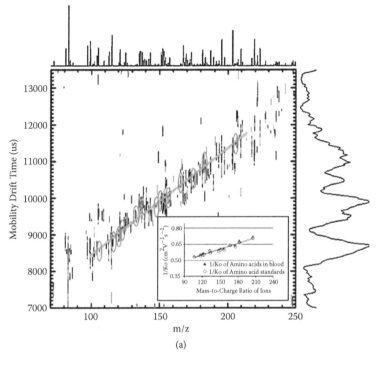

(a)

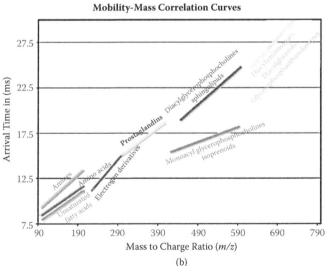

(b)

FIGURE 11.6 (a) MMC for amino acids detected in the blood extract. Peaks identified as amino acids in blood extract based on mass and reduced mobility data matched with that measured for standard solutions of amino acids. (b) MMCs for various classes of metabolites detected in the blood sample. Only protonated ions of metabolites constitute the MMCC (except for the sugars as sodium adducts). (Reprinted from Dwivedi, P.; Shultz, A.; & Hill, H. H. Jr. *Int. J. Mass Spectrom.* Accepted for publication. Copyright 2009, in press.)

known to determine the mathematical model. Another statistical approach is using an unsupervised method called principal component analysis (PCA). PCA is a statistical treatment used to pick out patterns in the data without knowing the classification of the sample. PCA was recently applied to IM-MS data from lymph fluid of fasting and fed rats shown in Figure 11.9.[46]

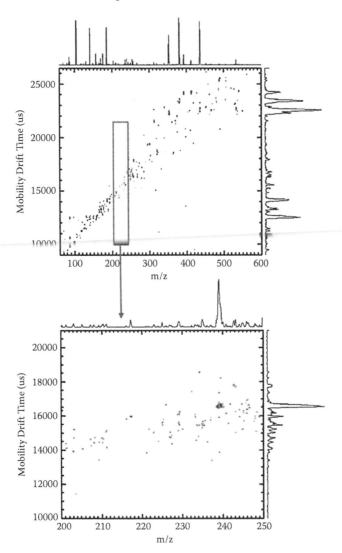

FIGURE 11.7   Top: Two-dimensional ESI-IM-MS spectra in the *m/z* range of 60–600 amu that shows metabolites detected in intracellular *E. coli* extract. Bottom: Enlarged 2D ESI-IM-MS spectra in the *m/z* range of 200–250 amu that illustrates detection and separation of approximately 100 metabolites in this mass range. (Reprinted from Dwivedi, P.; Wu, P.; Klopsch, S. J.; Puzon, G. J.; Xun, L.; Hill, H. H. Jr. *Metabolomics*. 2008, 4, 63–80. Copyright 2007, with permission from Springer Science and Business Media.)

In this study, lymph fluid samples were collected from rats that were fasting and then given a bolus (3-mL) injection of Ensure. Samples were taken in hourly intervals after the initial feeding. The PCA score plots pulled out the patterns detected in the IM-MS data. In both examples, there is a circular pattern indicating that the metabolome shifted after feeding and after six hours returned to the fasting condition.

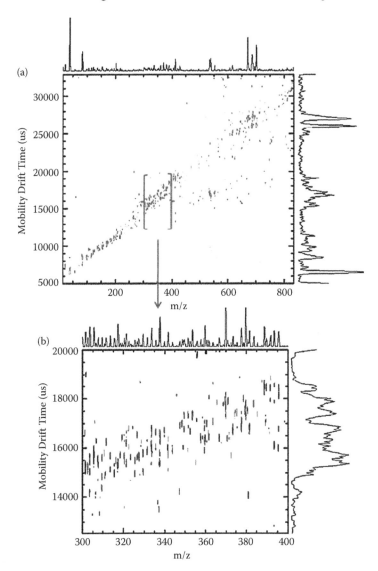

FIGURE 11.8 (a) Two-dimensional spectra of metabolic features measured in methanol extract of human blood. (b) A zoomed-in section of the IM-MS spectrum in the m/z range of 300–400 Da illustrating the peaks detected in the region with five ion counts or more along with separation of isomers and isobars. (Reprinted from Dwivedi, P.; Shultz, A.; & Hill, H. H. Jr. *Int. J. Mass Spectrom.* Accepted for publication. Copyright 2009, in press.)

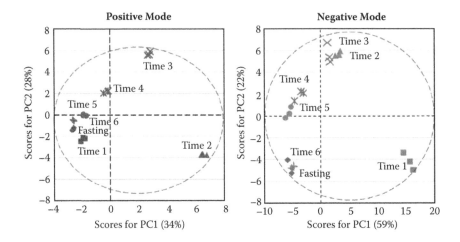

FIGURE 11.9 Principal component analysis score plots from IM-MS data collected in positive and negative mode for rat lymph samples collected in hourly intervals before and after feeding. (Reprinted from Kaplan, K; Dwivedi, P.; Davidson, S.; Yang, Q.; Tso, P.; Siems, W.; Hill, H. H. Jr. *Anal. Chem.* 2009, 81, 7944–7953. Copyright 2009, American Chemical Society.)

11.9   SUMMARY AND CONCLUSIONS

The simplicity of instrumentation, ease of interfacing with chromatography and mass spectrometry, high resolving powers, good sensitivity, and excellent reproducibility make ambient pressure ion mobility spectrometry a powerful analytical tool for metabolomics. Instrumentally, drift time IMS at ambient pressure requires no vacuum pumps, and due to the high collision frequency at ambient pressures, it can be housed in relatively small cells with short drift lengths. Because drift times in ambient pressure DTIMS are on the order of 1–100 ms, they fall into a time range that is intermediate to those required for separation in chromatography (1–100 minutes) and mass spectrometry (1–100 μs). Thus multiple IMS spectra can be obtained for each chromatogram while multiple mass spectra can be acquired for each ion mobility spectrum, producing a three-dimensional array of spectra. Resolving powers of DTIMS can be as high as those of capillary gas chromatography (>100,000 theoretical plates), which enables the separation of complex mixtures. One disadvantage of DTIMS is its duty cycle. At an IMS resolving power (t_d/w_h) of 100, only about 1% of the total ion current is used. When coupled to a mass spectrometer, however, IMS separates random electronic and chemical noise in mobility space, improving the signal-to-noise ratio and gaining back much of the sensitivity lost due to the duty cycle. Finally, the use of reduced mobility values provides a qualitative value that, unlike chromatographic retention times, are reproducible over time and among instruments and laboratories.

When applied to metabolomics, GC-IMS and IM-MS have proved to be promising novel analytical approaches. GC-IMS instruments provide a rapid method for diagnosing disease from the patterns of volatile metabolites in breath. Electrospray IM-MS monitors more complex nonvolatile metabolic changes in blood and other biological fluids. Specific applications that have been presented in this chapter include target metabolite analysis (drugs in hair), metabolite profiling (MMCs in rat brain and human blood), metabolomics (*E. coli* and blood samples), and metabonomics (lung cancer and fasting and fed rats). These initial applications demonstrate the potential breadth and depth of IMS as a complimentary analytical technique for metabolic analysis. Both GC-IMS and IM-MS instruments are now commercially available and their further application, as well as the development of an ion mobility database, should result in their routine use in the field of metabolomics.

REFERENCES

1. Duarte, N. C.; Becker, S. A.; Jamshidi, N.; Thiele, I.; Mo, M. L.; Vo, T. D.; et al. Global reconstruction of the human metabolic network based on genomic and bibliomic data. *Proc. Nat. Acad. Sci. U. S. A.* **2007**, 104, 1777-1782.
2. Hollywood, K.; Brison, D.; & Goodacre, R. Metabolomics: Current technologies and future trends. *Proteomics.* **2006**, 6, 4716–4723.
3. Goodacre, R.; Vaidyanathan, S.; Dunn, W. B.; Harrigan, G. G.; & Kell, D. B. Metabolomics by numbers: acquiring and understanding global metabolite data. *Trends Biotechnol.* **2004**, 22, 245–252.
4. Jiye, A; Trygg, J.; Gulleberg, J.; Johansson, A.; Jonsson, P.; Antti, H.; et al. Extraction and GC/MS analysis of the human blood plasma metabolome. *Anal. Chem.* **2005**, 24, 8086–8094.
5. Dunn, W. B. & Ellis, D. I. Metabolomics: Current analytical platforms and methodologies. *Trends Anal. Chem.* **2005**, 24, 285–294.
6. Wang, W.; Zhou, H.; Lin, H.; Roy, S.; Shaler, T.; Hill, L.; et al. Quantification of proteins and metabolites by mass spectrometry without isotopic labeling or spiked standard. *Anal. Chem.* **2003**, 75, 4818–4826.
7. Brindle, J. T.; Antti, H.; Holmes; E., Tranter; G., Nicholson; J. K., Bethell; H. W.; et al. Rapid and noninvasive diagnosis of the presence and severity of coronary heart disease using 1H-NMR-based metabonomics. *Nat. Med.* **2002**, 8, 1439–1444.
8. Odunsi, K.; Wollman, R. M.; Ambrosone, C. B.; Hutson, A.; McCann, S. E.; Tammela, J.; et al. Detection of epithelial ovarian cancer using 1H-NMR-based metabonomics. *Int. J. Cancer.* **2005**, 113, 782–788.
9. Shulaev, V. Metabolomics technology and bioinformatics. *Brief. Bioinform.* **2006**, 7, 128–139.
10. Ramautar, R.; Mayboroda, O. A.; Deelder, A. M.; Somsen, G. W.; & Jong, G. J. Metabolic analysis of body fluids by capillary electrophoresis using noncovalently coated capillaries. *J. Chromatogr. B.* **2008**, 871, 370–374.
11. Taibi, G. & Nicotra, C. Development and validation of a fast and sensitive chromatographic assay for all-trans-retinol and tocopherols in human serum and plasma using liquid–liquid extraction. *J. Chromatogr. B.* **2002**, 780, 261–267.
12. Want, E. J.; O'Maille, G.; Smith, C. A.; Brandon, T. R.; Uritboonthai, W.; Qin, C.; et al. Solvent-dependent metabolite distribution, clustering, and protein extraction for serum profiling with mass spectrometry. *Anal. Chem.* **2006**, 3, 743–752.
13. Wang, W.; Zhou, H.; Lin, H.; Roy, S.; Shaler, T.; Hill, L.; et. al. Quantification of proteins and metabolites by mass spectrometry without isotopic labeling or spiked standards. *Anal. Chem.* **2003**, 75, 4818–4826.

14. Ewing, R.; Atkinson, D.; Eiceman, G.; & Ewing, G. A critical review of ion mobility spectrometry for the detection of explosives and explosive related compounds. *Talanta.* **2001**, 54, 515–529.

15. Shvartsburg, A. A. & Smith, R. D. Fundamentals of traveling wave ion mobility spectrometry. *Anal. Chem.* **2008**, 80, 9689–9699.

16. Mukhopadhyay, R. IMS/MS: Its time has come. *Anal. Chem.* **2008**, 80, 7918–7920.

17. Shvartsburg, A. A.; Li, F.; Tang, K.; & Smith, R. D. High-resolution field asymmetric waveform ion mobility spectrometry using new planar geometry analyzers. *Anal. Chem.* **2006**, 78, 3706–3714.

18. Eiceman, G. A. Ion-mobility spectrometry as a fast monitor of chemical composition. *Trends Anal. Chem.* **2002**, 21, 259–275.

19. Guevremont, R. High-field asymmetric waveform ion mobility spectrometry (FAIMS). *Can. J. Anal. Sci. Spectroscopy.* **2004**, 46, 105–113.

20. Vautz, W. & Baumbach, J. I. Exemplar application of multi-capillary column ion mobility spectrometry for biological and medical purpose. *Int. J. Ion Mobil. Spec.* **2008**, 11, 35–41.

21. Stach, J. & Baumbach, J. I. Ion mobility spectrometry—Basic elements and applications. *Int. J. Ion Mobil. Spec.* **2002**, 5, 1–21.

22. Ruzsanyi, V.; Baumbach, J. I.; Sielemann, S.; Litterst, P.; Westhoff, M.; & Freitag, L. Detection of human metabolites using multi-capillary columns coupled to ion mobility spectrometers. *J. Chromatogr. A.* **2005**, 1084, 145–151.

23. Bader, S.; Urfer, W.; & Baumbach, J. I. Reduction of ion mobility spectrometry data by clustering characteristic peak parameters. *J. Chromatogr.* **2006**, 20, 128–135.

24. Kanu, A. B.; Dwivedi, P.; Tam, M.; Matz, L.; & Hill, H. H. Jr. Ion mobility-mass spectrometry. *J. Mass Spectrom.* **2008**, 43, 1–22.

25. Wu, C.; Siems, W. F.; Asbury, R.; & Hill, H. H. Jr. Electrospray ionization high-resolution ion mobility spectrometry-mass spectrometry. *Anal. Chem.* **1998**, 70, 4929–4938.

26. Beegle, L. W.; Kanik, I.; Matz, L.; & Hill, H. H. Jr. Electrospray ionization high-resolution ion mobility spectrometry for the detection of organic compounds, 1. Amino acids. *Anal. Chem.* **2001**, 73, 3028–3034.

27. Dwivedi, P.; Bendiak, B.; Clowers, B.; & Hill, H. H. Rapid resolution of carbohydrate isomers by electrospray ionization ambient pressure ion mobility spectrometry-time-of-flight mass spectrometry (ESI-APIMS-TOFMS). *J. Am. Soc. Mass Spectrom.* **2007**, 18, 1163–1175.

28. Clowers, B.; Dwivedi, P.; Steiner, W.; & Hill Jr., H. H. Separation of sodiated isobaric disaccharides and trisaccharides using electrospray ionization-atmospheric pressure ion mobility-time of flight mass spectrometry. *Am. Soc. Mass Spectrom.* **2005**, 16, 660–669.

29. Yamagaki, T. & Sato, A. Isomeric oligosaccharides analysis using negative-ion electrospray ionization ion mobility spectrometry combined with collision-induced dissociation MS/MS. *Anal. Sci.* **2009**, 25, 985–988.

30. Louis, R.; Hill, H. H.; & Eiceman, G. A. Ion mobility spectrometry in analytical chemistry. *Crit. Rev. Anal. Chem.* **1990**, 21, 321–355.

31. Jackson, S; Wang, H.-Y.; Woods, A. S.; Ugarov, M., Egan, T.; & Shultz, A. Direct tissue analysis of phospholipids in rat brain using MALDI-TOFMS and MALDI-ion mobility-TOFMS. *J. Am. Soc. Mass Spectrom.* **2005**, 16, 133–138.

32. Dwivedi, P.; Wu, P., Klopsch; S., Puzon, G.; Xun, L.; & Hill, H. H. Metabolic profiling by ion mobility mass spectrometry (IM-MS). *Metabolomics.* **2008**, 4, 63–80.

33. Valentine, S.; Kulchania, M.; Barnes, C.; & Clemmer, D. E. Multidimensional separations of complex peptide mixtures: A combined high-performance liquid chromatography/ion mobiility/time-of-flight mass spectrometry approach. *Int. J. Mass Spectrom.* **2001**, 212, 97–108.

34. Hill, H. H., Jr.; Graf, S.; Crawford, C. L.; Gonin, M.; Tanner, C.; and Davis, E. J. *Comparison of GCxGC-TOFMS with GC-IMSxTOFMS.* International Society of Capillary Chromatography and Electrophoresis, Portland, OR, May 2009.
35. Asbury, R. & Hill, H. H. Using different drift gases to change separation factors (α) in ion mobility spectrometry. *Anal. Chem.* **2000**, 72, 580–584.
36. Dwivedi, P. & Hill, H. H. Jr. A rapid analytical method for hair analysis using ambient pressure ion mobility mass spectrometry with electrospray ionization (ESI-IM-MS). *Int. J. Ion Mobil. Spec.* **2008**, 11, 61–69.
37. Chan, E. C. Y.; New, L. S.; Yap, C. W.; & Goh, L. T. Pharmaceutical metabolite profiling using quadrupole/ion mobility spectrometry/time-of-flight mass spectrometry. *Rapid Commun. Mass Spectrom.* **2009**, 23, 384–394.
38. Matz, L. & Hill, H. H. Jr. Evaluation of opiate separation by high-resolution electrospray ionization-ion mobility spectrometry/mass spectrometry. *Anal. Chem.* **2001**, 73, 1664–1669.
39. Matz, L.; Hill, H. H. Jr.; Beegle, L.; & Kanik, I. Investigation of drift gas selectivity in high-resolution ion mobility spectrometry with mass spectrometry detection. *J. Am. Soc. Mass Spectrom.* **2002**, 13, 300–307.
40. Steiner, W. E.; Clowers, B.; & Hill, H. H. Rapid separation of phenylthiohydantoin amino acids: Ambient pressure ion-mobility mass spectrometry (IM-MS). *Anal. Bioanal. Chem.* **2003**, 375, 99–102.
41. Karasek, F. W.; Kim, S. H.; & Rokushika, S. Plasma chromatography of alkyl amines. *Anal. Chem.* **1978**, 50, 2013–2016.
42. Karasek, F. W. & Tatone, O. S. Plasma chromatography of the mono-halogenated benzenes. *Anal. Chem.* **1972**, 44, 1758–1763.
43. Karasek, F.; Kilpatrick, W.; & Cohen, M. Qualitative studies of trace constituents by plasma chromatography. *Anal. Chem.* **1971**, 43, 1441–1447.
44. Woods, A.; Ugarov, M.; Egan, T.; Koomen, J.; Gillig, K.; Fuhrer, K.; Gonin, M.; & Schultz, J. A. Lipid/peptide/nucleotide separation with MALDI-ion mobility-TOF MS. *Anal. Chem.* **2004,** 76, 2187–2195.
45. Dwivedi, P.; Shultz, A.; & Hill, H. H. Jr. Metabolic profiling of human blood by high resolution ion mobility mass spectrometry (IM-MS). *Int. J. Mass Spectrom.* **2009.** (Accepted for publication)
46. Kaplan, K.; Dwivedi, P.; Davidson, S.; Yang, Q.; Tso, P.; Siems, W.; & Hill, H. H. Jr. Monitoring dynamic changes in lymph metabolome of fasting and fed rats by electrospray ionization-ion mobility mass spectrometry (ESI-IM-MS). *Anal. Chem.* **2009**, 81, 7944–7953.

12 Ion Mobility MALDI Mass Spectrometry and Its Applications

Amina S. Woods, J. Albert Schultz,
and Shelley N. Jackson

CONTENTS

12.1   INTRODUCTION

Two scientific developments that occurred in the early 1990s brought to the fore a new approach to solving biomolecular structures and enabled scientists in a wide variety of fields to make great strides in solving biological and pathophysiological problems. The first breakthrough was the completion of the Human Genome Project, which identified all the approximately 25,000 genes in human DNA and determined the sequences of the 3 billion chemical base pairs that make up human DNA, which were then translated into the proteins they code for and stored in huge databases that were made available to all scientists.[1–3] The other breakthrough was the development of two new mass spectrometric techniques: matrix-assisted laser desorption/ionization (MALDI) and electrospray ionization (ESI). The mass spectrometers were user friendly yet high performance, thus by combining these advances the scientific community was able to identify and characterize a wide variety of biomolecules, often from as little as a few femtomoles.[2,3] However, as technical progress is always on the march, one of the latest and most exciting breakthroughs was the coupling of ion mobility (IM) and mass spectrometry (MS),[4–6] as structure elucidation is an arduous task with a constant need for improved instrumentation and methodology in order to study lipids and proteins that are present in trace amounts, which is often the case for nonstructural proteins as cells are the most efficient entities in existence, thus proteins are only manufactured when needed.

In this chapter we present some of the applications of ion mobility MALDI. The ion mobility mass spectrometer used in these examples is an Ionwerks (Houston, TX) periodic focusing MALDI-IM-TOFMS (time-of-flight mass spectrometer) instrument that can be used in both positive and negative ion mode. The instrument has a mobility resolution of 30 (full width at half maximum [FWHM] of drift time) and a mass resolution of 3000 for mass-to-charge ratio (*m/z*) 1000, and uses an orthogonal time-of-flight mass spectrometer (o-TOFMS). Ions are generated in the source and are drawn into the mobility cell by a voltage between the sample and the exit of the helium-filled IM cell so that the ion's drift velocity is proportional to the density (surface area) of the ion. The length of the mobility cell is 15 cm. It is operated at 1700 V with 3.6 torr helium pressure. An X-Y sample stage provides 1-μm accuracy in beam positioning and sample scanning. An Nd:YLF (neodymium-doped yttrium lithium fluoride) ultraviolet (UV) laser ($\lambda = 349$ nm at 200 Hz) is used to generate ions in the source at the operating pressure of the mobility cell. Separation of isobaric ions, that is, ions with the same mass but different surface areas, is thus possible. Small, dense ions quickly traverse the IM cell and enter the mass spectrometer where a pulsed voltage applied to the orthogonal extraction region deflects the ion beam into a TOFMS equipped with a reflectron, which measures the ions' m/z directly and their ion mobility elution time indirectly. This indirect determination of the IM time is accomplished by "tagging" each such o-TOF mass spectrum with the elapsed time since the firing of laser pulse onto the sample. Ion drift times are typically under one millisecond for m/z 5000, yet the time of flight in the mass spectrometer is only a few tens of microseconds. Thus after each UV laser desorption pulse, several hundred mass spectra can be obtained from the eluted ions separated in the IM cell.

The data is presented as two-dimensional (2D) contour plots of ion intensity as a function of ion mobility drift time (*y*-axis) and m/z (*x*-axis). In the 2D contour plots of ion mobility versus mass, compounds that have the same molecular weight but different structures are observed along ion groupings that have different slopes, thus resulting in lipids, peptides, oligonucleotides, and small organic molecules to travel along familial trend lines. It was observed that oligonucleotides were the fastest and lipids the slowest, with peptides falling in between.[6,7] All contour plots were produced using IDL software.

12.2   SEPARATION OF LIPIDS BY ION MOBILITY MS

Lipidomics is a rapidly emerging field of study due to lipids' abundance in biomembranes, role in signal transduction, and function as an important reservoir of energy in organisms.[8-13] Two major groups of lipids in the brain are glycerophospholipids, consisting of phosphatidylcholines (PCs), phosphatidylethanolamines (PEs), phosphatidylinositols (PIs), phosphatidylserines (PSs), phosphatidylglycerol (PG), etc., and sphingolipids, consisting of cerebrosides, sphingomyelin (SM), sulfatides (STs), and gangliosides. Glycerophospholipids and sphingolipids represent large and very diverse groups of lipids that vary due to their head groups and radyl substituents. Previous studies have used MALDI-IM MS[14] and ESI-IM MS[15] to analyze complex mixtures of glycerophospholipids and sphingolipids.

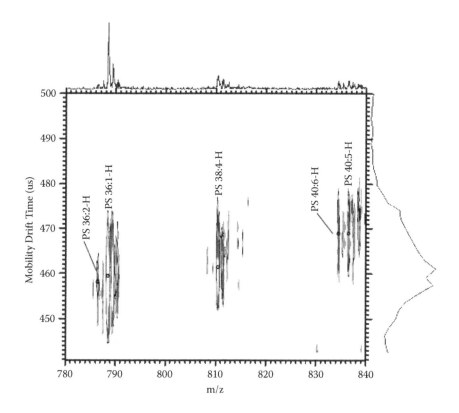

FIGURE 12.1 MALDI-IM 2D spectrum of 200 pmol of brain PS extract with DHA matrix in negative ion mode.

Figure 12.1 illustrates a MALDI-IM MS spectrum of 200 pmol of porcine brain PS extract in negative ion mode with DHA (2,6-dihydroxyacetophenone) matrix. In this spectrum, five PS species are observed. As shown in Figure 12.1, as the degree of unsaturation (number of double bonds) on the acyl chain increases, the ion mobility drift time decreases. This is most likely due to the fact that the presence of a double bond causes the acyl chain to bend, making it more compact and thus causing a smaller ion collision cross section and a decrease in mobility time. This result was previously observed in positive ion mode for phospholipids by MALDI-IM MS[14] and ESI-IM MS.[15]

PCs are one of the most abundant glycerophospholipid classes. In addition to the basic diacyl species of glycerophospholipids, in which two acyl groups are attached to the sn-1 and sn-2 positions of the glycerol backbone, PCs also contain plasmalogen species, in which a vinyl ether group instead of an acyl group is attached to the sn-1 position. Plasmalogen species are a major constituent of heart PC, accounting for approximately 40% of PC in the bovine heart. Figure 12.2 shows a MALDI-IM MS spectrum of 50 pmol of bovine heart extract in positive ion mode with DHA matrix. The mass range in Figure 12.2 contains two plasmalogen species (PC 34:2p and PC 34:1p) and two diacyl species (PC 34:2a and PC 34:1a). As shown in this figure, the loss of one double bond in the fatty acid chains of the PC species causes a greater

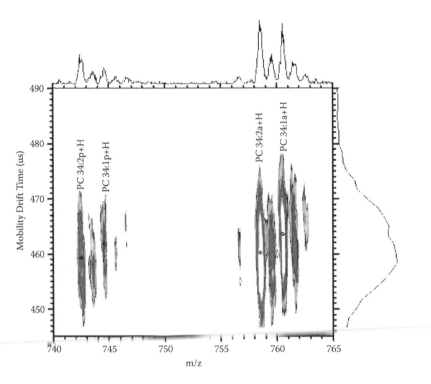

FIGURE 12.2 MALDI-IM 2D spectrum of 50 pmol of heart PC extract with DHA matrix in positive ion mode.

TABLE 12.1
Effects of Head Group upon Mobility Drift Time

Mass	Peak Mass (*m/z*)	Mobility Time (μs)[a]
PC18:1/18:1+Na	808.6	479.2
PE18:1/18:1+Na	766.5	466.4
PI18:1/18:1+Na	885.4	488.0
PS18:1/18:1+Na	810.4	470.1

[a] Mobility time measurements normalized against an internal standard C_{60}.

increase in mobility time compared to the difference between plasmalogen and diacyl species with the same fatty acid chains. This result makes it easy to identify plasmalogen and diacyl species in complex mixtures and was previously observed for brain PE species using MALDI-IM MS.[14]

Glycerophospholipids are divided into classes due to their different polar head groups. Table 12.1 summarizes the effects of head group upon mobility drift time. In this study, 100 pmol of five glycerophospholipids (PC 18:1/18:1, PE 18:1/18:1, PI 18:1/18:1, PS 18:1/18:1) were analyzed individually with an internal C_{60} standard in positive ion mode with DHB matrix.

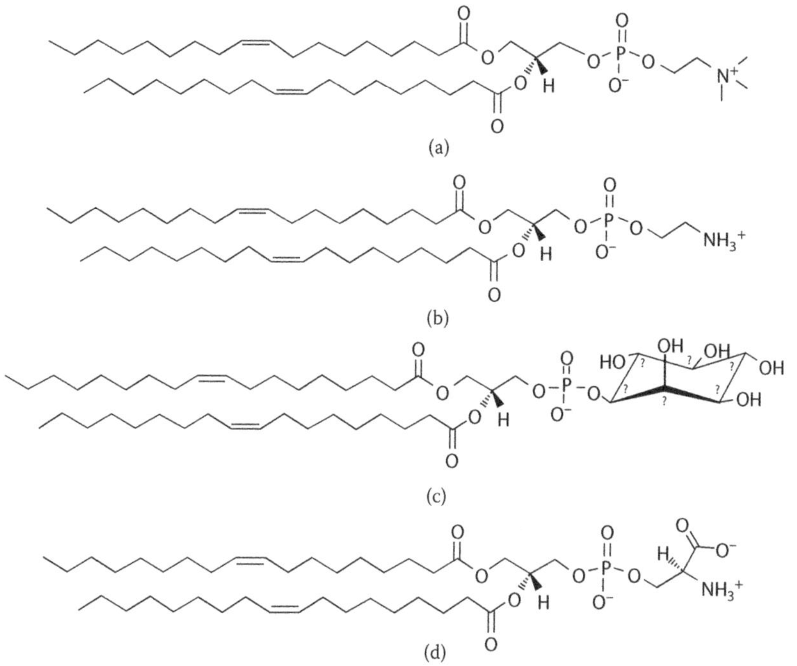

FIGURE 12.3   Basic structure for (a) PC 18:1/18:1, (b) PE 18:1/18:1, (c) PI 18:1/18:1, and (d) PS 18:1/18:1.

The mobility time for sodium adducts with each lipid species normalized against the C_{60} internal standard is listed in Table 12.1. The mobility time observed for head group from high to low is PI > PC > PS > PE. As shown in Table 12.1, this result is not due only to the m/z value, i.e., higher m/z value results in higher mobility time, but is also due to the chemical structure of the head group as can be seen in Figure 12.3. The difference in head group and radyl substituents can be used to separate lipids in complex mixtures onto class trend lines.

Figure 12.4 illustrates a MALDI-IM spectrum of 50 pmol each of porcine brain PS extract and brain cerebroside extract with DHA matrix in positive ion mode. As shown in Figure 12.4, the cerebroside species have a higher mobility time compared to the PS species, which allows for the two different classes of lipids to have different trend lines. The zoom view in the figure demonstrates the utility of ion mobility to separate species with similar/same m/z values, which are not easily identified with just MS analysis.

12.3   IM IMAGING

We have used MALDI IM not just for direct tissue profiling, but also for the imaging of lipids in the brain, as MALDI IM is a versatile analytical technique that allows us to map and image the distribution of biomolecules and elucidation of their molecular structure with minimal preparation.[16] Figure 12.5 illustrates the imaging of a brain

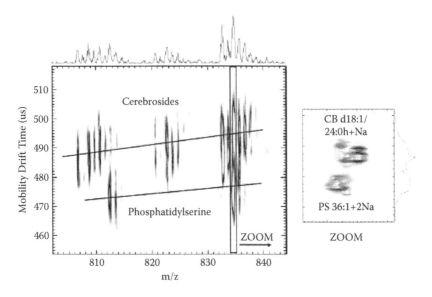

FIGURE 12.4 MALDI-IM 2D spectrum of 50 pmol each of brain PS extract and brain cerebroside extract with DHA matrix in positive ion mode.

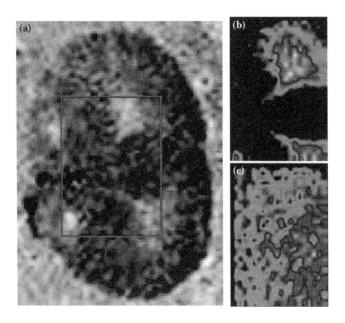

FIGURE 12.5 Negative ion mode imaging of a brain tissue section. The DHB matrix, 200-μm resolution, 4.5 × 6-mm image, 2-hour acquisition. (a) is the photograph of the brain slice. (b) is the distribution of hydroxyl sulfatide 24:1 in white matter. (c) is the distribution of phosphatidyl inositol.

tissue section using MALDI IM in negative ion mode in order to image the distribution and localization of hydroxyl sulfatide 24:1 in white matter and the distribution of phosphatidyl inositol 38:4, which is more widely distributed.

12.4   A STUDY OF NONCOVALENT COMPLEXES

IM-MALDI MS is also a very useful technique for studying noncovalent complexes. We have used IM-MALDI MS to understand and clarify the key role electrostatic interactions play in determining the conformation of interacting proteins or peptides. These interactions involve the guanidinium groups of adjacent arginines (Arg) on one peptide or protein and a phosphorylated residue or the carboxyl groups of adjacent glutamate or aspartate on the other peptide or protein involved in the interaction.[17–23] Figure 12.6a shows the interaction between a fragment of a neuropeptide dynorphin 1 to 8 YGGFL**RR**I (ppt1, MH$^+$ 981.6), which contains adjacent Arg, and an epitope of the cannabinoide receptor SVSTD**pTpS**AE (ppt2, MH$^+$ 1056.3), which contains two phosphorylated residues, a serine (Ser) and a threonine (Thr). Both peptides as well as their noncovalent complex are seen; one complex is between ppt1 and ppt2 (MH$^+$ 2036.9), and the other is between two ppt1 and one ppt2 (MH$^+$ 3017.5) as this peptide has two phosphates available for interaction.

As we previously showed with MALDI and ESI the noncovalent bond between the guanidinium groups of Args and the phosphate was maintained while the covalent bond between the phosphate and Ser/Thr was broken.[20–22] Thus in Figure 12.6b (2D) and 12.6c (3D), there is a peak at 1070.6 representing YGGFL**RR**I+H$_3$PO$_4$ and just below the contour plot for this peak there is another one at 1078.3 representing SVSTD**pTpS**AE+Na$^+$. Although these two molecular ions differ by 1.3 amu, their mobility differs by about 21 μsec. So the mobility makes detection of such fragments much easier.

Our first demonstration of phosphate-quaternary amines noncovalent complex formation was done using IM-MALDI.[23] Phosphorylated angiotensin II (DRV**pY**IHPF; molecular weight [MW] = 1126.2) formed a noncovalent complex with acetylcholine (MW = 146.2) while the nonphosphorylated form (DRVYIHPF; MW = 1046.2) did not. We noted that the phosphorylated angiotensin MH$^+$ and its fragments align along a trend line with a lower slope than the trend line for the nonphosphorylated ions, indicating that the phosphorylated ions have a faster drift time through the mobility cell than nonphosphorylated ions of the same mass. This difference in mobility is especially obvious when comparing two fragments, one phosphorylated and the other not, at nominal mass 756, in which we detected two contours, at mass 756, although their mobility drift time differed by 40 μsec. Investigation of the fragments gave the following two possibilities: mass 756.3 (y5 from phosphorylated angiotensin) and mass 756.4 (a6 from nonphosphorylated angiotensin), confirming that phosphorylated peptides and their fragments do indeed have a faster mobility than their nonphosphorylated counterparts.

Sphingomyelin (Figure 12.7a) is a lipid found in high concentration in the brain. It has a phosphate group, a quaternary amine, and heterogeneous fatty acid content. Chlorisondamine (Figure 12.7a) forms a noncovalent complex with sphingomyelin, a cation-π interaction occurs between the aromatic ring of chlorisondamine and the

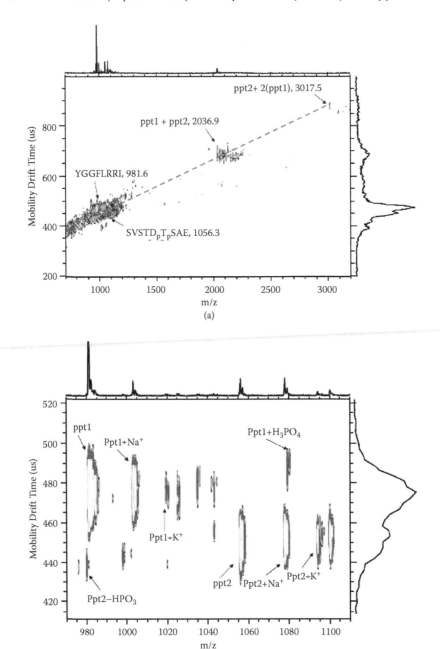

FIGURE 12.6 (a) The 2D contour plots of ion intensity of the noncovalent complex formed between a fragment of the neuropeptide Dyn (1-8) YGGFLRRI (ppt1) and a phosphorylated epitope of the cannabinoide receptor SVSTDpTpSAE (ppt2). (b) The 2D contour plots of ion intensity of the neuropeptide Dyn (1-8) YGGFLRRI (ppt1) and a phosphorylated epitope of the cannabinoide receptor SVSTDpTpSAE (ppt2).

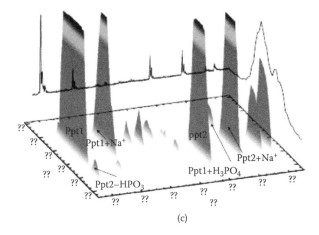

(c)

FIGURE 12.6 (c) The 3D contour plots of ion intensity of the neuropeptide Dyn (1-8) YGGFL**RR**I (ppt1) and a phosphorylated epitope of the cannabinoide receptor SVSTD**pTpS**AE (ppt2).

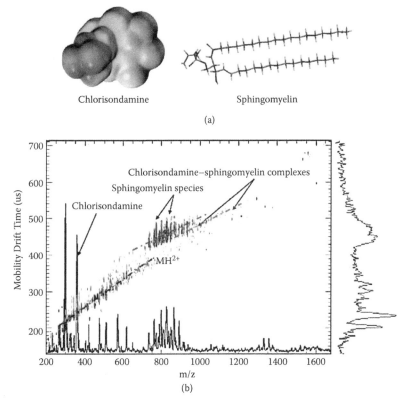

FIGURE 12.7 (a) Hartree-Fock 6-31G optimized structure of chlorisondamine showing its surface electrostatic potential. Next to it is the structure of sphingomyelin. (b) 2D and 1D spectra of the interaction of chlorisondamine and sphingomyelin showing the formation of noncovalent complexes between the lipids and the chlorisondamine.

quaternaryamine of sphingomyelin, and an electrostatic interaction occurs between the quaternaryamines in chlorisondamine and the phosphate in sphingomyelin,[24,25] resulting in very stable noncovalent complexes (Figure 12.7b). The 2D spectrum also shows that the complexes have a faster mobility than the sphingomyelins.

12.5   CONCLUSION

In conclusion, IM-MALDI MS is truly a wonderful tool as with a single glance one can separate isobaric molecules and even suggest as to what biochemical family the various molecules belong and whether or not they are related, depending on the fact that they are located along the same trend line or not.

REFERENCES

1. International Human Genome Sequencing Consortium. Initial Sequencing and Analysis of the Human Genome. *Nature* **409**, 860–921 (2001).
2. Jensen ON. Interpreting the Protein Language Using Proteomics. *Nature Rev.* **7**, 391–403 (2006).
3. Domon B and Aebersold R. Mass Spectrometry and Protein Analysis. *Science* **312**, 212–217 (2006).
4. Gillig KJ, Ruotolo BT, Stone EG, Russell DH, Fuhrer K, Gonin M, and Schultz JA. Coupling High-Pressure MALDI with Ion Mobility/Orthogonal Time-of-Flight Mass Spectrometry. *Anal. Chem.* **72**, 3965–3971 (2000).
5. Hoaglund CS, Valentine SJ, Sporleder CR, Reilly JP, and Clemmer DE. Three-Dimensional Ion Mobility/TOFMS Analysis of Electrosprayed Biomolecules. *Anal. Chem.* **70**, 2236–2242 (1998).
6. Woods AS, Fuhrer K, Egan T, Ugarov M, Koomen J, Gonin M, Gillig KJ, and Schultz JA. Lipid/Peptide/Nucleotide Separation with MALDI-Ion Mobility-TOFMS. *Anal. Chem.* **76**, 2187–2195 (2004).
7. Gidden J, Bushnell JE, and Bowers MT. Gas-Phase Conformations and Folding Energetics of Oligonucleotides: dTG⁻ and dGT⁻ *J. Am. Chem. Soc.* **123**, 5610–5611 (2001).
8. Holthuis JC and Levine TP. Lipid Traffic: Floppy Drives and a Superhighway. *Nature Rev. Mol. Cell Biol.* **6** 209–2020 (2005).
9. Watson AD. Thematic Review Series: Systems Biology Approaches to Metabolic and Cardiovascular Disorders. Lipidomics: A Global Approach to Lipid Analysis in Biological Systems. *J. Lipid Res.* **47**, 2101–2111 (2006).
10. van Meer G. Cellular Lipidomics. *EMBO J.* **24**, 3159–3165 (2005).
11. Wenk MR. The Emerging Field of Lipidomics. *Nature Rev. Drug Discov.* **7**, 594–610 (2005).
12. Piomelli D, Astarita G, and Rapaka R. A Neuroscientist guide to lipidomics. *Nature Rev. Neurosci.* **8**, 743–754 (2007).
13. Han X. *Front. Biosci.* **12**, 2601–2607 (2007).
14. Jackson SN, Ugarov M, Post J, Egan T, Langlais D, Schultz JA, and Woods AS. A Study of Phospholipids by Ion Mobility TOFMS. *J. Am. Soc. Mass Spectrom.* **19**, 1655–1662 (2008).
15. Trimpin S, Tan B, Bohrer BC, O'Dell DK, Merenbloom SI, Pazos MX, Clemmer DE, and Walker JM. Profiling of Phospholipids and Related Structures Using Multidimensional Ion Mobility-Spectrometry Mass Spectrometry. *Int. J. Mass Spectrom.* In press (2009).

16. Jackson SN, Ugarov M, Egan T, Post JD, Langlais D, Schultz JA, and Woods AS. MALDI-Ion Mobility-TOFMS Imaging of Lipids in Rat Brain Tissue. *J. Mass Spectrom.* **42**, 1093–1098 (2007).

17. Woods AS and Huestis MA. A Study of Peptide–Peptide Interaction by MALDI. *J. Am. Soc. Mass Spectrom.* **12**, 88–96 (2001).

18. Woods AS. The Mighty Arginine, the Stable Quaternary Amines, the Powerful Aromatics and the Aggressive Phosphate: Their Role in the Noncovalent Minuet. *J. Proteome Res.* **3**, 478–484 (2004).

19. Woods AS, Koomen J, Ruotolo B, Gillig KJ, Russell DH, Fuhrer K, Gonin M, Egan T, and Schultz JA. A Study of Peptide–Peptide Interactions Using MALDI Ion Mobility o-TOF and ESI-TOF Mass Spectrometry. *J. Am. Soc. Mass Spectrom.* **13**, 166–169 (2002).

20. Woods AS and Ferré S. Amazing Stability of the Arginine–Phosphate Electrostatic Interaction. *J. Proteome Res.* **4,** 1397–1402 (2005).

21. Jackson SN, Wang HY, and Woods AS. Study of the Fragmentation Patterns of the Phosphate–Arginine Noncovalent Bond. *J. Proteome Res.* **4**, 2360–2363 (2005).

22. Jackson SN, Moyer SC, and Woods AS. The Role of Phosphorylated Residues in Peptide–Peptide Noncovalent Complexes Formation. *J. Am. Soc. Mass Spectrom.* **19,** 1535–1541 (2008).

23. Woods AS, Fuhrer K, Gonin M, Egan T, Ugarov M, Gillig KJ, and Schultz JA. Angiotensin II/Acetylcholine Non-Covalent Complexes Analyzed with MALDI-Ion Mobility-TOFMS. *J. Biomol. Tech.* **14**, 1–8 (2003).

24. Woods AS, Moyer SC, Wang HY, and Wise RA. Interaction of Chlorisondamine with the Neuronal Nicotinic Acetylcholine Receptor. *J. Proteome Res.* **2**, 207–212 (2003).

25. Woods AS, Moyer SC, and Jackson SN. The Amazing Stability of Phosphate–Quaternary Amine Interactions. *J. Proteome Res.* **7**, 3423–3427 (2008).

13 Profiling and Imaging of Tissues by Imaging Ion Mobility–Mass Spectrometry

Whitney B. Parson and Richard M. Caprioli

CONTENTS

13.1   INTRODUCTION

MALDI (matrix-assisted laser desorption/ionization) IMS (imaging mass spectrometry) can provide the spatial distribution, relative abundance, and molecular identity of thousands of endogenous analytes of biological species (lipids, metabolites, peptides, proteins, and drugs) directly from tissue sections.[1–3] Mass spectra can be acquired at multiple *x-y* coordinates across a thin tissue section, and the intensities of mass-to-charge (*m/z*) values are used to reconstruct two-dimensional (2D) ion density maps. IMS is currently used in clinical and medical research where it has been successfully applied to the molecular analysis of diseases such as cancer and to aid in the assessment of clinical diagnosis and prognosis.[4–7] This technology has also been utilized to study normal biological processes to understand underlying molecular mechanisms. For instance, proteomic and lipidomic molecular maps have been generated to study early developmental stages of mouse embryo implantation.[8,9] Additionally, IMS has been used to determine the spatial distribution of exogenous pharmaceuticals and metabolites in animal organs and whole rat tissue sections.[10–12]

In the past decade, advances in MS instrumentation have allowed more effective analysis of tissue sections by MALDI IMS. In particular, the addition of a rapid post-ionization gas-phase ion mobility separation allows an ion fractionation process to be integrated into the imaging experiment. MALDI imaging ion mobility MS separates analytes first by collision cross section (CCS) in the mobility cell and second by m/z in the MS analyzer.[13–17] The ion separation step before MS analysis provides a rapid (μs-ms) separation of complex samples without additional sample preparation or a significant increase in analysis time. For the imaging experiment in particular, the additional ion fractionation step provides the capability of resolving two isobaric analytes by structure alone, producing independent images.[18] Ion mobility can separate different classes of biomolecules based on their intrinsic gas-phase packing efficiencies along CCS *versus* m/z trend-lines.[19] The point at which a given analyte falls along these trend-lines provides qualitative information of the molecular species without the need for MS/MS analysis (Figure 13.1). As visualized in a 2D conformation plot in Figure 13.1a, lipids have larger CCS (Å^2) values than peptides of the same m/z. A benefit of the ion mobility separation for the analysis of complex samples, such as tissue sections, is illustrated in Figure 13.1b where peptides are separated and exported apart from the more abundant endogenous phospholipids. Ion mobility separation additionally enhances the IMS experiment by separating analytes of interest from endogenous chemical noise and provides the capability of selectively analyzing or imaging one class of biomolecules in the presence of another. Ion mobility can also be used to target peptides containing post-translational modifications that deviate from the peptide trend-line and simultaneously fragment all species after ion mobility separation, which is especially applicable to high-throughput pharmaceutical imaging.[18,20–22]

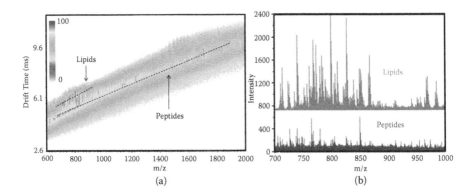

(a)   (b)

FIGURE 13.1 MALDI ion mobility MS analysis of a grade II human astrocytoma tissue section. (a) The 2D conformation plot with predicted phospholipid and peptide trend-lines are indicated by dashed lines. Ion mobility MS signal intensity is indicated by false coloring (scale displayed). (b) Signals from the phospholipid and peptide trend-lines were exported separately and plotted as m/z to intensity.

13.2 THE TISSUE IMAGING EXPERIMENT

13.2.1 SAMPLE PREPARATION

Typically, a flash-frozen tissue sample is prepared for MALDI IMS by preparing a 12-μm-thick section using a cryostat (~ –15°C) and thaw-mounting it onto a MALDI target. After thaw-mounting, tissue sections may be stored in a desiccator. Targets may be either gold-coated stainless steel, microscope slides, or ITO (indium-tin oxide)-coated glass slides.[6,23] While gold-coated targets offer better light contrast for observation of specific tissue features, ITO-coated glass slides (Delta-Technologies, Stillwater, MN) and microscope slides provide the capability of histologically staining the tissue section before or after MS analysis. ITO-coated glass slides have an electrically conductive surface that is compatible with the high voltage sources used in MALDI-TOF (time-of-flight) MS instruments.[6]

MALDI matrix is typically a small organic acid that co-crystallizes with analytes extracted from the tissue section to allow absorption of energy of the pulsed laser. Matrix and analytes are ablated and ionized from the tissue surface with subsequent MS analysis.[24,25] Common matrixes used for the analysis of peptides, lipids, and small drug molecules are 2,5-dihydroxybenzoic acid (DHB) and α-cyano-4-hydroxycinnamic acid (CHCA) while sinapinic acid (SA) is most often used for protein analysis.[10,11,23,26,27] An organic solvent such as acetonitrile or methanol can be used for extraction of small molecules from tissue sections prior to matrix co-crystallization.[10]

Mass spectrometric analysis of tissue sections can be performed in either a profiling or imaging mode. Profiling consists of applying the matrix in a low density array to specific areas of the tissue and can be performed in a histology-directed fashion on small areas or groups of cells within the tissue (Figure 13.2, left). Imaging is performed to map an entire tissue section or areas within a tissue section in one or more molecular dimensions (Figure 13.2, right).[6] The profiling strategy is applicable to high-throughput sample analysis with the use of an automatic reagent spotter and MS acquisition.[28] For a histology-directed approach, the tissue section for MS analysis or a serial section is stained and features of interest are selected and marked for MS analysis. Matrix is applied to the specific features marked using a robotic reagent spotter. Typically, these spotters create a matrix spot size of 100–200 μm in diameter. The x-y coordinates of the matrix spots are then transferred to the MS.[29]

For MS analysis of tissue sections, matrix application can be performed by spraying the matrix solution directly onto the section to achieve a homogeneous coating either manually using a thin-layer chromatography (TLC) nebulizer or by commercial automatic sprayers such as the ImagePrep (Bruker Daltonics, Bremen, Germany) or other spray devices.[30,31] Depending on the matrix and preparation, the size of the matrix crystals is typically less than 20 μm. With a homogeneous coating of matrix across the tissue section, the spatial resolution of the image is dependent on the laser spot size on the target. However, with spray techniques, one must take care to prevent delocalization of analytes caused by over wetting the tissue section. Robotic reagent spotters such as the Portrait 630 (Labcyte, Sunnyvale, CA) or the piezoelectric-based chemical ink-jet printer (ChIP-1000, Shimadzu Co., Kyoto, Japan) apply matrix in a high-density array (~100–250-μm center to center

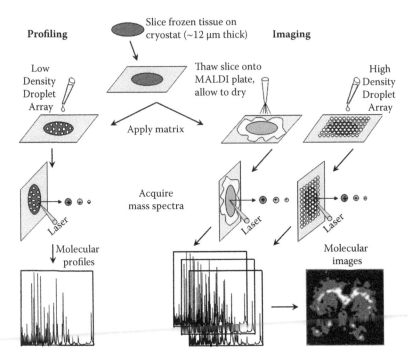

FIGURE 13.2 General profiling and imaging workflow schemes for the analysis of biological tissue sections using MALDI MS. Fresh-frozen tissue is sectioned by a cryostat. Matrix is applied in either a low-density droplet array (profiling) or in a high-density array or homogeneous coating (imaging). MS analysis is performed at predefined x-y coordinates, and ion density maps are reconstructed from each m/z analyzed. (From P. Chaurand, J. L. Norris, D. S. Cornett, J. A. Mobley, and R. M. Caprioli. *J. Proteome Res.* 5(11), 2889–2900, 2006. With permission.)

spacing).[27,28,32,33] By applying the matrix in an array, analytes are contained within the diameter of the matrix spot, ensuring limited analyte delocalization. Further, signal intensities are typically higher and more reproducible when the matrix is applied in an automated fashion.

13.2.2 Mass Spectrometric Analysis

Data acquisition is accomplished by a raster of the laser across a tissue section at predefined x-y coordinates (pixels). A UV laser (N_2, 337 nm; or frequency-tripled Nd:YAG [neodymium-doped yttrium aluminum garnet], 355 nm) has a laser spot size focused between ~25 and 150 μm and are typically operated between 200–1000 Hz.[6] Newer frequency-tripled Nd:YLF (neodymium-doped yttrium lithium fluoride, 349 nm) UV lasers have a high repetition rate (up to 20 kHz) but typically perform at repetition rates of 5 kHz or below.[34] Recently, high lateral resolution has been achieved by decreasing the laser irradiation spot to the average size of most mammalian cells (~10 μm) using optics with a coaxial illumination gridless two-stage ion source and N_2 laser. This configuration was able to image proteins up to m/z 27,000 directly from a 12-μm-thick mouse epididymis tissue section at ~7 μm spatial resolution.[35] High-resolution imaging has also been achieved utilizing magnifying ion optics configured as an ion microscope to preserve the spatial arrangement of the analytes on the sample (stigmatic imaging) using a position-sensitive detector. Features within an area irradiated by a ~200-μm laser beam are spatially resolved at about 4 μm for a selected *m/z* value. Stigmatic imaging was used to determine the localization of neuropeptides within rat and human pituitary tissue sections.[36] Oversampling of the laser can be performed to increase lateral resolution by setting a smaller laser raster increment than the laser beam width, although it is critical that all material is ablated to prevent signal carryover.[37] This approach was successfully applied to image neuropeptides from *Aplysia* nerve cells (~80–100 μm) at a lateral resolution of 25 μm.

Imaging ion mobility MS has currently been reported on two different platforms: a MALDI drift cell ion mobility–TOFMS instrument (Ionwerks, Inc., Houston, TX) and traveling-wave ion mobility–TOFMS (SYNAPT HDMS System, Waters Corporation, Manchester, UK).[3,18,20,38,39] The MALDI drift cell ion mobility instrument utilizes a solid-state Nd:YLF (349 nm) high repetition rate laser routinely operated between 300 and 1000 Hz. Ions are injected into a drift cell about 15 cm in length that is maintained at 3–5 Torr He to produce a mobility resolution (ΔT/T) of ~30–50. MALDI is performed at the same pressure of the drift cell offering collisional cooling of the ions. After mobility separation, ions are directed through a differentially pumped region and analyzed by orthogonal reflectron TOF MS. The arrival time of an ion to the pulsar in front of the orthogonal TOF analyzer is commonly referred to as the arrival time distribution (ATD) or drift time. The mobility separation is completed on a μs-ms time scale, and flight times in the TOF analyzer are on the order of tens of μs so that several mass spectra are recorded for each mobility separation. Due to the large data file size (from both the ion mobility drift cell and TOF analyzer) for each pixel, MALDI imaging ion mobility MS is generally performed by selecting a 4-μs drift time window for each imaging experiment. This approach increases the TOF duty cycle while maintaining a manageable data file size (less than 2 Gb for most tissue sections). Mass spectra are only recorded from a region of the lipid or peptide trend-line providing biomolecular-class-specific images devoid of isobaric chemical noise. This is especially useful when analyzing peptides that fall within the lipid-rich region of the spectrum (between *m/z* 700–900). The 2D conformation space displaying the ATD or CCS is plotted as a function of *m/z* (as in Figure 13.1a) and images are reconstructed using customized software.[18,20]

In the case of the MALDI traveling-wave ion mobility instrument, ions are guided to a quadrupole where they are either transmitted or isolated for subsequent tandem MS analysis. After exiting the quadrupole, ions are accumulated in a trap region before being released into a traveling-wave ion mobility device (ms drift times) and sequentially analyzed by an orthogonal TOF analyzer. For this particular instrument platform, 200 mass spectra are taken per each ion mobility event. Further, all drift time and *m/z* data is collected for the entire image, creating files greater than 20 Gb for 10-mm² sections imaged with a 200-μm spatial resolution. Class-specific imaging is performed by collecting all data from the ion mobility and TOF analyzer during the imaging experiment and selecting signals of interest from the resulting 2D conformation plot for image reconstruction. Spectra are visualized using MassLynx software (Waters Corp.) and images are converted to the Analyze file format using MALDI Imaging Converter Software (Waters Corp.) and visualized using Biomap (Novartis, Basel, Switzerland).[3]

The traveling-wave technology consists of a series of stacked ring electrodes in which a continuous wave of dc pulses on adjacent electrodes guides ions through N_2 gas (~10^{-1} Torr in the cell). Two parameters that influence the mobility resolution are the dc wave height (typically 7.5–12 V but can be set as high as 25 V) and the dc wave velocity (between 300 and 600 m/z). The traveling waves of dc pulses provide a higher sensitivity due to the radial refocusing of ions, but typically have a lower mobility resolution (~10–15) than drift cell ion mobility instruments.[39] The difference in mobility resolution is compensated to an extent by the high resolution of the orthogonal reflectron TOF analyzer, which can be operated in either V mode (~15,000 R) or W mode (~20,000 R). Since traveling-wave technology does not utilize a continuous electrostatic field, the drift time of analytes is not inversely proportional to the absolute CCS, and therefore CCS values cannot be directly calculated as with the drift cell technology. However, several labs have successfully estimated CCS values using calibrants of known CCS measured using drift cell ion mobility-MS instruments.[40–42]

13.3   APPLICATIONS OF TISSUE IMAGING

The ability to characterize molecular components of specific cell populations within tissue sections and between healthy and diseased tissue samples has made MS a valuable tool for both proteomic- and lipidomic-based research.[1,43,44] The high-throughput nature of MALDI MS provides the potential to analyze large numbers of tissue samples rapidly in a spatially resolved manner from a small amount of sample. For example, MALDI IMS provides the capability to analyze specific subpopulations of different grades and malignancies of tumor cells within a heterogeneous sample providing a spatially resolved molecular snapshot. This technology has been widely applied to probe the molecular composition of multiple diseased tissues including human gliomas, breast tumors, and non-small-cell lung tumors along with normal tissue samples to further understand underlying molecular mechanisms.[5,7–9,29]

Currently, the major method of diagnosis, prognosis, and predicting effective therapies for human diseases, specifically cancer, is histological/pathological evaluation. While pathology is critical for timely disease evaluation and prognosis, additional molecular information is expected to greatly enhance individual patient care. Of specific clinical interest is the determination of biomarkers for reliable early disease detection, therapy response prediction, and elucidation of the specific molecular events found in and adjacent to diseased tissues. Figure 13.3 illustrates MALDI MS analysis of a human renal cell carcinoma biopsy, showing the presence of tumor and adjacent noncancerous tissue.[45] For protein analysis by MALDI MS, matrix (25 mg/mL sinapinic acid in 50/50/0.1 of acetonitrile/H_2O/trifluoroacetic acid [TFA]) was automatically applied by a robotic spotter every 250 μm in rows across the tumor margin. The resulting spectra (m/z range 4–20 kDa) were combined and displayed as a heat map with spatial and m/z information clearly displaying several signal intensity

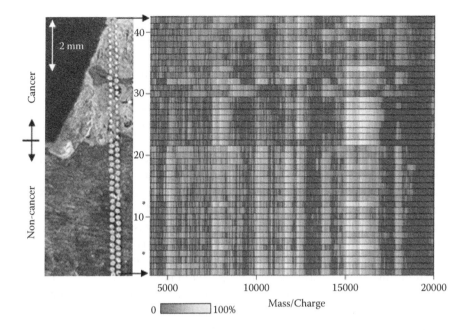

FIGURE 13.3 MALDI MS analysis of a human renal cell carcinoma (optical image, top) with attached nontumor (optical image, bottom) tissue biopsy section. The protein profiles are displayed (right) as a heat map illustrating the benefit of analyzing heterogeneous tissue samples in a spatially resolved manner. (From P. Chaurand, J. L. Norris, D. S. Cornett, J. A. Mobley, and R. M. Caprioli. *J. Proteome Res.* 5(11), 2889–2900, 2006. With permission.)

differences between the cancerous area (Figure 13.3, displayed at the top of the optical section) and the adjacent noncancerous tissue (Figure 13.3, shown at the bottom of the optical section). The molecular patterns that were observed within the cancerous tissue itself are consistent with the heterogeneous cellular content of the tumor. Spectra acquired from within the noncancerous area (Figure 13.3, spectra 5 and 12, which are highlighted with asterisks) contain some features similar to those in spectra acquired from the cancerous area (traces 22–29). It is likely that these molecular similarities between the noncancerous tissue and the cancerous tissue represent the beginning of cellular transformation in cells near the tumor and may indeed be early signs of precancerous growth.[45]

Protein identification is a major focus of many studies and is of particular interest in tissue image analysis. In classic proteomic studies, protein identification schemes consist of tissue homogenization and separation followed by enzymatic cleavage before peptide sequencing by MS/MS and database searching.[5,8] Generally, this is time-consuming, requires relatively large amounts of tissue, and can be insensitive with respect to low levels of protein expression. Recently, methods have been optimized to complete protease digestion directly onto the tissue section without the need for tissue homogenization.[27,46] In situ enzymatic digestion is performed by spotting trypsin onto the tissue section before matrix application and directly analyzing the resulting tryptic peptides (m/z 500–4000) by MALDI IMS. This technique provides the ability to rapidly map and sequence tryptic peptides directly from the tissue section and search protein databases to identify proteins with high confidence. The image analysis of tryptic peptides enables the analysis of higher molecular weight proteins not normally desorbed and ionized. Several studies using in situ tryptic digestion of fresh-frozen tissue sections has demonstrated the localization and identification of neurogranin, several isoforms of rat brain myelin basic protein and the 71 kDa synapsin I protein.[27] Additionally, this protocol can be adapted to use other in situ chemistries directly on tissue sections.

Until recently, MALDI IMS has been applied to the analysis of fresh-frozen tissue sections since formalin fixation alters protein molecular weight and structure. Nevertheless, FFPE (formalin-fixed, paraffin-embedded) tissues represent an extensive archive of diseased tissues with most having patient histories and known outcomes. Formalin fixation stabilizes proteins by chemical cross-linking throughout a tissue section, preventing postmortem enzymatic proteolysis while maintaining the cellular histology and allows tissues to be stored at room temperature for extended periods of time.[47] Protocols have been developed to minimize the effects of formalin-fixation (referred to as antigen retrieval) for use with genomic and proteomic techniques, such as immunohistochemistry (IHC).[48–50] The development of in situ chemistry prior to MALDI IMS analysis has made it possible to analyze enzymatically cleaved peptides directly from antigen-retrieved FFPE tissues, allowing a vast number of archived tissues available for rapid analysis by MALDI IMS for proteomic studies.[46,51,52] For example, MALDI IMS of FFPE tissue sections has been applied to the analysis of the 6-hydroxydopamine (6-OHDA) Parkinson's disease animal model. [52] In this model system, 6-OHDA was administered directly to the rat brain where it is toxic to the dopaminergic receptors, providing similar effects as Parkinson's disease. MALDI IMS analysis of the FFPE rat brain sections validated and confirmed previous biomarkers of Parkinson's disease including lower expression levels of ubiquitin, *trans*-elongation factor 1, neurofilament-M, and hexokinase while F1 ATPase, α-enolase, and peroxiredoxin 6 are up-regulated.[52] MALDI IMS has also been applied to the direct analysis of FFPE tissue microarrays (TMAs).[51] A TMA consists of an array of as many as several hundred core tumor biopsies arranged in a paraffin block. TMAs were developed for high-throughput screening and are routinely analyzed for DNA, mRNA, and proteins by fluorescence in situ hybridization, in situ hybridization, and IHC, respectively.[53–55] By utilizing antigen retrieval and in situ tryptic digestion, TMAs are easily analyzed by MALDI IMS for high-throughput screening of hundreds of individual tumor samples in a single imaging experiment. TMA analysis by MALDI IMS has been applied to studying various types of human lung cancer such as squamous cell carcinoma, adenocarcinoma, and bronchioloalveolar carcinoma and additional noncancerous tissue from matched individuals.[51] From analysis of these tumors, three tryptic peptides from heat shock protein β-1 were almost exclusively localized to a subset of the squamous cell carcinoma tissue biopsies. This particular example illustrates the concept that molecular analysis of these tumors may determine specific subsets beyond histological classification schemes.

13.4 APPLICATIONS OF TISSUE IMAGING BY ION MOBILITY–MASS SPECTROMETRY

The addition of an ion mobility separation step prior to the MS analyzer enhances MALDI IMS analysis of tissue sections by providing molecular fractionation based on the apparent surface area of a molecule. The rapid 2D ion mobility MS separation is beneficial for analyzing complex samples having many nominal isobaric molecular species. For instance, the analysis of a single matrix spot from in situ enzymatic digested tissue sections produces thousands of distinct peaks with a signal-to-noise ratio greater than three.[27] Ion mobility MS may also provide qualitative information without the need for MS/MS analysis by resolving different classes of biomolecules based on their intrinsic gas-phase packing efficiencies along trend-lines.[56,57] This is especially useful in the peptide and lipid mass range below *m/z* 1000 where abundant lipid species from the tissue section may mask peptide signals. The drift time or gas-phase CCS of tryptic peptides provides information on peptide secondary structure. Deviations from the peptide trend-line indicate that post-translational modifications such as phosphorylation and glycosylation may be present.[17,58,59]

The specificity of ion mobility MS makes this technology a powerful tool for 'omics studies. The potential to simultaneously analyze and image several biomolecular species provides a larger snapshot of the composition of tissue without additional analysis or sample preparation. For example, Figure 13.1a illustrates the separation of phospholipids and tryptic peptides from a grade II human glioma tissue section. By exporting the peptide and lipid trend-lines separately into two different mass spectra (Figure 13.1b), signals from each molecular class can be interrogated separately. Figure 13.4 illustrates separation and imaging of nominally isobaric lipid and peptide species deposited on a 12-μm mouse liver section.[18] For this experiment, DHB matrix (40 mg/mL in 70% methanol and 0.1% TFA) was spiked with either 1 mg/mL of the peptide RPPGFSP or 2 mg/mL of a phosphatidylcholine (PC) extract. Spiked matrices were deposited onto the tissue section using an automatic reagent spotter (Portrait 630 Reagent Multispotter) in an array pattern of an "X" where each line was deposited as either the peptide RPPGFSP ("/") or PC ("\") (Figure 13.4a, optical image). A representative 2D ATD *versus m/z* conformation plot for a mixture of the peptide and PC standards is shown in Figure 13.4b in which the peptide, RPPGFSP, and an isobaric lipid (PC 34:2) are baseline resolved by ion mobility alone with ATD peak maxima at 449 and 504 μs, respectively. Ion density maps for RPPGFSP and PC 34:2 are illustrated in Figure 13.4c (left and middle, respectively). The same 1-Da window (*m/z* 759–760) is interrogated for all three images but at different ATD drift times. The peptide, RPPGFSP, and PC 34:2 display two different images while the overlayed ion density map (Figure 13.4c, right) illustrates normal MALDI MS–only imaging. This figure depicts the advantages of eliminating chemical noise on the basis of structural separation. For example, if the targeted biomolecule were the lipid species, PC 34:2, then chemical noise (i.e., the isobaric peptide) would result in the erroneous image of this lipid.[18] Endogenous lipid selective imaging of thin tissue sections has been performed by MALDI imaging ion mobility MS.[18,20]

Lipidomics is a rapidly growing research field in part because of the role of lipids in disease processes. Phospholipids are a diverse group of molecules that are

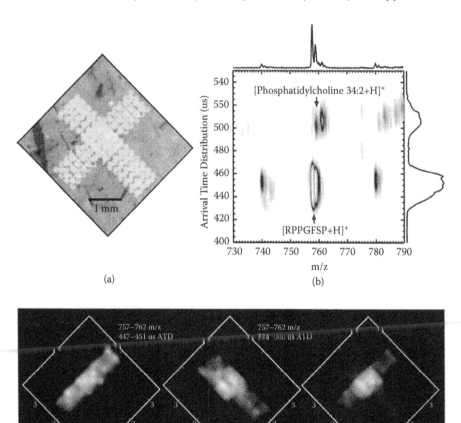

(a)   (b)

(c)

FIGURE 13.4   MALDI imaging ion mobility MS of an isobaric peptide (RPPGFSP) and phospholipid (PC 34:2). (a) An optical image of RPPGFSP and a phosphatidylcholine (PC) extract spiked into the DHB matrix and deposited separately on a mouse liver tissue section in the pattern of an "X". The "/" and "\" lines are where RPPGFSP and PC were deposited, respectively. (b) A representative 2D conformation plot resulting from the analysis of RPPGFSP and PC 34:2, illustrating the ability of ion mobility MS to resolve these two nominal species. (c) Extracted ion density maps for RPPGFSP (left, green), PC 34:2 (middle, blue), and an overlayed image of the two ion density maps (right). (From J. A. McLean, W. B. Ridenour, and R. M. Caprioli. *J. Mass Spectrom.* 42(8), 1099–1105, 2007. With permission.)

the structural components of cell membranes and are involved in signaling pathways, protein sorting, and energy storage. Phospholipids also serve as precursors for ceramides, eicosanoids, inositol phosphates, and lysophosphatidic acids, which in turn activate various cellular responses including proliferation, angiogenesis, and apoptosis.[60] Altered levels in lipid composition are found in many pathological conditions such as Alzheimer's disease, Down syndrome, diabetes, and cancer as well as many other diseases.[61–63] Analysis by MALDI IMS has shown, for example, that differential intensities of specific lipid species have been revealed across tumor margins of human renal cell carcinoma (S. Puolitaival, unpublished data). The two main groups of phospholipids observed by MALDI-MS in positive mode are PC and SM (sphingomyelin). While these two biomolecular species are similar, they differ in their carbon chain backbone. MALDI ion mobility MS has the specificity to resolve these two species based on their gas-phase packing efficiencies in addition to m/z specificity, with SM species having longer mobility drift times and larger CCS.[44,56] Ion mobility can selectively image phospholipid species directly from tissue sections separately from all other endogenous chemical components.[18,20] Thus a specific phospholipid image can be obtained without isobaric endogenous molecules of a different molecular class contributing to the image. For example, selective MALDI imaging ion mobility MS analysis of phospholipids was performed by a MALDI drift cell instrument of a human clear cell renal cell carcinoma (ccRCC) tissue section with adjacent normal cortex (Figure 13.5). Several phospholipid species were found to have different intensities across the tissue section including m/z 741

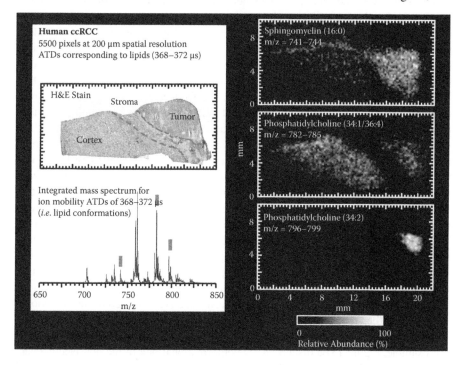

FIGURE 13.5 Species-selective MALDI imaging ion mobility MS of phospholipid species analyzed directly from a grade I human clear cell renal cell carcinoma (ccRCC) tissue biopsy section with an adjacent nontumor cortex region. Top left: An H&E stained serial section with outlined normal cortex, stroma, and tumor. Bottom left: An average mass spectrum of the image. Right: The resulting phospholipid images of SM (16:0), PC (34:1/36:0), and PC (34:2) illustrating ion density differences across the tumor margin.

identified by MS/MS as [SM 16:0+H]$^+$ that is more abundant in the tumor than in the surrounding histologically normal tissue. Additionally, m/z 782 and 796 were found to be differentially localized and were identified as [PC 34:1+Na]$^+$/[PC 36:4+H]$^+$ and [PC 34:2+K]$^+$, respectively. Interestingly, one can observe a drastic difference in the distribution of [PC 34:1+Na]$^+$ and [PC 34:2+K]$^+$ even though there is only one degree of saturation difference in the fatty acid carbon chain.

13.5   PHARMACEUTICAL APPLICATIONS

MALDI imaging ion mobility MS has been shown to be an effective analytical tool for the analysis of exogenous pharmaceuticals analyzed directly from tissue sections. Mapping the localization of drugs and their metabolites to determine the pharmacokinetics or absorption, distribution, metabolism, and excretion (ADME) is essential in order to obtain U.S. Food and Drug Administration (FDA) approval. Current pharmaceutical imaging technologies, such as whole body autoradiography (WBA) or fluorescence imaging, detect the distribution of a label throughout dosed animals, but it is not known if one is measuring the original drug, a metabolite, or other species resulting from the transfer of the label. Therefore, further analysis is required to obtain additional information on the nature of the monitored species.[64]

MALDI IMS provides information on the spatial distribution and molecular identity of drugs and drug metabolites in tissue at therapeutic doses.[10–12] Typically a single reaction monitoring (SRM) strategy is employed in drug IMS experiments because of the high abundance of endogenous analytes and matrix interference below m/z 1000. When imaging is performed in this fashion, the intensity of the transition of a precursor ion to a major fragment is monitored, and the fragment ion intensity is used to reconstruct the ion density map.[11] To date, several studies have analyzed pharmaceuticals and metabolites by MALDI IMS, e.g., antipsychotics such as clozapine (m/z 327 → 192) in rat brain and olanzapine (m/z 313 → 256) and its two first-pass metabolites, N-desmethyl-olanzapine (m/z 299 → 256) and 2-hydroxymethyl-olazapine (m/z 329 → 272) in whole animal sections.[10,65] The antitumor drug SCH 22637 (m/z 695 → 228) was also mapped in mouse tumor tissue using the SRM strategy to eliminate interference by ions from the matrix that are at the same m/z of the drug.[11]

The analysis of dosed tissue sections by MALDI imaging ion mobility MS provides several advantages including the ability to separate pharmaceuticals from endogenous chemical noise before fragmentation and the possibility of simultaneous fragmentation of all species following ion mobility separation increasing throughput. Fragment ions correlate with the drift time of the parent molecule and are separated in conformational space from other species of the same m/z.[21,22] Therefore, each drug and metabolite in a mixture is able to be fragmented and analyzed in a single imaging experiment. For example, Figure 13.6a displays a 2D conformational plot of the nonselective beta blocker, propranolol (m/z 260) with fragmentation occurring after ion mobility drift cell separation. The fragments of m/z 218, 182, 157, and 116 all have the same drift time as the parent molecule and are also resolved from a matrix-derived analyte.[19] MALDI imaging ion mobility MS of pharmaceuticals using collision-induced dissociation (CID) after mobility separation has been applied to the analysis of vinblastine (VLB), an antitumor agent, in whole body tissue sections (Figure 13.6b).[3] After one hour of

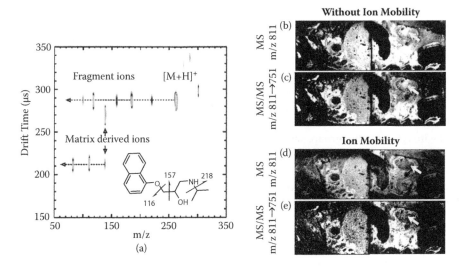

FIGURE 13.6 MALDI imaging ion mobility MS analysis of exogenous pharmaceuticals. (a) A 2D conformation plot of MALDI IM-MS analysis of the drug propranolol (*m/z* 260, structure shown). Since CID was performed after ion mobility separation, fragment ions correlate to the drift time of the parent ion. (From L. S. Fenn, and J. A. McLean. *Anal. Bioanal. Chem.* 391(3), 905–909, 2008. With permission.) (b) MALDI IMS of vinblastine (VLB, *m/z* 811). (c) MALDI IMS performed using an SRM strategy to monitor the intensity of the transition *m/z* 811 → 751. The incorrect representation of VLB located to the kidney is due to an isobaric phospholipid that shares the fragment *m/z* 751. (d) MALDI imaging ion mobility MS of VLB. (e) MALDI imaging ion mobility MS/MS of VLB (*m/z* 811 → 751). Notice VLB does not appear to localize to the kidney when ion mobility separation is performed before fragmentation due to the separation of VLB fragment at *m/z* 751 from an isobaric phospholipid that also contains a fragment at *m/z* 751, which contributes to the analyte distribution in (b) and (c). (From P. J. Trim, C. M. Henson, J. L. Avery, A. McEwen, M. F. Snel, E. Claude, P. S. Marshall, A. West, A. P. Princivalle, and M. R. Clench. *Anal. Chem.* 80(22), 8628–8634, 2008. With permission.)

administrating 6 mg/kg VLB intravenously (IV), rats were sacrificed and sectioned for analysis. Matrix (25 mg/mL CHCA in 70% ethanol and 0.1% TFA) was applied by a manual spray system. Whole rat animal sections were analyzed by MALDI MS/MS using MALDI imaging traveling-wave ion mobility MS (Waters, Corp.). VLB (*m/z* 811) is fragmented to *m/z* 751, 542, 524, 719, 649, and 691. All the reconstructed images for each fragment indicate that the distribution of VLB is localized to the liver, renal cortex, and in tissues surrounding the gastrointestinal (GI) tract, in agreement with WBA imaging. On the other hand, the distribution of the parent ion at *m/z* 811 and the transition of *m/z* 811 → 751 indicated VLB is additionally localized to the kidney (Figure 13.6b and c). When whole animal sections were analyzed by MALDI imaging ion mobility MS/MS (fragmentation occurs after ion mobility separation), the distribution of VLB for the transition *m/z* 811 → 751 at the drift time of VLB was not localized to the kidney but had a distribution more similar with the other VLB transitions and when MALDI imaging ion mobility MS was performed (Figure 13.6d and e). The localization to the kidney area when analyzed without utilizing an ion mobility separation was determined to arise from an interfering isobaric endogenous lipid also containing a fragment at m/z 751. Thus, prefractionation using ion mobility allows VLB and its fragments to be resolved from interfering endogenous lipid species. These data illustrate the power of the additional, post-ionization separation dimension in such studies.

The limitation for simultaneous fragmentation of all species is the ability to resolve molecules of interest, and this strategy is especially applicable in cases where structurally different metabolites or other modifications of differing structure are being monitored. Since several metabolites only differ from the parent molecule by a hydroxyl or methyl group, separation may be difficult in these cases, but strategies utilizing more polar drift gases such as CO_2 may be used and have shown promise for the analysis of drugs and other small molecules.[66,67] MALDI imaging ion mobility MS will also be useful in the analysis of the spatial distribution of drugs after administration of multiple drugs.

13.6   CONCLUSIONS AND PERSPECTIVES

MALDI IMS provides a high-throughput technology for the analysis of tissue sections amendable for the analysis of human tumor samples and rapid determination of prognosis and tissue-based diagnosis. Further instrumental advances open possibilities to determine a proteomic profile of a tissue section within a timeframe currently used for intraoperative histological examination. MALDI IMS provides the possibility of not only determining tumor grading but also determining a therapeutic strategy likely to give a positive response.[6] The additional dimension of separation provided by ion mobility enhances current imaging protocols by performing a rapid (μs–ms) structural separation before m/z selectivity without the need for further sample preparation. Ion mobility is currently able to resolve and image nominal isobaric species by structural differences and is sufficiently sensitive to analyze endogenous and exogenous species directly from thin tissue sections.[3,18,20] The additional separation step enables imaging to be performed in a biomolecular class-selective mode by reconstructing images selectively from signals that correlate to drift times of species of interest. Further instrumentation advances in ion mobility to provide higher sensitivity, higher mobility resolutions and an increased mass range will provide still greater capabilities for analysis of discrete molecular species in complex mixtures.

Pharmaceutical analysis of dosed tissue sections by MALDI imaging ion mobility MS provides the means to resolve isobaric species before fragmentation is preformed.[3] The ability to simultaneously fragment and image exogenous drugs provides a spatially resolved, molecular-specific, high-throughput analysis of the distribution of a drug molecule and related metabolites.[21,22] MALDI imaging ion mobility MS technology does and will continue to provide a new dimension to the spatially resolved analysis of molecular components of tissue samples.

REFERENCES

1. Chaurand, P.; Schwartz, S. A.; Caprioli, R. M., "Imaging mass spectrometry: a new tool to investigate the spatial organization of peptides and proteins in mammalian tissue sections", *Curr. Opin. Chem. Biol.* 2002, *6*, 676–681.

2. Cornett, D. S.; Reyzer, M. L.; Chaurand, P.; Caprioli, R. M., "MALDI imaging mass spectrometry: molecular snapshots of biochemical systems", *Nature Meth.* 2007, *4*, 828–833.

3. Trim, P. J.; Atkinson, S. J.; Princivalle, A. P.; Marshall, P. S.; West, A.; Clench, M. R., "Matrix-assisted laser desorption/ionisation mass spectrometry imaging of lipids in rat brain tissue with integrated unsupervised and supervised multivariant statistical analysis", *Rapid Commun. Mass Spectrom.* 2008, *22*, 1503–1509.

4. Xu, B. J.; Li, J.; Beauchamp, R. D.; Shyr, Y.; Li, M.; Washington, M. K.; Yeatman, T. J.; Whitehead, R. H.; Coffey, R. J.; Caprioli, R. M., "Identification of early intestinal neoplasia protein biomarkers using laser capture microdissection and MALDI MS", *Mol. Cell Proteomics* 2009, *8*, 936–945.

5. Schwartz, S. A.; Weil, R. J.; Thompson, R. C.; Shyr, Y.; Moore, J. H.; Toms, S. A.; Johnson, M. D.; Caprioli, R. M., "Proteomic-based prognosis of brain tumor patients using direct-tissue matrix-assisted laser desorption ionization mass spectrometry", *Cancer Res.* 2005, *65*, 7674–7681.

6. Chaurand, P.; Cornett, D. S.; Caprioli, R. M., "Molecular imaging of thin mammalian tissue sections by mass spectrometry", *Curr. Opin. Biotechnol.* 2006, *17*, 431–436.

7. Yanagisawa, K.; Shyr, Y.; Xu, B. J.; Massion, P. P.; Larsen, P. H.; White, B. C.; Roberts, J. R.; Edgerton, M.; Gonzalez, A.; Nadaf, S.; Moore, J. H.; Caprioli, R. M.; Carbone, D. P., "Proteomic patterns of tumour subsets in non-small-cell lung cancer", *Lancet* 2003, *362*, 433–439.

8. Burnum, K. E.; Tranguch, S.; Mi, D.; Daikoku, T.; Dey, S. K.; Caprioli, R. M., "Imaging mass spectrometry reveals unique protein profiles during embryo implantation", *Endocrinology* 2008, *149*, 3274–3278.

9. Burnum, K. E.; Cornett, D. S.; Puolitaival, S. M.; Milne, S. B.; Myers, D. S.; Tranguch, S.; Brown, H. A.; Dey, S. K.; Caprioli, R. M., "Spatial and temporal alterations of phospholipids determined by mass spectrometry during mouse embryo implantation", *J. Lipid Res.* 2009, *50*, 2290–2298.

10. Khatib-Shahidi, S.; Andersson, M.; Herman, J. L.; Gillespie, T. A.; Caprioli, R. M., "Direct molecular analysis of whole-body animal tissue sections by imaging MALDI mass spectrometry", *Anal. Chem.* 2006, *78*, 6448–6456.

11. Reyzer, M. L.; Hsieh, Y.; Ng, K.; Korfmacher, W. A.; Caprioli, R. M., "Direct analysis of drug candidates in tissue by matrix-assisted laser desorption/ionization mass spectrometry", *J. Mass Spectrom.* 2003, *38*, 1081–1092.

12. Atkinson, S. J.; Loadman, P. M.; Sutton, C.; Patterson, L. H.; Clench, M. R., "Examination of the distribution of the bioreductive drug AQ4N and its active metabolite AQ4 in solid tumours by imaging matrix-assisted laser desorption/ionisation mass spectrometry", *Rapid Comm. Mass Spectrom.* 2007, *21*, 1271–1276.

13. Kanu, A. B.; Dwivedi, P.; Tam, M.; Matz, L.; Hill, H. H., Jr., "Ion mobility–mass spectrometry", *J. Mass Spectrom.* 2008, *43*, 1–22.

14. St. Louis, R. H.; Hill, H. H., "Ion mobility spectrometry in analytical chemistry", *Crit. Rev. Anal. Chem.* 1990, *21*, 321–355.

15. Clemmer, D. E.; Jarrold, M. F., "Ion mobility measurements and their applications to clusters and biomolecules", *J. Mass Spectrom.* 1997, *32*, 577–592.

16. Wyttenbach, T.; Bowers, M. T. In *Modern mass spectrometry*; Springer-Verlag Berlin: Berlin, 2003; Vol. 225, pp 207–232.

17. McLean, J. A.; Ruotolo, B. T.; Gillig, K. J.; Russell, D. H., "Ion mobility-mass spectrometry: a new paradigm for proteomics", *Int. J. Mass Spectrom.* 2005, *240*, 301–315.

18. McLean, J. A.; Ridenour, W. B.; Caprioli, R. M., "Profiling and imaging of tissues by imaging ion mobility-mass spectrometry", *J. Mass Spectrom.* 2007, *42*, 1099–1105.

19. Fenn, L. S.; McLean, J. A., "Biomolecular structural separations by ion mobility–mass spectrometry", *Anal. Bioanal. Chem.* 2008, *391*, 905–909.

20. Jackson, S. N.; Ugarov, M.; Egan, T.; Post, J. D.; Langlais, D.; Schultz, J. A.; Woods, A. S., "MALDI-ion mobility-TOFMS imaging of lipids in rat brain tissue", *J. Mass Spectrom.* 2007, *42*, 1093–1098.

21. Hoaglund-Hyzer, C. S.; Clemmer, D. E., "Ion trap/ion mobility/quadrupole/time-of-flight mass spectrometry for peptide mixture analysis", *Anal. Chem.* 2001, *73*, 177–184.

22. Baker, E. S.; Tang, K.; Danielson, W. F., 3rd; Prior, D. C.; Smith, R. D., "Simultaneous fragmentation of multiple ions using IMS drift time dependent collision energies", *J. Am. Soc. Mass Spectrom.* 2008, *19*, 411–419.

23. Schwartz, S. A.; Reyzer, M. L.; Caprioli, R. M., "Direct tissue analysis using matrix-assisted laser desorption/ionization mass spectrometry: practical aspects of sample preparation", *J. Mass Spectrom.* 2003, *38*, 699–708.

24. Gluckmann, M.; Pfenninger, A.; Kruger, R.; Thierolf, M.; Karas, M.; Horneffer, V.; Hillenkamp, F.; Strupat, K., "Mechanisms in MALDI analysis: surface interaction or incorporation of analytes?", *Int. J. Mass Spectrom.* 2001, *210*, 121–132.

25. Dreisewerd, K., "The desorption process in MALDI", *Chem. Rev.* 2003, *103*, 395–426.

26. Kruse, R.; Sweedler, J. V., "Spatial profiling invertebrate ganglia using MALDI MS", *J. Am. Soc. Mass Spectrom.* 2003, *14*, 752–759.

27. Groseclose, M. R.; Andersson, M.; Hardesty, W. M.; Caprioli, R. M., "Identification of proteins directly from tissue: in situ tryptic digestions coupled with imaging mass spectrometry", *J. Mass Spectrom.* 2007, *42*, 254–262.

28. Aerni, H. R.; Cornett, D. S.; Caprioli, R. M., "Automated acoustic matrix deposition for MALDI sample preparation", *Anal. Chem.* 2006, *78*, 827–834.

29. Cornett, D. S.; Mobley, J. A.; Dias, E. C.; Andersson, M.; Arteaga, C. L.; Sanders, M. E.; Caprioli, R. M., "A novel histology-directed strategy for MALDI-MS tissue profiling that improves throughput and cellular specificity in human breast cancer", *Mol. Cell. Proteomics* 2006, *5*, 1975–1983.

30. Deininger, S.; Ebert, M. P.; Futterer, A.; Gerhard, M.; Rocken, C., "MALDI imaging combined with hierarchical clustering as a new tool for the interpretation of complex human cancers", *J. Proteome Res.* 2008, *7*, 5230–5236.

31. Walch, A.; Rauser, S.; Deininger, S. O.; Hofler, H., "MALDI imaging mass spectrometry for direct tissue analysis: a new frontier for molecular histology", *Histochem. Cell Biol.* 2008, *130*, 421–434.

32. Baluya, D. L.; Garrett, T. J.; Yost, R. A., "Automated MALDI matrix deposition method with inkjet printing for imaging mass spectrometry", *Anal. Chem.* 2007, *79*, 6862–6867.

33. Drexler, D. M.; Garrett, T. J.; Cantone, J. L.; Diters, R. W.; Mitroka, J. G.; Prieto Conaway, M. C.; Adams, S. P.; Yost, R. A.; Sanders, M., "Utility of imaging mass spectrometry (IMS) by matrix-assisted laser desorption ionization (MALDI) on an ion trap mass spectrometer in the analysis of drugs and metabolites in biological tissues", *J. Pharmacol. Toxicol. Methods* 2007, *55*, 279–288.

34. Spengler, B.; Hubert, M., "Scanning microprobe matrix-assisted laser desorption ionization (SMALDI) mass spectrometry: instrumentation for sub-micrometer resolved LDI and MALDI surface analysis", *J. Am. Soc. Mass Spectrom.* 2002, *13*, 735–748.

35. Chaurand, P.; Schriver, K. E.; Caprioli, R. M., "Instrument design and characterization for high resolution MALDI-MS imaging of tissue sections", *J. Mass Spectrom.* 2007, *42*, 476–489.

36. Altelaar, A. F. M.; Taban, I. M.; McDonnell, L. A.; Verhaert, P.; de Lange, R. P. J.; Adan, R. A. H.; Mooi, W. J.; Heeren, R. M. A.; Piersma, S. R., "High-resolution MALDI imaging mass spectrometry allows localization of peptide distributions at cellular length scales in pituitary tissue sections", *Int. J. Mass Spectrom.* 2007, *260*, 203–211.

37. Jurchen, J. C.; Rubakhin, S. S.; Sweedler, J. V., "MALDI-MS imaging of features smaller than the size of the laser beam", *J. Am. Soc Mass Spectrom.* 2005, *16*, 1654–1659.

38. Pringle, S. D.; Giles, K.; Wildgoose, J. L.; Williams, J. P.; Slade, S. E.; Thalassinos, K.; Bateman, R. H.; Bowers, M. T.; Scrivens, J. H., "An investigation of the mobility separation of some peptide and protein ions using a new hybrid quadrupole/travelling wave IMS/oa-ToF instrument", *Int. J. Mass Spectrom.* 2007, *261*, 1–12.

39. Giles, K.; Pringle, S. D.; Worthington, K. R.; Little, D.; Wildgoose, J. L.; Bateman, R. H., "Applications of a travelling wave-based radio-frequency only stacked ring ion guide", *Rapid Comm. Mass Spectrom.* 2004, *18*, 2401–2414.

40. Ruotolo, B. T.; Benesch, J. L.; Sandercock, A. M.; Hyung, S. J.; Robinson, C. V., "Ion mobility-mass spectrometry analysis of large protein complexes", *Nat. Protoc.* 2008, *3*, 1139–1152.

41. Thalassinos, K.; Grabenauer, M.; Slade, S. E.; Hilton, G. R.; Bowers, M. T.; Scrivens, J. H., "Characterization of phosphorylated peptides using traveling wave-based and drift cell ion mobility mass spectrometry", *Anal. Chem.* 2009, *81*, 248–254.

42. Scarff, C. A.; Thalassinos, K.; Hilton, G. R.; Scrivens, J. H., "Travelling wave ion mobility mass spectrometry studies of protein structure: biological significance and comparison with x-ray crystallography and nuclear magnetic resonance spectroscopy measurements", *Rapid Commun. Mass Spectrom.* 2008, *22*, 3297–3304.

43. Fournier, I.; Wisztorski, M.; Salzet, M., "Tissue imaging using MALDI-MS: a new frontier of histopathology proteomics", *Expert Rev. Proteomics* 2008, *5*, 413–424.

44. Jackson, S. N.; Wang, H. Y.; Woods, A. S., "Direct profiling of lipid distribution in brain tissue using MALDI-TOF MS", *Anal. Chem.* 2005, *77*, 4523–4527.

45. Chaurand, P.; Norris, J. L.; Cornett, D. S.; Mobley, J. A.; Caprioli, R. M., "New developments in profiling and imaging of proteins from tissue sections by MALDI mass spectrometry", *J. Proteome Res.* 2006, *5*, 2889–2900.

46. Lemaire, R.; Desmons, A.; Tabet, J. C.; Day, R.; Salzet, M.; Fournier, I., "Direct analysis and MALDI imaging of formalin-fixed, paraffin-embedded tissue sections", *J. Proteome Res.* 2007, *6*, 1295–1305.

47. Fox, C. H.; Johnson, F. B.; Whiting, J.; Roller, P. P., "Formaldehyde fixation", *J. Histochem. Cytochem.* 1985, *33*, 845–853.

48. Sompuram, S. R.; Vani, K.; Messana, E.; Bogen, S. A., "A molecular mechanism of formalin fixation and antigen retrieval", *Am. J. Clin. Pathol.* 2004, *121*, 190–199.

49. Jiang, X.; Jiang, X.; Feng, S.; Tian, R.; Ye, M.; Zou, H., "Development of efficient protein extraction methods for shotgun proteome analysis of formalin-fixed tissues", *J. Proteome Res.* 2007, *6*, 1038–1047.

50. Hwang, S. I.; Thumar, J.; Lundgren, D. H.; Rezaul, K.; Mayya, V.; Wu, L.; Eng, J.; Wright, M. E.; Han, D. K., "Direct cancer tissue proteomics: a method to identify candidate cancer biomarkers from formalin-fixed paraffin-embedded archival tissues", *Oncogene* 2007, *26*, 65–76.

51. Groseclose, M. R.; Massion, P. P.; Chaurand, P.; Caprioli, R. M., "High-throughput proteomic analysis of formalin-fixed paraffin-embedded tissue microarrays using MALDI imaging mass spectrometry", *Proteomics* 2008, *8*, 3715–3724.

52. Stauber, J.; Lemaire, R.; Franck, J.; Bonnel, D.; Croix, D.; Day, R.; Wisztorski, M.; Fournier, I.; Salzet, M., "MALDI imaging of formalin-fixed paraffin-embedded tissues: application to model animals of Parkinson disease for biomarker hunting", *J. Proteome Res.* 2008, *7*, 969–978.

53. Kononen, J.; Bubendorf, L.; Kallioniemi, A.; Barlund, M.; Schraml, P.; Leighton, S.; Torhorst, J.; Mihatsch, M. J.; Sauter, G.; Kallioniemi, O. P., "Tissue microarrays for high-throughput molecular profiling of tumor specimens", *Nat. Med.* 1998, *4*, 844–847.

54. Battifora, H., "The multitumor (sausage) tissue block: novel method for immunohistochemical antibody testing", *Lab. Invest.* 1986, *55*, 244–248.

55. Battifora, H.; Mehta, P., "The checkerboard tissue block. An improved multitissue control block", *Lab. Invest.* 1990, *63*, 722–724.

56. Fenn, L. S.; Kliman, M.; Mahsut, A.; Zhao, S. R.; McLean, J. A., "Characterizing ion mobility-mass spectrometry conformation space for the analysis of complex biological samples", *Anal. Bioanal. Chem.* 2009, *394*, 235–244.

57. Woods, A. S.; Ugarov, M.; Egan, T.; Koomen, J.; Gillig, K. J.; Fuhrer, K.; Gonin, M.; Schultz, J. A., "Lipid/peptide/nucleotide separation with MALDI-ion mobility-TOF MS", *Anal. Chem.* 2004, *76*, 2187–2195.

58. Ruotolo, B. T.; Verbeck, G. F.; Thomson, L. M.; Gillig, K. J.; Russell, D. H., "Observation of conserved solution-phase secondary structure in gas-phase tryptic peptides", *J. Am. Chem. Soc.* 2002, *124*, 4214–4215.

59. von Helden, G.; Wyttenbach, T.; Bowers, M. T., "Conformation of macromolecules in the gas phase: use of matrix-assisted laser desorption methods in ion chromatography", *Science* 1995, *267*, 1483–1485.

60. Fahy, E.; Subramaniam, S.; Brown, H. A.; Glass, C. K.; Merrill, A. H., Jr.; Murphy, R. C.; Raetz, C. R.; Russell, D. W.; Seyama, Y.; Shaw, W.; Shimizu, T.; Spener, F.; van Meer, G.; VanNieuwenhze, M. S.; White, S. H.; Witztum, J. L.; Dennis, E. A., "A comprehensive classification system for lipids", *J. Lipid Res.* 2005, *46*, 839–861.

61. Murphy, E. J.; Schapiro, M. B.; Rapoport, S. I.; Shetty, H. U., "Phospholipid composition and levels are altered in Down syndrome brain", *Brain Res.* 2000, *867*, 9–18.

62. Han, X.; Holtzman, D. M.; McKeel, D. W., Jr., "Plasmalogen deficiency in early Alzheimer's disease subjects and in animal models: molecular characterization using electrospray ionization mass spectrometry", *J. Neurochem.* 2001, *77*, 1168–1180.

63. Han, X.; Yang, J.; Yang, K.; Zhao, Z.; Abendschein, D. R.; Gross, R. W., "Alterations in myocardial cardiolipin content and composition occur at the very earliest stages of diabetes: a shotgun lipidomics study", *Biochemistry* 2007, *46*, 6417–6428.

64. Solon, E. G.; Balani, S. K.; Lee, F. W., "Whole-body autoradiography in drug discovery", *Curr. Drug Metab.* 2002, *3*, 451–462.

65. Hsieh, Y.; Casale, R.; Fukuda, E.; Chen, J.; Knemeyer, I.; Wingate, J.; Morrison, R.; Korfmacher, W., "Matrix-assisted laser desorption/ionization imaging mass spectrometry for direct measurement of clozapine in rat brain tissue", *Rapid Commun. Mass Spectrom.* 2006, *20*, 965–972.

66. Asbury, G. R.; Hill, H. H., Jr., "Using different drift gases to change separation factors (alpha) in ion mobility spectrometry", *Anal. Chem.* 2000, *72*, 580–584.

67. Matz, L. M.; Hill, H. H., Jr.; Beegle, L. W.; Kanik, I., "Investigation of drift gas selectivity in high resolution ion mobility spectrometry with mass spectrometry detection", *J. Am. Soc. Mass Spectrom.* 2002, *13*, 300–307.

14 Deciphering Carbohydrate Structures by IMS-MS

Applications to Biological Features Related to Carbohydrate Chemistry and Biology

Sergey Y. Vakhrushev and Jasna Peter-Katalinić

CONTENTS

14.1   INTRODUCTION TO CARBOHYDRATE STRUCTURE: FROM MONO- TO POLYSACCHARIDES

Carbohydrates are ubiquitous constituents of plants, animals, and microorganisms. Structure and conformation of carbohydrates strongly influence their physical properties, their chemical reactivity, and their biological functions. Monosaccharides are predominantly six- or five- membered cyclic compounds, formed via a semiacetal bond between a hydroxyl and a carbonyl group. The isobaric aldoses or ketoses differ within their subgroup by absolute configuration on chiral carbon atoms giving rise to 2^n possible diastereomers. The stereochemistry on the anomeric carbon at the reducing end is fixed in an α- or β-position if its hydroxy group is substituted by another monosaccharide in a glycosidic linkage or by an aglycon. The carbohydrate chain can be further extended in a linear way and/or by branching, since the glycosidic linkages can be closed between monosaccharides and/or oligosaccharides in different positions. The majority of carbohydrates, as optically active compounds found on earth, belong to the D-series, of which D-glucose (Glc) is the most abundant. Biologically highly abundant carbohydrates also include D-mannose (Man), D-galactose (Gal), two pentoses, D-xylose (Xyl), and D-ribose (Rib), as well as a nonasaccharide N-acetyl-D-neuraminic acid (NeuAc). High-molecular-weight D-glucose polysaccharides like glycogen, cellulose and dextrans, and inulin containing D-fructose, are homopolymers with a defined linkage pattern of the main chain and branching. Carrageenans from seaweeds contain mainly galactose, which is also present in the modified forms, by lack of hydroxy groups at 3 and 6 positions and by sulfation at position 4. Hyaluronic acid, a major component of synovial fluid and connective tissue, composed of the disaccharide repeating unit D-glucuronic acid-N-acetyl-D-glucosamine (GlcA-GlcNAc), can be prolonged to contain up to 10,000 monosaccharide units in the chain. Glycan chains attached to proteins via an N- or O-glycosidic linkage may contain 1–30 monosaccharide units and are structurally conserved. For all N-glycan types a pentasaccharide core structure is represented by two N-acetyl-D-glycosamines and three D-mannoses, where for O-glycans there are eight defined core structures, which all have an N-acetyl-D-galactosamine core attachment to a serin or a threonin in the protein chain in common (Figure 14.1). A high number of statistically possible variables of glycosylation is somewhat reduced according to defined biosynthetic rules, by which glycosyl transferases as primary gene products select the substrates for attachment of additional monosaccharides following their enzymatic specificity.

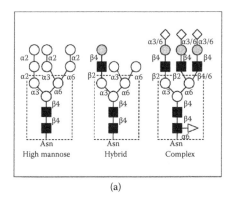

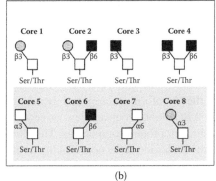

(a)

(b)

FIGURE 14.1 Basic architecture of *N*- and *O*-linked glycans. (a) The different types of *N*-glycan: complex, hybrid, and high mannose. The pentasaccharide "core" common to all *N*-glycans is shown by dashed lines. (b) *O*-glycan core structure. Core types 1–4 are the most common. The core structures can be further processed in the golgi. Symbols: empty circle, mannose; filled circle, galactose; empty square, *N*-acetyl galactosamine; filled square, *N*-acetyl glycosamine; empty diamond, *N*-acetylneuraminic acid; empty triangle, fucose.

14.2 INTRODUCTION TO GLYCOBIOLOGY

The expression "glycobiology," referring to a biology of glycoconjugates, was coined about 30 years ago as an idiom following a rapid growth of interest in glycoconjugates by academic and industrial research and development communities in the last part of the 20th century.[1,2] Although the chemical and biochemical aspects of glycoconjugates have been research topics of interest since 18th century, a number of instrumental methods developed in the second half of the 20th century for analysis of complex carbohydrate structure, like high field nuclear magnetic resonance (NMR) and soft ionization mass spectrometry (MS), which contributed significantly to an increase of the speed of discovery, providing a platform for detailed analytical studies of functionally interesting carbohydrate structures obtained by chemical synthesis or biosynthesis. In glycobiological studies it could be shown that different glycoconjugates represent key molecular players in cell biology, development, and interactions. Carbohydrates were shown to play a crucial role in signal transduction, cell adhesion, and immune recognition. It has been estimated that more than half of all proteins are glycosylated.[3] In some cases oligosaccharides were found to be responsible for the protein function, and in many cases carbohydrates are modifying chemical and physical properties of proteins, but in most cases their role is still unknown. Because of the nontemplate glycan structure, even a single glycosylation site of the protein may result in a wide heterogeneity with many different glycoforms. Besides, glycans of the same compositional constitution may exist as different isomeric structures (anomericity, branching distribution, linkage differentiation).

The interest of the biotechnology industry to develop glycoproteins as new human therapeutic products gave a strong impulse to produce a number of glycosylated recombinant proteins with distinct functions, like immunoglobulin G (IgG), tissue plasminogen activator (TPA), erythropoietin (EPO), and γ-interferon (γ-IFN). In IgG, TPA, and EPO, an accurate glycosylation pattern is an essential structural requirement for the efficiency of the therapy and for minimum or no side effects. According to definitions of regulatory agencies, the glycosylation of proteins produced in nonhuman cells has to meet stringent rules concerning the immunogenicity of the glycoforms present and their heterogeneity.

ABO blood group antigens on erythrocyte membranes are carbohydrates attached to lipid or protein components. The blood group specificity is defined in the terminal tetrasaccharide at the nonreducing terminus, in which antigens are defined by immunodominant sugars, GalNAc for A, Gal for B, attached to the H-antigen, carrying an L-fucose (Fuc) attached to an N-acetyl-lactosamine (LacNAc) epitope.[4] More complex structures exist in subgroups of A and B antigens. Mass spectrometry was shown to play an important role in monitoring the expression of potential immunogenic carbohydrate structures directly and to encompass the heterogeneity of protein and lipid glycosylation since then. Several types of workflows were probed in order to meet high demands on accurate sets of analytical data, which can provide the confidence necessary for batch-to-batch reproducibility.[5] The time scale of such workflows has been shown to take weeks and months, in which a purification of the sample was the speed-determining step.

14.3 COMPLEXITY OF CARBOHYDRATE STRUCTURE IN BIOLOGICAL SYSTEMS

Biosynthesis of the complex carbohydrate chains of glycoproteins, glycolipids, and proteoglycans is regulated by specifity and cell topology of glycosyltransferases. Over 300 glycosyltransferases expressing more than 100 glycogenes exist in mammalian tissues. In the brush border membrane-enriched fraction from epithelial cells the complexity of N-glycosylation is reflected by mapping of glycoforms in terms of their composition, where around 100 different molecular ions of different intensity were detected.[6] Considering the possibility of isoforms, this number could be significantly higher.

Polysialic acid, a homopolymer of α2,8-glycosidically linked sialic acid residues, is a major regulator of cell–cell interactions in the developing nervous system and a key factor in maintaining neural plasticity and exerts different glycopatterns in the brain regarding the time scale during brain development.[7]

According to the recent study on position isomeric glycoconjugates in α2,3- versus α2,6-sialylation in human granulocyte glycolipids, the heterogeneity of the mixture revealed more than 180 molecular structures of different quantity.[8]

Abberant glycosylation in proteins and lipids has been associated with defect glycogens and studied in relation to the pathophysiology of the disease. The goal of such studies is to define primarily the "healthy" and the "diseased" glycopatterns and to measure qualitative and quantitative differences in their respective expression profiles. Usually such profiles contain different glycoforms representing a ladder of homologous structures generated by anabolic or catabolic processes. The prelude task is the analytical one: to dissect the glycoprofiles starting with sequencing of all single components in complex mixtures and to quantify those that appear to be relevant. This information will represent a molecular basis for functional studies of single glycoforms in glycoproteins, glycolipids, and proteoglycans.

The goal to approach biologically relevant complex mixtures could be followed more clearly only since 1999, when the term "glycomics" was coined, as described below.

14.4 MASS SPECTROMETRY IN GLYCOMICS

14.4.1 GLYCOMICS

The term *glycome* encompasses a full set of carbohydrate structures in a cell or in an organism. Due to the high diversity of oligosaccharides, the glycome may be the most complex essence in nature, more complex than the genome and the proteome. Thus glycomics is a discipline that deals with all aspects related to the structure and function of the entire population of glycans within single organs and organisms. Presently the most challenging objectives of glycomics are to establish potent protocols for structure elucidation of all types of carbohydrates, which can serve to explore their functions in living organisms.

A major breakthrough in glycomics is yet to be achieved. Presently significant efforts are being invested in improving the efficiency and accuracy of molecular sequencing.[9–11] Combined with bioinformatics these data will be the main contributors to enhancing the computational organization and interpretation of sequence data.[12–14]

14.4.2 GENERAL ASPECTS OF MASS SPECTROMETRY IN GLYCOMICS

For the sequence analysis of glycoconjugates the most common analytical tools are NMR and MS. In spite of the ability to fully characterize a purified glycan by using NMR spectroscopy with little or no assistance from other methods, the use of NMR is limited by the fact that even the most sensitive instruments require a minimum of 1–10 nmol of purified oligosaccharides. In comparison to NMR the sensitivity of MS is in general significantly higher, namely in the femto- to zeptomole range. Besides, another significant feature of MS technology is the ability to analyze very complex mixture samples, frequently containing more than 100 components. Thus for structural analysis of oligosaccharide mixtures obtained from native sources, MS was shown to represent a versatile and most efficient final analyzer tool. The basic experiments to be performed are mapping and sequencing of all components from the mixture with high sensitivity and accuracy of identification.[15–20] In complex mixtures high precision can be achieved if during the sample admission great care is dedicated to keeping the polar analytes intact, by ensuring a low degree of in-source fragmentation. To perform such experiments, much attention should be focused toward the conditions in the ion source. For electrospray ionization (ESI), the electric field and the cone voltage should be adapted in such a way that the labile complex carbohydrate structure elements, NeuAc, Fuc, and sulfates (S), present in most biological mixtures are preserved during the ionization process and can enter the MS for analysis as intact species. For matrix-assisted laser desorption/ionization (MALDI) the main factors in preserving the integrity of the carbohydrate sample are the proper choice of organic matrix, the pH of the matrix and solvents, and the laser fluence.[15,21] The use of reflectrons in time-of-flight (TOF) analyzers is in most cases possible. In proteome analysis, characterization of glycosylation as a major co- and post-translational modification in proteolytic digests is rather difficult due to the lower ionization efficiency of carbohydrates in comparison to that of nonmodified peptides, resulting in ion suppression. Besides, different types and sizes of glycan chains can modify a single glycosylation site in the protein sequence, contributing to the microheterogeneity. Since the glycosylation patterns of proteins cannot be predicted from the nucleotide sequence as the protein core structure, it is a priori not clear how many and in which dynamic ratio different glycoforms from the same glycosylation site are present and how they contribute to the biological function.

Both ionization methods, MALDI and ESI, are adequate for the analysis of low- and high-molecular-weight molecular ions. Data provided by mass-to-charge ratio (m/z) determination can in some cases directly depict a compositional assignment of the molecule under investigation if the resolution and mass accuracy are sufficient, as in quadrupole TOF (QTOF) analyzers, or high, as in ion cyclotron resonance (ICR) and Orbitrap analyzers.

14.4.3 FRAGMENTATION AND MS/MS DATA INTERPRETATION

The fragmentation process of complex glycans is dependent on a number of parameters, which have been under investigation since 1988.[22] For sequencing of oligomeric components different methods of tandem mass spectrometry (MS/MS) were employed, like low energy CID, TOF-TOF fragmentation, sustained off-resonance irradiation (SORI), infrared multiphoton dissociation (IRMPD), and electron capture dissociation (ECD), basically attempting to obtain diagnostic fragment ions specific or unique to single structures.[23–28] Different techniques were probed for sequencing of components in carbohydrate mixtures[16,29–38] Additionally, the assignment of a carbohydrate structure from a single experiment still represents a challenging task.

Six series of fragment ions, A, B, C and X, Y, Z are necessary to encompass the full sequence[39] produced by the cleavage of glycosidic bonds, yielding fragment ions from reducing and nonreducing ends, according to nomenclature established by Domon and Costello.[40] Ring cleavage ions of A and X types are frequently accompanied by double cross-ring cleavages containing either the reducing or the nonreducing end and an additional glycosidic cleavage to obtain internal fragments. The combinations between these series, such as triple glycosidic or glycosidic and cross-ring dissociation, are frequently observed.[15–17,32,33,38,41–45] The first three series represent major tools for sequencing, while cross-ring fragmentation serves to establish branching patterns (Figure 14.2).

The possibility to obtain distinct fragmentation patterns under controlled collision energy and gas conditions along with the ability of automatic switching between MS and MS/MS mode by an electrospray QTOF instrument under low-energy collision-induced dissociation (CID) represents a powerful option for high-throughput

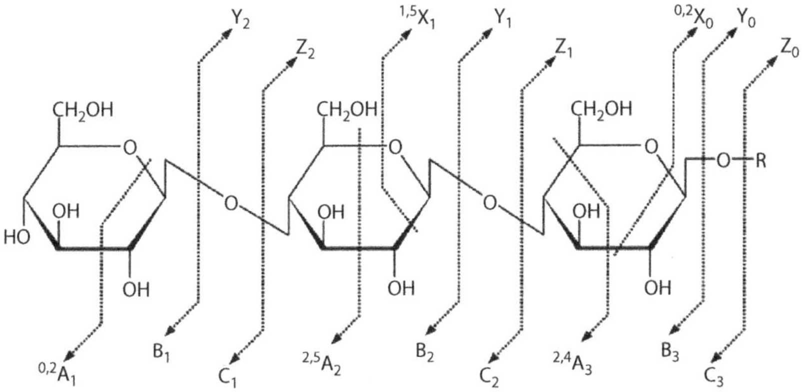

FIGURE 14.2 Nomenclature of oligosaccharide fragmentation pattern suggested by Domon and Costello. (From Domon, B.; Costello, C. E. *Glycoconjugate Journal* 5, 397–409, 1988. With permission.)

analysis of complex carbohydrate mixtures. The ESI followed by formation of fragment ions by CID allows sensitive mapping and sequencing of oligosaccharides for their identification.

14.4.4 MASS SPECTROMETRIC ANALYSIS OF COMPLEX GLYCOCONJUGATE MIXTURES, DIRECT INLET, AND PRESEPARATION BY LIQUID CHROMATOGRAPHY AND CAPILLARY ELECTROPHORESIS: POTENTIALS AND LIMITS

Since the majority of glycoconjugate species can exist at different isomeric forms, and therefore have the same m/z values, the direct mass spectrometric analysis of such samples represents a highly challenging task, which can in some cases be solved by using the mass analyzers of high resolving power (ICR, Orbitrap), but the aspect of differentiation of isomeric carbohydrate structures cannot be solved. By coupling separation devices like liquid chromatographs (LC) or capillary electrophoresis (CE) to an MS analyzer, the sample complexity in a time window can be largely reduced, improving the level of structural characterization significantly. Theoretically, positional and linkage isomers as well as structural isobars could be characterized separately, if the choice of the column and separation conditions are properly selected. However, in many cases this ideal LC separation cannot be achieved and a reasonable compromise has to be found in the case of simultaneous analysis of biomolecules of different nature (e.g., peptides and glycans or lipids and glycans). Another limiting factor of the LC-MS approach is the timeline, which can for an adequate sample separation reach several hours. In addition to that aspect, the fragmentation analysis in the LC-MS/MS approach has to be performed within a very short acquisition time, which is sometimes inadequate for collecting all diagnostic ions for assignment. In this context IMS MS could significantly contribute to the analysis of complex glycoconjugate mixtures, providing the ability to probe a gas-phase separation of different classes of biomolecules, as well as isomers and isobars, during the MS experiment.

14.5   ION MOBILITY MASS SPECTROMETRY OF CARBOHYDRATES

14.5.1   BRIEF INTRODUCTION TO ION MOBILITY MASS SPECTROMETRY

In a few recently published reviews related to the introduction of a combination of ion mobility spectrometry and mass spectrometry (IMS-MS), IMS was shown to add significantly to the efficiency of MS analysis. Upon separation of ionized components by size, shape, and mass, ionic species of different charge states can be isolated in different ion arrival distributions, avoiding the overlap.[46] The gas-phase ions accelerated by the linear electric field, E, perform a constant drift velocity (v_d) after collision with a counter flow of neutral gas. The ratio between the drift velocity and the strength of electric field is defined as the mobility of an ion, $K = v_d/E$, valid at low electric field strength (<1000 V/cm).[47] Ion mobility depends on the ion charge, the number density of buffer gas, the reduced mass of ion-neutral gas complex, the absolute gas temperature, and the average collision cross section, Ω_d. Upon adjustment of all parameters to constant values, except Ω_d, the ion mobility constant will only be dependent on the average cross section, $K \sim 1/\Omega_d$.[48–50] By coupling IMS to the time-of-flight (TOF) MS analyzers, ions pulsed into the MS will finally be separated in two dimensions: by their mobility and by their m/z ratios.[51–55] Presently, custom-built and commercial IMS-MS instruments can be distinguished according to the basic principle of ion mobility separation underlying each distinct method: drift-time ion mobility, aspiration ion mobility, differential ion mobility, and traveling-wave ion mobility.[56]

Although already tested for proteomics projects, the application of IMS-MS to the analysis of glycoconjugates is still in its early stages. Here we summarize the current state of the art of IMS-MS for carbohydrate analysis and application to glycomics.

14.5.2   IMS-MS FOR CARBOHYDRATE ANALYSIS:
MONO- AND OLIGOSACCHARIDES

In the first report on the IMS analysis of carbohydrates the aspect of isomer separation had been followed on the samples of raffinose [α-D-Galp-(1-6)-α-D-Glcp-(1-2)-β-D-Fruf] and melezitose [α-D-Glcp-(1-3)-β-D-Fruf-(2-1)-α-D-Glcp] as isomeric trisaccharides. The in-house constructed instrument had an electrospray ion source linked to an injection ion mobility drift tube equipped with a quadrupole MS. Slightly different drift times of 2.135 ms for melezitose and 2.162 ms for raffinose in the negative ion mode were measured. For investigation of differences in their fragmentation patterns, an injection energy of 1000 eV was applied, followed by scanning the fragment ions coming from the drift tube by the quadrupole to monitor unique fragment ions. Experiments with α-, β-, and γ-cyclodextrins, which represent cyclic α-glucose homologues as 6-, 7-, and 8-membered rings, respectively, were designed to measure the drift times of fragment ions generated at 1000 eV injection energy. High abundant A and X cross-ring ions from α-cyclodextrin were compared to that of the drift times for the parent ions. Mobility distributions for α-cyclodextrin fragment ions revealed the peak splitting for the fragment ion at m/z 545 assigned to $^{0,2}X_3$, which might be explained by the existence of different conformers.[57,58]

Carbohydrate ions generated by matrix-assisted laser desorption/ionization (MALDI) were analyzed by IMS on a modified double-focusing sector instrument to probe the feasibility of a heatable ion chromatography cell containing He gas at 2–5 Torr. Commercial tetra- and hexaose oligosaccharides were ionized as sodium adducts in 2,5-dihydroxybenzoic acid (DHB) as a matrix and measured in the positive ion mode. Cross-section data of oligosaccharides determined by ion mobility studies were compared to predictions calculated by molecular mechanics/molecular dynamics. The arrival times of three tetraoses—cellotetraose, maltotetraose, and isomaltotetraose—and those of the two hexaoses, α-cyclodextrin and maltohexaose, were very similar indicating that their cross sections may be similar, which was in accord with calculated values. The measurement of temperature dependence between 200 and 600 K revealed the inverse relation to the formation of cross section for all five oligosaccharide standards.[59]

In another investigation on the feasibility of ESI IMS for carbohydrate analysis, and as a detection method for efficiency of a high-performance liquid chromatography (HPLC) separation, 21 commercially available carbohydrates in the range of molecular masses from 180 to 1153 were chosen, including monosaccharides, oligosaccharides up to heptasaccharide, sugar alcohols, and amino sugars. The mobility constants of isomers D-glucose, D-fructose, and D-galactose, with the molecular mass of 180.16 Da, were measured at 1.25, 1.31, and 1.28 ms, respectively. Mobility constants of isomeric disaccharides with the free reducing end, D-cellobiose, β-D-maltose, and lactose, were at 0.68, 0.70, and 1.37 ms, respectively, while that without the free reducing end, sucrose, was at 1.00 ms. A common trend for dependency between the reduced mobility and the respective molecular size could be defined, but no clear explanation of the ion mobility behavior concerning isomeric structures was found.[60]

Although the chirality of biomolecules plays an important role in biological systems, there are only a few methods for analysis. In "omics"-related MS applications this factor is mostly neglected, unless the MS analysis is performed on a prior separation of optical isomers by chiral chromatography or capillary electrophoresis. [61–65] Alternatively, as a stand-alone MS approach for the distinction of enantiomers, an ion-molecule reaction between the enantiomeric mixture with a common chiral reagent followed by MS/MS fragmentation analysis for potential distinction could be performed.[66,67] The concept of chiral ion mobility spectrometry (CIMS) was recently described as a method by which the gas-phase ions are separated upon their stereospecific interaction with a chiral gas. Using CIMS, a racemic mixture of D- and L-methyl α-glucopyranosides[68] was analyzed according to the drift time of their sodiated adducts to be at 25.24 ms and 27.56 ms, respectively. The efficiency of CIMS for oligosaccharides with multiple chiral centers has not yet been tested.

The potential of the ambient pressure ion mobility time-of-flight mass spectrometry (AP IMS TOF MS) has been tested for the separation of different anomeric methyl glycosides derived from D-mannose, D-galactose, and D-glucose with the same anomeric configuration on C1. To study the separation effect, Ag^+, Ca^{2+}, Cu^{2+}, Hg^{2+}, Pb^{2+} acetates, and Co^{2+} and Pb^{2+} acetylacetonates were added to hexoses and their glycosides under different drift gas regimens (He, N_2, Ar, and CO_2). It could be shown that, whereas methyl glycosides were detected in a single IMS peak, mono-saccharides with the free reducing end gave more than one IMS peak, reflecting the potential presence of different anomers and ring forms in the gas phase. Regardless of the ambiguous influence of metal cations involved in ion formation and the type of drift gas on the isomer separation, it was clearly determined that β-anomers were of higher ion mobility (shorter drift time) than their α counterparts.[69]

To investigate the proof-of-principle for the separation of positional carbohydrate isomers by AP IMS TOF MS simple mixtures of reduced disaccharides, melibiitol and cellobiitol, and nonreduced trisaccharides, melezitose, raffinose, and isomaltotri-ose, were selected. For a mixture of two positional isomers, α-D-GalNAc-(1-3)-D-GalNAc-ol and α-D-GalNAc-(1-6)-D-GalNAc-ol, as sodiated adducts for a baseline separation has been achieved, as well as others, such as β-D-GlcNAc-(1-3)-D-GalNAc-ol and β-D-GlcNAc-(1-6)-D-GalNAc-ol. In both cases, disaccharides with 1-6 glycosidic linkage, either α or β, had longer mobility drift times in comparison to their (1-3) positional isomers, as could be expected due to their higher conforma-tional freedom.[70]

Ion mobility MS and the density functional theory have been applied in con-formational studies of Zn-ligand hexose diastereomers. The metal-carbohydrate complexes play an important role in biological processes, and accordingly the rel-evance of such studies could be significant. In this experiment it could be shown that a single diastereomeric carbohydrate/metal complex can be separated to pro-duce unique diagnostic fragmentation patterns upon collision-induced dissociation. This approach was applicable to hexoses, hexosamines, and N-acetyl hexosamines. In comparing the three biologically most common monosaccharides, D-mannose, D-glucose, and D-galactose, conjugated with Zn(II)/diethylenetriamine, it was deter-mined that the mannose complex is the more compact than that of galactose and that of galactose is more compact than that of glucose, according to the measured cross section for mannose at 96.3 Å, for galactose at 100.0 Å, and for glucose metal-ligand complexes at 101.2 Å.[71]

To investigate potentials of high-field asymmetric waveform ion mobility spec-trometry (FAIMS) for gas-phase separation and differentiation of carbohydrate ano-mer, linkage, and position isomers, an electrospray instrument built in house at the University of Alberta linked to the quadrupole analyzer was tested for disaccharide isomers, derivatized at the reducing end by methyl esters of aliphatic carbonic acids or aliphatic hydrocarbons. The investigated samples were chosen as pairs, differing in the anomericity, linkage, or attachment site. In these investigations molecular ion formation was assisted by metal salts for the positive and by anorganic or organic acids for the negative ion mode analysis. In all cases disaccharide derivatives could be partially or fully separated; however, they were not analyzed by CID for diagnos-tic differences in fragmentation processes.[72]

The ability of the custom-built dual gate/ion mobility quadrupole ion trap mass spectrometer equipped with electrospray ionization (ESI-DG-IM-QIT-MS) to per-form mobility-filtered ion selection prior to mass analysis has been evaluated with peptides in the positive ion mode and carbohydrates in the negative ion mode. The separation of a disaccharide melibiose [α-D-Gal-(1-6)-D-Glc] and a trisaccharide raffinose [α-D-Gal-(1-6)-α-D-Glc-(1-2)-β-D-Fru] as chlorine adducts in the negative ion mode was successful according to their respective drift times at 22.15 and 26.95 ms, respectively, followed by MS/MS analysis for identification.[73]

14.6 IMS-MS APPLICATIONS TO GLYCOCHEMISTRY AND GLYCOBIOLOGY

14.6.1 APPLICATION TO FLAVONOID DIGLYCOSIDE ISOMERS

Flavonoids are organic molecules that are extensively involved in plant tissue regulation and metabolism. They appear to be nontoxic compared to other plant compounds like alkaloids, and display antiallergic, anti-inflammatory, antimicrobial, and anticancer activity, playing an important role in the human diet. In flavonoids, carbohydrate moieties are linked to the carbon skeleton via the C-glycosidic bond or via hydroxyl groups as the O-linked glycosides. As for other glycoconjugates, one of the relevant questions to be addressed in addition to the MS determination of molecular weight and substitution patterns is the determination of the stereochemistry, e.g., the presence of enantiomers and/or diastereomers. IMS MS has been introduced to address these aspects using a combination of atmospheric pressure–dual gate–ion mobility spectrometer and the quadrupole ion trap mass analyzer (ESI-DG-IM-QIT-MS).[74] Individual flavonoid diglycosides have been investigated to determine the difference in their cross sections. Commercially available isobaric flavonoid diglycosides with molecular weight (MW) 610.2 (hesperidin, neohesperidin, and rutin) and those with MW 580.2 (naringin and narirutin) were measured sequentially and in mixture under enhanced ionization efficiency as silver, sodium, and potassium adducts. An effective separation was observed for all sets of isobaric flavonoids. In the sodiated and potassiated forms two cross sections were observed for the isobaric flavonoid glycosides with MW 610.2, but not for those with MW 580.2. This finding could be rationalized by the existence of an additional conformer. The silver adducts pair of MW 610.2 was well separated but partially overlapped with those of MW 580.2; consequently, specific fragmentation patterns could not be obtained for all components from the QIT analyzer. The gas-phase separation was less efficient in the negative ion mode, which could be explained by the fact that more than one negative-charge-bearing functional group in a single molecular ion was probably formed. Silver cationization was found to be the most efficient under the chosen IM-QIT MS conditions.

14.6.2 APPLICATION TO GLYCOSAMINOGLYCANS

Glycosaminoglycans (GAGs) belong to a biologically highly relevant but structurally least investigated class of glycoconjugates. In proteoglycans, GAG chains are covalently linked to proteins on serins or threonins via a tetrasaccharide core structure, GlcA-Gal-Gal-Xyl. By extension of the carbohydrate backbone, repeating uronic acids and glucosamine disaccharide units form a linear skeleton, modified by different degrees and different attachment positions of sulfates. Structural identification of GAGs requires a precise determination of monosaccharide identity and their sequence and distribution of sulfate groups all over the molecule, where a high degree of sulfation and the epimerization of GlcA to IduA are postulated to decrease the oligosaccharide conformational freedom. According to the high similarity of their chemical composition, but differences in positional attachments and absolute configuration, it is a challenging task to identify biologically distinct components, which may be represented as isobaric species at the same m/z ratios. In the analysis of highly sulfated heparan sulfate (HS)/heparin glycosaminoglycans, obtained by enzymatic digestion of a low-molecular-weight heparin, a disaccharide [ΔUA(2S)-GlcNS(6S)-Na$_3$] and three tetrasaccharides [ΔUA(2S)-GlcNS(6S)-IdoA(2S)-GlcNS(6S)-Na$_6$, ΔUA(2S)-GlcNS(6S)-IdoA(2S)-GlcNS-Na$_6$, and ΔUA(2S)-GlcNS(6S)-GlcA-GlcNS(6S)-Na$_5$] were investigated by IMS as negatively charged sodium adduct ions, which appear to be extensively involved in the stabilization of ions in the gas phase.[75] The IMS instrument was the same as in the citations.[59,76] A relatively high level of oligomerization of disaccharide ions was observed under the working conditions in the negative ion mode, indicating that most probably a high concentration of substrates and inefficient desalting may play a major role in the formation of compact structures generated by oligomerization in the gas phase. The most significant result in this study was the fact that already slight differences in sulfate density and the position of sulfate attachment play a significant role in the conformations of oligosaccharides. Followed by molecular modeling, the theoretical cross sections of oligosaccharides, calculated as average cross sections of the 10% lowest energy structures, were found to be in good agreement with those obtained in IMS experiments.

14.6.3 Application to Glycosylation Analysis of Proteins

Ion mobility mass spectrometry in combination with molecular modeling calculation was recently applied as a potential approach for the glycomics analysis of N-glycan mixtures derived from glycoproteins by PNGase F cleavage, as shown for ovalbumin[77] and for those in human serum associated with liver cancer and cirrhosis.[78]

To investigate the presence of potential human serum markers a total mixture of N-glycans released enzymatically upon treatment by exoglycosidase PNGase F was submitted to IMS analysis for mapping of possible changes between the normal and diseased human plasma. Serum samples of 30 healthy persons, 30 persons diagnosed with hepatocellular carcinoma, and 30 persons with liver cirrhosis at different stages of disease were investigated to define their specific glycoprofiles. Several stable ionic species that could be assigned to different conformers of the same component were observed, indicating the presence of multiple mixtures of conformers or mixtures of conformers and isomers according to their different cross-section profiles. Clustering of data obtained after collection of the total drift time acquisition for a supervised principal component analysis (PCA) provided the opportunity to compare single glycoforms that might be specific for such disease markers, which are already in use in immunochemical diagnostic protocols. For potential clinical use, the level of confidence for cancer diagnosis is not sufficient, but could be improved by introducing fragmentation analysis for the identification of specific glycoforms in human serum.[78]

In another IMS study of *N*-glycans from ovalbumin, a glycoprotein of high glycosylation heterogeneity, the release of *N*-glycans upon the treatment by PNGase F was followed by chemical derivatization, where all –OH groups were methylated.[77] This *N*-glycan mixture, when submitted to IMS, revealed the presence of fully and partially methylated glycan structures. Information from the IMS "nested" distribution was based on the fact that fully and partially methylated glycan structures can be separated according to their drift time versus *m/z* plot. Extracting profiles along the diagonal plot line, smaller structures such as the Man3GlcNAc2, the common core structure for all *N*-glycans, appear to be well separated. IMS distributions that may represent multiple isomers exhibited more complicated behavior. Hybrid-type glycans, known to be present in ovalbumin in different isomeric forms, could not be discerned. In such cases an optimization of the gas-phase IMS separation followed by MS/MS analysis for identification is expected.

In a top-down analysis of intact glycoproteins, the measurement of the glycosylation profile in an intact protein containing a single glycosylation site was successfully achieved in the case of an IgG monoclonal antibody. The experiments were supported by additional data obtained by a novel approach using the LC/MS method for direct glycopeptide screening of the IgG tryptic digest, using reversed-phase (RP) solid phase under different gradient regimens.[79]

14.6.4 APPLICATION TO LIPOPOLYSACCHARIDES

Bacterial lipopolysaccharides (LPSs) are known to exert a high structural heterogeneity in three structurally and functionally distinct parts: lipid A, oligosaccharide core, and the *O*-side chain. The structural variety of the oligosaccharide core with truncations and possible phosphate and/or phosphoethanolamine modifications and the lipid A portion, decorated with up to eight fatty acid moieties, contributes to the high structural diversity of the intact molecule. An elegant way to couple FAIMS to CE-MS was applied to the analysis of LPS preparation from *Haemophilus influenzae* strain 375.[80] In this study it was shown that the majority of singly and multiply charged ions of LPS glycoforms can be detected in operating FAIMS at a few discrete values of compensation voltage (CV). Due to the narrow CE peaks, running FAIMS under these conditions in contrast to scanning CV across the wide range promotes the optimal coupling between CE and FAIMS. The chemical noise reduction was very helpful with respect to significantly improved LPS detection limit (~70 amol of the major glycoforms). It was demonstrated that two LPS species of $Hex_4Hep_3PCho_1PEtn_2P_1KDO_1lipidA\text{-}OH$ differing in a single phosphoethanolamine unit attachment can be separated and identified: The isomer with the lower electrophoretic mobility was shown to contain a KDO-PPEtn moiety, whereas the second one only KDO-P.

14.6.5 APPLICATION TO A MIXTURE CONSTITUTED FROM BIOMOLECULES OF DIFFERENT CLASSES

In glycomics the analytical task is most frequently defined as an effort to detect and quantify all glycoconjugate species in the presence of other biopolymers, such as proteins, peptides, and lipids. In the classical sense, this task is related to single or multiple preseparation and cleaning steps to achieve the glycoconjugate preconcentration. All these manipulations increase the experimental time and reduce sample quantity, which is usually available in very limited amounts. Moreover, during such steps minor components that might be crucial for the biological activity might be lost. Presently, it can be postulated that the IM-MS protocols can in most cases overcome these difficulties and facilitate analysis in terms of costs and time by providing opportunities to analyze a mixture of different biomolecules simultaneously. In one example using a mock mixture of different classes of oligomers, a large dataset of the collision cross sections was obtained for singly charged species of oligonucleotides (ca. 100), carbohydrates (ca. 200), lipids (ca. 60),[81] and peptides (ca. 600).[82] Analyzing the relationship between collision cross sections and m/z ratios for different classes of biomolecules allows for the construction of the order of relative collision cross sections at certain m/z values as follows: oligonucleotides < carbohydrates < peptides < lipids. A good spatial distribution on the two-dimensional (2D)-plot of MALDI IMS MS was obtained for the simultaneous analysis of peptides and glycans obtained after sequential digestion of ribonuclease B by the trypsin and PNGase F cleavage.[83,84]

To enhance the potential for IM separation between different biomolecules an ion mobility shift reagent strategy was developed. Briefly, the ion mobility shift reagent was selected to derivatize molecules in order to induce a structural shift in the conformation area above or below the predicted region. Two structurally different boronic acids (BAs) were chosen as shift reagents: ferrocene boronic acid (FBA) and 4-[(2′, 6′-diisopropylphenoxy)methyl]phenyl boronic acid (PBA). FBA and PBA served as high- and low-density anchors, respectively. For the series of carbohydrates lacto-N-fucopentaose 1 (LNFP1), lacto-N-fucopentaose 2 (LNFP2), N-acetyl D-lactoseamine, maltose, lactose, and synthetic tri- and tetra-saccharides, the desired shift in conformation space was achieved for the most species. Besides, structurally based isobaric separation was demonstrated for four pairs of isomeric carbohydrates.[83]

14.7 IMS IN GLYCOURINOMICS

In urine, both N- and O-linked-type free glycans as well as aglycon-linked species of unknown constitution and glycan type are present. De novo identification of full-size, or truncated, oligosaccharides and glycopeptides from glycourinome mixtures and their ratios by a direct nanoESI QTOF MS analysis is jeopardized by a high density of ions and overlapping of singly and multiply charged ions. This is a severe obstacle for a potential automated compositional assignment of molecular ions on one hand and their identification by fragmentation mass spectrometry on the other. In human glycourinome a large heterogenic population of glycoconjugates has been postulated, but only a fraction of structures has been characterized, where single fractions containing molecular entities differing in size, branching patterns, and degree of modification by sialic acid, fucose, phosphate, sulphate, and amino acids need to be fully characterized.

An ion mobility separation device combined with a QTOF mass spectrometer has been probed for the development of high-throughput identification of glycoconjugates present in complex urine samples, GlycoUrinomics. In particular, patients suffering from congenital disorders of glycosylation (CDGs) excrete high amounts of glycoconjugates of high structural complexity in their urine. Analysis of these complex biological samples by mass spectrometry holds significant promise for the detection of markers for inherited disorders and the discovery of potential biomarkers to monitor the efficacy of the treatment. In this IMS QTOF approach, complex mixtures of free glycans, glycosylated amino acids, and glycopeptides are analyzed. To match the speed and accuracy of this improved screening a computer-assisted approach for identification from either MS and/or MS/MS data using an algorithm has been designed. Ions separated by their mobility are subsequently detected in the TOF mass analyzer, according to their drift time versus their *m/z* value. The inspection of this ratio allows a discrimination of distinct trend areas. In this protocol, selected areas are used for generating extracted ion current (XIC) chromatograms with retained drift time. From the mass spectra domains, obtained by averaging over XIC chromatograms, singly, doubly, and triply charged ionic species populations can be separately detected. Besides, upon the filtering of ions by their mobility, the signal-to-noise ratio is increased, improving the chance for automated compositional assignment at high accuracy. A negative ion detection is a preferential analysis mode for naturally occuring glycoconjugates, since a majority of them contain a preformed negative charge, located either in sialic acid building block moieties, and/or from modifications by phosphates or sulphates.[85,86] Samples from urine prepared by twofold gel permeation chromatography prior to purification by graphitized carbon cartridges and desalting by AG50 (H⁺) resins were used.

In one fraction of glycoconjugates extracted from the urine of a patient suffering from a CDG analyzed in the negative ion mode ESI QTOF MS, over all mobility drift times more than 170 singly, doubly, and triply charged distinct ionic species were detected (Figure 14.3a). The majority of singly charged and multiply charged ions in different ratios were below 30% of relative intensity, and all ions were highly overlapping. In the inset of Figure 14.3a, a mass range window of *m/z* 1007–1012 acquired over all mobility drift times is presented, where partially overlapping ionic species at *m/z* 1007.340 and 1008.837 were detectable as doubly and triply charged ions, respectively. The plot of the ionic species drift time versus their *m/z* value shows the distribution of ions achieved after the activation of the IMS cell (Figure 14.3b). The inspection of this plot allows a discrimination of three distinct trend areas labeled as A, B, and C. Those areas are used for generating XIC chromatograms with retained drift time. From the mass spectra domains, obtained by averaging over XIC chromatograms, singly, doubly, and triply charged ionic species were separately detected (Figure 14.4).

Upon ion mobility separation, in the mass range of *m/z* 1007–1012 doubly charged ions at *m/z* 1008.827 are not overlapping any more, allowing a direct computer-assisted compositional assignment to the dodecasaccharide $[NeuAc_2Hex_5HexNAc_3-2H]^{2-}$, since ionic species of higher charge states superposed with this precursor have been successfully filtered out. The triply charged ions at *m/z* 1007.340 are now well detectable and directly assigned by computer assistance to the composition of the tetradecasaccharide glycan $[NeuAc_3dHexHex_5HexNAc_5-3H]^{3-}$ (Figure 14.5a–c).

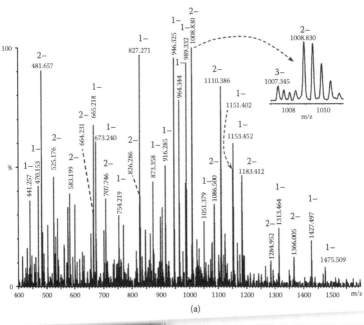

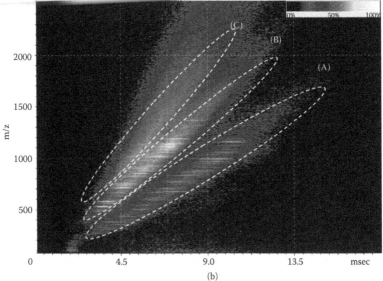

FIGURE 14.3 (a) Negative ion mode nanoESI QTOF MS of the fraction M5 obtained from urine of the patient KL after gel permeation chromatography. Spectrum was acquired from the TIC chromatogram over all mobility drift times. Inset: Expansion of the mass range at *m/z* 1007–1012. (b) Plot of the drift time versus *m/z* values for the negative ion mode nanoESI Q-IMS-TOF MS analysis of KLM5 sample. Ions have been distributed onto areas of predominantly singly (A), doubly (B), and triply (C) charged ionic species (selected by dashed lines). XIC chromatograms A, B, and C were generated from the corresponding area. (Reprinted from Vakhrushev, S. Y. et al. *Anal. Chem.* 80, 2506–2513, 2008. With permission. © 2008 American Chemical Society.)

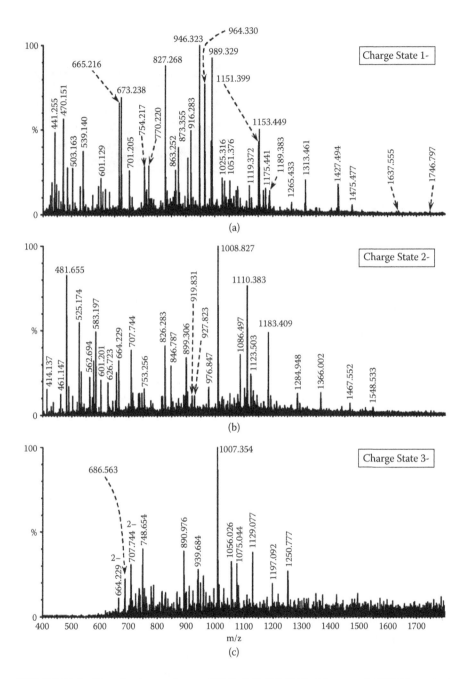

FIGURE 14.4 Negative ion mode nanoESI Q-IMS-TOF MS of the sample KLM5 acquired over (a) XIC chromatogram A, (b) XIC chromatogram B, and (c) XIC chromatogram C (Figure 14.3b). (Reprinted from Vakhrushev, S. Y. et al. *Anal. Chem.* 80, 2506–2513, 2008. With permission. © 2008 American Chemical Society.)

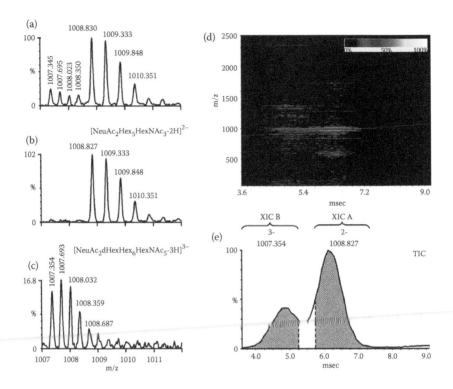

FIGURE 14.5 (a) Expansion of the mass range at *m/z* 1007–1012 of the negative ion mode nanoESI Q-TOF MS of the fraction KLM5, acquired from TIC chromatogram over all mobility drift times. (b) Expansion of the mass range at *m/z* 1007–1012 of the spectrum acquired from XIC chromatogram of doubly charged distributed area after IM separation. (c) Expansion of the mass range at *m/z* 1007–1012 of the spectrum acquired from XIC chromatogram of triply charged distributed area after IM separation. (d) Plot of fragment ions obtained by CID of overlapped precursor ions at *m/z* 1007.354 (left) and 1008.827 (right) versus *m/z* values. (e) Total ion current chromatogram with retained drift time of overlapped precursor ions at *m/z* 1007.354 (left) and 1008.827 (right). Selected areas indicate extracted ion current chromatogram A for the precursor ions at *m/z* 1008.827 and chromatogram B for the precursor ions at *m/z* 1007.354. (Reprinted from Vakhrushev, S. Y. et al. *Anal. Chem.* 80, 2506–2513, 2008. With permission. © 2008 American Chemical Society.)

As already discussed in cases from the literature, only the nonoverlapped precursor ions can render meaningful fragmentation patterns upon MS/MS analysis. In the mass range of *m/z* 1007–1012, upon their IMS separation in the drift tube according to their mobility, all cleavage ions will accordingly have the same mobility as their precursor ions; they will be detected by mass analyzer as two separate series of fragment ions (Figure 14.5d). According to the total ion current (TIC) profile, fragment ions originated from the IMS-separated doubly and triply charged precursors are located in the XIC peaks, respectively (Figure 14.5e).

The fragmentation pattern obtained by MS/MS of the doubly charged ionic species at m/z 1008.837 assigned to the composition of $[NeuAc_2Hex_5HexNAc_3\text{-}2H]^{2-}$ reflects the glycan sequence derived from fragment ions detected in the spectrum and following the biosynthetic rules of glycan assembly (Figure 14.6). The structure of the truncated sialylated biantennary N-glycan NeuAc(α2-3/6)Gal(β1-4)GlcNAc(β1-2) Man(α1-6)[NeuAc(α2-3/6)Gal(β1-4)GlcNAc(β1-2)Man(α1-3)]Man(β1-4)GlcNAc, previously found in the urine of a patient suffering from sialidosis,[87] with 85% sequence assignment coverage, as calculated by in silico fragmentation of this structural candidate. According to the fragmentation pattern of the triply charged precursor ions at m/z 1007.340 assigned to the composition $[NeuAc_3dHexHex_5HexNAc_5\text{-}3H]^{3-}$ (Figure 14.7), the structure of the sialylated triantennary N-glycan with core fucose was proposed. Also, applying computational assignment to the mass spectrum generated from the appropriate XIC chromatogram allowed for the estimation of compositions of approximately 67% of the singly charged detected ions.

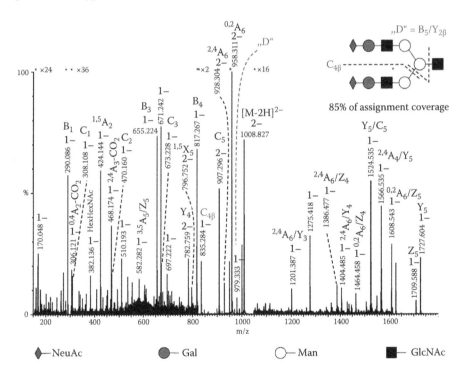

FIGURE 14.6 Negative ion mode nanoESI IMS QTOF MS/MS of the precursor ions at m/z 1008.827 (Figure 14.4b). Fragmentation spectrum obtained by averaging XIC chromatogram A (Figure 14.5e) after IMS separation. Inset: Proposed structure—truncated sialylated biantennary N-glycan. Greek letters representing branched structures were used for labeling fragment ions diagnostic for the antennae distribution, but not for cleavages, which are identical for all antennae. Diagnostic ions are highlighted by red. (Reprinted from Vakhrushev, S. Y. et al. *Anal. Chem.* 80, 2506–2513, 2008. With permission. © 2008 American Chemical Society.)

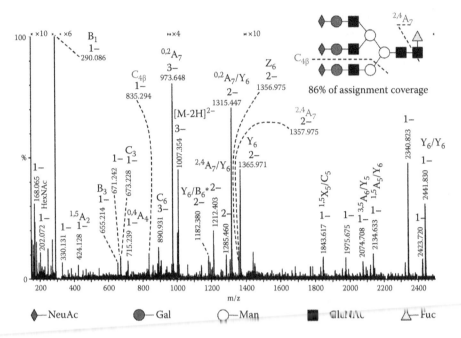

FIGURE 14.7 Negative ion mode nanoESI IMS QTOF MS/MS of the precursor ions at *m/z* 1007.354 (Figure 14.4c). Fragmentation spectrum obtained by averaging XIC chromatogram B (Figure 14.5e) after IM separation. Inset: Proposed structure—sialylated triantennary *N*-glycan with core fucose. Greek letters representing branched structures were used for labeling fragment ions diagnostic for the antennae distribution, but not for cleavages, which are identical for all antennae. Diagnostic ions are highlighted by red. Asterisk (*) denotes the fragment ion, which can be equally assigned to the Z_6/C_6 internal cleavage ion. (Reprinted from Vakhrushev, S. Y. et al. *Anal. Chem.* 80, 2506–2513, 2008. With permission. © 2008 American Chemical Society.)

14.8 SUMMARY

Introducing a novel approach of IMS-MS for analysis of complex biological samples the overlapping of the crowded carbohydrate mass range landscape could be highly reduced. Using this strategy, highly efficient identification by sequencing glycan precursor ions following their separation by mobility, or drift time, through the traveling-wave IMS cell can be directly achieved. The quality of fragmentation spectra obtained is significantly higher, and upon the inspection of the sequence coverage unique ions typical for the determination of antennarity and bisecting GlcNAc can be found. Submitting this type of data to a previously proposed automatization procedure, a recently developed software platform for automated glycomics,[88] an efficient novel high-quality combined tool for glycomics is created that can be applied to clinical samples.[86] Different levels of modifications such as sialylation and fucosylation as common features in biological samples can be dissected in a more accurate way to obtain an overview of possible specific carbohydrate epitopes present.

Upon IMS analysis followed by a direct and automatic data assignment, the newly assigned structures can be correlated with those already deposited in databases. In this approach a high chance for discovery of novel structures and their quantification is given.[86] Through the use of IMS the overlapping carbohydrate m/z landscape was highly reduced in a given window, allowing the development of a highly efficient identification procedure by sequencing precursor ions using transfer CID fragmentation after separation by IMS. Along with the chip technology[23,89] and hyphenation by LC and CE,[8,90–96] IMS technology is opening new perspectives for advanced glycomics.

REFERENCES

1. Dwek, R. A. Glycobiology: Toward understanding the function of sugars. *Chemical Reviews* 1996, *96*, 683–720.
2. Rademacher, T. W.; Parekh, R. B.; Dwek, R. A. Glycobiology. *Annual Review of Biochemistry* 1988, *57*, 785–838.
3. Apweiler, R.; Hermjakob, H.; Sharon, N. On the frequency of protein glycosylation, as deduced from analysis of the SWISS-PROT database. *Biochimica Et Biophysica Acta-General Subjects* 1999, *1473*, 4–8.
4. Oriol, R., ABO, Hh, Lewis and secretion—Serology, genetics and tissue distribution. In *Blood cell biochemistry*, Cartron, J. P.; Rouger, P., Eds. Plenum Press: New York, 1995; Vol. 6, pp 37–73.
5. Grabenhorst, E.; Schlenke, P.; Pohl, S.; Nimtz, M.; Conradt, H. S. Genetic engineering of recombinant glycoproteins and the glycosylation pathway in mammalian host cells. *Glycoconjugate Journal* 1999, *16*, 81–97.
6. Morelle, W.; Stechly, L.; Andre, S.; Van Seuningen, I.; Porchet, N.; Gabius, H. J.; Michalski, J. C.; Huet, G. Glycosylation pattern of brush border-associated glycoproteins in enterocyte-like cells: Involvement of complex-type N-glycans in apical trafficking. *Biological Chemistry* 2009, *390*, 529–544.
7. Muhlenhoff, M.; Oltmann-Norden, I.; Weinhold, B.; Hildebrandt, H.; Gerardy-Schahn, R. Brain development needs sugar: The role of polysialic acid in controlling NCAM functions. *Biological Chemistry* 2009, *390*, 567–574.
8. Kirsch, S.; Muthing, J.; Peter-Katalinić, J.; Bindila, L. On-line nano-HPLC/ESI QTOF MS monitoring of alpha 2-3 and alpha 2-6 sialylation in granulocyte glycosphingolipidome. *Biological Chemistry* 2009, *390*, 657–672.
9. Harvey, D. J.; Dwek, R. A.; Rudd, P. M. Determining the structure of glycan moieties by mass spectrometry. *Current Protocols in Protein Science* 2006, Chapter 12, Unit 12 7.
10. Mutenda, K. E.; Matthiesen, R. Analysis of carbohydrates by mass spectrometry. *Methods in Molecular Biology* 2007, *367*, 289–301.
11. Peter-Katalinić, J. Methods in enzymology: O-glycosylation of proteins. *Methods in Enzymology* 2005, *405*, 139–171.
12. Packer, N. H.; von der Lieth, C. W.; Aoki-Kinoshita, K. F.; Lebrilla, C. B.; Paulson, J. C.; Raman, R.; Rudd, P.; Sasisekharan, R.; Taniguchi, N.; York, W. S. Frontiers in glycomics: Bioinformatics and biomarkers in disease—An NIH white paper prepared from discussions by the focus groups at a workshop on the NIH campus, Bethesda, MD (September 11–13, 2006). *Proteomics* 2008, *8*, 8–20.
13. von der Lieth, C. W.; Bohne-Lang, A.; Lohmann, K. K.; Frank, M. Bioinformatics for glycomics: Status, methods, requirements and perspectives. *Briefings in Bioinformatics* 2004, *5*, 164–178.

14. von der Lieth, C. W.; Lutteke, T.; Frank, M. The role of informatics in glycobiology research with special emphasis on automatic interpretation of MS spectra. *Biochimica et Biophysica Acta* 2006, *1760*, 568–577.

15. Harvey, D. J. Matrix-assisted laser desorption/ionization mass spectrometry of carbohydrates. *Mass Spectrometry Reviews* 1999, *18*, 349–450.

16. Harvey, D. J. Electrospray mass spectrometry and fragmentation of N-linked carbohydrates derivatized at the reducing terminus. *Journal of the American Society of Mass Spectrometry* 2000, *11*, 900–915.

17. Harvey, D. J. Identification of protein-bound carbohydrates by mass spectrometry. *Proteomics* 2001, *1*, 311–328.

18. Park, Y.; Lebrilla, C. B. Application of Fourier transform ion cyclotron resonance mass spectrometry to oligosaccharides. *Mass Spectrometry Reviews* 2005, *24*, 232–264.

19. Peter-Katalinić, J. Analysis of glycoconjugates by fast atom bombardment mass spectrometry and related MS techniques. *Mass Spectrometry Reviews* 1994, *13*, 77–90.

20. Zaia, J. Mass spectrometry of oligosaccharides. *Mass Spectrometry Reviews* 2004, *23*, 161–227.

21. Harvey, D. J. Analysis of carbohydrates and glycoconjugates by matrix-assisted laser desorption/ionization mass spectrometry: an update for 2003–2004. *Mass Spectrometry Reviews* 2009, *28*, 273–361.

22. Domon, B.; Mueller, D. R.; Richter, W. J. Determination of interglycosidic linkages in disaccharides by high performance tandem mass spectrometry. *International Journal of Mass Spectrometry Ion Processes* 1990, *100*, 301–311.

23. Bindila, L.; Peter-Katalinić, J. Chip-mass spectrometry for glycomic studies. *Mass Spectrometry Reviews* 2009, *28*, 223–253.

24. Froesch, M.; Bindila, L.; Zamfir, A.; Peter-Katalinić, J. Sialylation analysis of O-glycosylated sialylated peptides from urine of patients suffering from Schindler's disease by Fourier transform ion cyclotron resonance mass spectrometry and sustained off-resonance irradiation collision-induced dissociation. *Rapid Communications in Mass Spectrometry* 2003, *17*, 2822–2832.

25. Froesch, M.; Bindila, L. M.; Baykut, G.; Allen, M.; Peter-Katalinić, J.; Zamfir, A. D. Coupling of fully automated chip electrospray to Fourier transform ion cyclotron resonance mass spectrometry for high-performance glycoscreening and sequencing. *Rapid Communications in Mass Spectrometry* 2004, *18*, 3084–3092.

26. Mormann, M.; Paulsen, H.; Peter-Katalinić, J. Electron capture dissociation of O-glycosylated peptides: Radical site-induced fragmentation of glycosidic bonds. *European Journal of Mass Spectrometry* 2005, *11*, 497–511.

27. Vakhrushev, S. Y.; Snel, M. F.; Langridge, J.; Peter-Katalinić, J. MALDI-QTOFMS/MS identification of glycoforms from the urine of a CDG patient. *Carbohydrate Research* 2008, *343*, 2172–2183.

28. Vakhrushev, S. Y.; Zamfir, A.; Peter-Katalinić, J. $^{0,2}A_n$ cross-ring cleavage as a general diagnostic tool for glycan assignment in glycoconjugate mixtures. *Journal of the American Society for Mass Spectrometry* 2004, *15*, 1863–1868.

29. Butler, M.; Quelhas, D.; Critchley, A. J.; Carchon, H.; Hebestreit, H. F.; Hibbert, R. G.; Vilarinho, L.; Teles, E.; Matthijs, G.; Schollen, E.; Argibay, P.; Harvey, D. J.; Dwek, R. A.; Jaeken, J.; Rudd, P. M. Detailed glycan analysis of serum glycoproteins of patients with congenital disorders of glycosylation indicates the specific defective glycan processing step and provides an insight into pathogenesis. *Glycobiology* 2003, *13*, 601–622.

30. Chai, W.; Piskarev, V.; Lawson, A. M. Negative-ion electrospray mass spectrometry of neutral underivatized oligosaccharides. *Analytical Chemistry* 2001, *73*, 651–657.

31. Chai, W.; Piskarev, V.; Lawson, A. M. Branching pattern and sequence analysis of underivatized oligosaccharides by combined MS/MS of singly and doubly charged molecular ions in negative-ion electrospray mass spectrometry. *Journal of the American Society for Mass Spectrometry* 2002, *13*, 670–679.

32. Harvey, D. J. Collision-induced fragmentation of underivatized N-linked carbohydrates ionized by electrospray. *Journal of Mass Spectrometry* 2000, *35*, 1178–1190.

33. Harvey, D. J. Ionization and collision-induced fragmentation of N-linked and related carbohydrates using divalent cations. *Journal of the American Society for Mass Spectrometry* 2001, *12*, 926–937.

34. Pfenninger, A.; Karas, M.; Finke, B.; Stahl, B. Structural analysis of underivatized neutral human milk oligosaccharides in the negative ion mode by nano-electrospray MS(n) (part 1: methodology). *Journal of the American Society for Mass Spectrometry* 2002, *13*, 1331–1340.

35. Que, A. H.; Novotny, M. V. Structural characterization of neutral oligosaccharide mixtures through a combination of capillary electrochromatography and ion trap tandem mass spectrometry. *Analytical and Bioanalytical Chemistry* 2003, *375*, 599–608.

36. Quemener, B.; Cabrera Pino, J. C.; Ralet, M. C.; Bonnin, E.; Thibault, J. F. Assignment of acetyl groups to O-2 and/or O-3 of pectic oligogalacturonides using negative electrospray ionization ion trap mass spectrometry. *Journal of Mass Spectrometry* 2003, *38*, 641–648.

37. Robbe, C.; Capon, C.; Coddeville, B.; Michalski, J. C. Diagnostic ions for the rapid analysis by nano-electrospray ionization quadrupole time-of-flight mass spectrometry of O-glycans from human mucins. *Rapid Communications in Mass Spectrometry* 2004, *18*, 412–420.

38. Šagi, D.; Peter-Katalinić, J.; Conradt, H. S.; Nimtz, M. Sequencing of tri- and tetraantennary N-glycans containing sialic acid by negative mode ESI QTOF tandem MS. *Journal of the American Society for Mass Spectrometry* 2002, *13*, 1138–1148.

39. de Hoffmann, E., Stroobant, V. *Mass spectrometry. Principles and applications;* John Wiley & Sons, Ltd: Chichester, West Sussex, England, 2002; p 407.

40. Domon, B.; Costello, C. E. A systematic nomenclature for carbohydrate fragmentation in FAB MS/MS of glycoconjugates. *Glycoconjugate Journal* 1988, *5*, 397–409.

41. Bindila, L.; Froesch, M.; Lion, N.; Vukelić, Z.; Rossier, J. S.; Girault, H. H.; Peter-Katalinić, J.; Zamfir, A. D. A thin chip microsprayer system coupled to Fourier transform ion cyclotron resonance mass spectrometry for glycopeptide screening. *Rapid Communications in Mass Spectrometry* 2004, *18*, 2913–2920.

42. Froesch, M.; Bindila, L.; Zamfir, A.; Peter-Katalinić, J. Sialylation analysis of O-glycosylated sialylated peptides from urine of patients suffering from Schindler's disease by Fourier transform ion cyclotron resonance mass spectrometry and sustained off-resonance irradiation collision-induced dissociation. *Rapid Communications in Mass Spectrometry* 2003, *17*, 2822–2832.

43. Vakhrushev, S. Y., Mormann, M., Peter-Katalinić, J. Identification of complex glycoconjugates related to congenital disorders of glycosylation by FT-ICR MS and MS/MS in combination with computational algorithm. Proceedings of the 52nd ASMS Conference on Mass Spectrometry and Allied Topics, Nashville, Tennessee, 2004; p WPC047.

44. Vakhrushev, S. Y.; Zamfir, A.; Peter-Katalinić, J. $^{0,2}A_n$ cross-ring cleavage as a general diagnostic tool for glycan assignment in glycoconjugate mixtures. *Journal of the American Society for Mass Spectrometry* 2004, *15*, 1863–1868.

45. Zamfir, A.; Vakhrushev, S.; Sterling, A.; Niebel, H. J.; Allen, M.; Peter-Katalinić, J. Fully automated chip-based mass spectrometry for complex carbohydrate system analysis. *Analytical Chemistry* 2004, *76*, 2046–54.

46. Clemmer, D. E.; Jarrold, M. F. Ion mobility measurements and their applications to clusters and biomolecules. *Journal of Mass Spectrometry* 1997, *32*, 577–592.

47. Collins, D. C.; Lee, M. L. Developments in ion mobility spectrometry-mass spectrometry. *Analytical and Bioanalytical Chemistry* 2002, *372*, 66–73.
48. Bowers, M. T.; Marshall, A. G.; McLafferty, F. W. Mass spectrometry: Recent advances and future directions. *Journal of Physical Chemistry* 1996, *100*, 12897–12910.
49. Henderson, S. C.; Valentine, S. J.; Counterman, A. E.; Clemmer, D. E. ESI/ion trap/ion mobility/time-of-flight mass spectrometry for rapid and sensitive analysis of biomolecular mixtures. *Analytical Chemistry* 1999, *71*, 291–301.
50. Verbeck, G. F.; Ruotolo, B. T.; Sawyer, H. A.; Gillig, K. J.; Russell, D. H. A fundamental introduction to ion mobility mass spectrometry applied to the analysis of biomolecules. *Journal of Biomolecular Techniques* 2002, *13*, 56–61.
51. Guevremont, R.; Siu, K. W. M.; Wang, J. Y.; Ding, L. Y. Combined ion mobility time-of-flight mass spectrometry study of electrospray-generated ions. *Analytical Chemistry* 1997, *69*, 3959–3965.
52. Hoaglund, C. S.; Valentine, S. J.; Sporleder, C. R.; Reilly, J. P.; Clemmer, D. E. Three-dimensional ion mobility TOFMS analysis of electrosprayed biomolecules. *Analytical Chemistry* 1998, *70*, 2236–2242.
53. Lee, Y. J., Hoaglund-Hyzera, C. S., Srebalus Barnes, C. A., Hilderbrand, A. E., Valentine, S. J., Clemmer, D. E. Development of high-throughput liquid chromatography injected ion mobility quadrupole time-of-flight techniques for analysis of complex peptide mixtures. *Journal of Chromatography B* 2002, *785*, 342–351.
54. Myung, S.; Lee, Y. J.; Moon, M. H.; Taraszka, J.; Sowell, R.; Koeniger, S.; Hilderbrand, A. E.; Valentine, S. J.; Cherbas, L.; Cherbas, P.; Kaufmann, T. C.; Miller, D. F.; Mechref, Y.; Novotny, M. V.; Ewing, M. A.; Sporleder, C. R.; Clemmer, D. E. Development of high-sensitivity ion trap ion mobility spectrometry time-of-flight techniques: A high-throughput nano-LC-IMS-TOF separation of peptides arising from a *Drosophila* protein extract. *Analytical Chemistry* 2003, *75*, 5137–5145.
55. Srebalus, C. A.; Li, J. W.; Marshall, W. S.; Clemmer, D. E. Gas phase separations of electrosprayed peptide libraries. *Analytical Chemistry* 1999, *71*, 3918–3927.
56. Kanu, A. B.; Dwivedi, P.; Tam, M.; Matz, L.; Hill, H. H. Ion mobility-mass spectrometry. *Journal of Mass Spectrometry* 2008, *43*, 1–22.
57. Liu, Y. S.; Clemmer, D. E. Characterizing oligosaccharides using injected-ion mobility mass spectrometry. *Analytical Chemistry* 1997, *69*, 2504–2509.
58. Liu, Y. S.; Valentine, S. J.; Counterman, A. E.; Hoaglund, C. S.; Clemmer, D. E. Injected-ion mobility analysis of biomolecules. *Analytical Chemistry* 1997, *69*, A728–A735.
59. Lee, S.; Wyttenbach, T.; Bowers, M. T. Gas phase structures of sodiated oligosaccharides by ion mobility ion chromatography methods. *International Journal of Mass Spectrometry* 1997, *167*, 605–614.
60. Lee, D. S.; Wu, C.; Hill, H. H. Detection of carbohydrates by electrospray ionization ion mobility spectrometry following microbore high-performance liquid chromatography. *Journal of Chromatography A* 1998, *822*, 1–9.
61. Gubitz, G. Separation of drug enantiomers by HPLC using chiral stationary phases—A selective review. *Chromatographia* 1990, *30*, 555–564.
62. Tanaka, M.; Yamazaki, H.; Hakusui, H. Direct HPLC separation of enantiomers of pantoprazole and other benzimidazole sulfoxides using cellulose-based chiral stationary phases in reversed-phase mode. *Chirality* 1995, *7*, 612–615.
63. Nassar, A. E. F.; Stuart, J. D. Separations of chiral compounds and proteins by CE-MS. *Abstracts of Papers of the American Chemical Society* 1996, *211*, 30-Anyl.
64. Fanali, S. Chiral separations by CE employing CDs. *Electrophoresis* 2009, *30 Suppl 1*, S203–S210.
65. Rocco, A.; Fanali, S. Enantiomeric separation of acidic compounds by nano-liquid chromatography with methylated-beta-cyclodextrin as a mobile phase additive. *Journal of Separation Science* 2009, *32*, 1696–1703.

66. Wu, L. M.; Meurer, E. C.; Cooks, R. G. Chiral morphing and enantiomeric quantification in mixtures by mass spectrometry. *Analytical Chemistry* 2004, *76*, 663–671.

67. Grigorean, G.; Ramirez, J.; Ahn, S. H.; Lebrilla, C. B. A mass spectrometry method for the determination of enantiomeric excess in mixtures of D,L-amino acids. *Analytical Chemistry* 2000, *72*, 4275–4281.

68. Dwivedi, P.; Wu, C.; Matz, L. M.; Clowers, B. H.; Siems, W. F.; Hill, H. H. Gas-phase chiral separations by ion mobility spectrometry. *Analytical Chemistry* 2006, *78*, 8200–8206.

69. Dwivedi, P.; Bendiak, B.; Clowers, B. H.; Hill, H. H. Rapid resolution of carbohydrate isomers by electrospray ionization ambient pressure ion mobility spectrometry-time-of-flight mass spectrometry (ESI-APIMS-TOFMS). *Journal of the American Society for Mass Spectrometry* 2007, *18*, 1163–1175.

70. Clowers, B. H.; Dwivedi, P.; Steiner, W. E.; Hill, H. H.; Bendiak, B. Separation of sodiated isobaric disaccharides and trisaccharides using electrospray ionization-atmospheric pressure ion mobility-time of flight mass spectrometry. *Journal of the American Society for Mass Spectrometry* 2005, *16*, 660–669.

71. Leavell, M. D.; Gaucher, S. P.; Leary, J. A.; Taraszka, J. A.; Clemmer, D. E. Conformational studies of Zn-ligand-hexose diastereomers using ion mobility measurements and density functional theory calculations. *Journal of the American Society for Mass Spectrometry* 2002, *13*, 284–293.

72. Gabryelski, W.; Froese, K. L. Rapid and sensitive differentiation of anomers, linkage, and position isomers of disaccharides using high-field asymmetric waveform ion mobility spectrometry (FAIMS). *Journal of the American Society Mass Spectrometry* 2003, *14*, 265–277.

73. Clowers, B. H.; Hill, H. H. Mass analysis of mobility-selected ion populations using dual gate, ion mobility, quadrupole ion trap mass spectrometry. *Analytical Chemistry* 2005, *77*, 5877–5885.

74. Clowers, B. H.; Hill, H. H. Influence of cation adduction on the separation characteristics of flavonoid diglycoside isomers using dual gate-ion mobility-quadrupole ion trap mass spectrometry. *Journal of Mass Spectrometry* 2006, *41*, 339–351.

75. Jin, L.; Barran, P. E.; Deakin, J. A.; Lyon, M.; Uhrin, D. Conformation of glycosaminoglycans by ion mobility mass spectrometry and molecular modelling. *Physical Chemistry Chemical Physics* 2005, *7*, 3464–3471.

76. Wyttenbach, T.; Kemper, P. R.; Bowers, M. T. Design of a new electrospray ion mobility mass spectrometer. *International Journal of Mass Spectrometry* 2001, *212*, 13–23.

77. Plasencia, M. D.; Isailovic, D.; Merenbloom, S. I.; Mechref, Y.; Clemmer, D. E. Resolving and assigning N-linked glycan structural isomers from ovalbumin by IMS-MS. *Journal of the American Society for Mass Spectrometry* 2008, *19*, 1706–1715.

78. Isailovic, D.; Kurulugama, R. T.; Plasencia, M. D.; Stokes, S. T.; Kyselova, Z.; Goldman, R.; Mechref, Y.; Novotny, M. V.; Clemmer, D. E. Profiling of human serum glycans associated with liver cancer and cirrhosis by IMS-MS. *Journal of Proteome Research* 2008, *7*, 1109–1117.

79. Olivova, P.; Chen, W. B.; Chakraborty, A. B.; Gebler, J. C. Determination of N-glycosylation sites and site heterogeneity in a monoclonal antibody by electrospray quadrupole ion-mobility time-of-flight mass spectrometry. *Rapid Communications in Mass Spectrometry* 2008, *22*, 29–40.

80. Li, J. J.; Purves, R. W.; Richards, J. C. Coupling capillary electrophoresis and high-field asymmetric waveform ion mobility spectrometry mass spectrometry for the analysis of complex lipopolysaccharides. *Analytical Chemistry* 2004, *76*, 4676–4683.

81. Fenn, L. S.; Kliman, M.; Mahsut, A.; Zhao, S. R.; McLean, J. A. Characterizing ion mobility-mass spectrometry conformation space for the analysis of complex biological samples. *Analytical and Bioanalytical Chemistry* 2009, *394*, 235–244.

82. Tao, L.; McLean, J. R.; McLean, J. A.; Russell, D. H. A collision cross-section database of singly-charged peptide ions (vol 18, pg 1232, 2007). *Journal of the American Society for Mass Spectrometry* 2007, *18,* 1727–1728.

83. Fenn, L. S.; McLean, J. A. Enhanced carbohydrate structural selectivity in ion mobility-mass spectrometry analyses by boronic acid derivatization. *Chemical Communications* 2008, 5505–5507.

84. Fenn, L. S.; McLean, J. A. Biomolecular structural separations by ion mobility-mass spectrometry. *Analytical and Bioanalytical Chemistry* 2008, *391,* 905–909.

85. Vakhrushev, S. Y.; Langridge, J.; Campuzano, I.; Hughes, C.; Peter-Katalinić, J. Ion mobility mass spectrometry analysis of human glycourinome. *Analytical Chemistry* 2008, *80,* 2506–2513.

86. Vakhrushev, S. Y.; Langridge, J.; Campuzano, I.; Hughes, C.; Peter-Katalinić, J. Identification of monosialylated N-glycoforms in the CDG urinome by ion mobility tandem mass spectrometry: The potential for clinical applications. *Journal of Clinical Proteomics* 2008, *4,* 47–57.

87. Van Pelt, J.; Kamerling, J. P.; Bakker, H. D.; Vliegenthart, J. F. A comparative study of sialyloligosaccharides isolated from sialidosis and galactosialidosis urine. *Journal of Inherited Metabolic Disease* 1991, *14,* 730–740.

88. Vakhrushev, S. Y.; Dadimov, D.; Peter-Katalinić, J. Software platform for high throughput glycomics. *Analytical Chemistry* 2009, *81,* 3252–3260.

89. Zamfir, A.; Vakhrushev, S.; Sterling, A.; Niebel, H. J.; Allen, M.; Peter-Katalinić, J. Fully automated chip-based mass spectrometry for complex carbohydrate system analysis. *Analytical Chemistry* 2004, *76,* 2046–2054.

90. Bindila, L.; Peter-Katalinić, J.; Zamfir, A. Sheathless reverse-polarity capillary electrophoresis-electrospray-mass spectrometry for analysis of underivatized glycoconjugates. *Electrophoresis* 2005, *26,* 1488–1499.

91. Zamfir, A.; Peter-Katalinić, J. Glycoscreening by on-line sheathless capillary electrophoresis/electrospray ionization-quadrupole time of flight-tandem mass spectrometry. *Electrophoresis* 2001, *22,* 2448–2457.

92. Zamfir, A. D.; Bindila, L.; Lion, N.; Allen, M.; Girault, H. H.; Peter-Katalinić, J. Chip electrospray mass spectrometry for carbohydrate analysis. *Electrophoresis* 2005, *26,* 3650–3673.

93. Ruhaak, L. R.; Deelder, A. M.; Wuhrer, M. Oligosaccharide analysis by graphitized carbon liquid chromatography-mass spectrometry. *Analytical and Bioanalytical Chemistry* 2009, *394,* 163–174.

94. Wuhrer, M.; de Boer, A. R.; Deelder, A. M. Structural glycomics using hydrophilic interaction chromatography (hilic) with mass spectrometry. *Mass Spectrometry Reviews* 2009, *28,* 192–206.

95. Mechref, Y.; Novotny, M. V. Glycomic analysis by capillary electrophoresis-mass spectrometry. *Mass Spectrometry Reviews* 2009, *28,* 207–222.

96. Zaia, J. On-line separations combined with MS for analysis of glycosaminoglycans. *Mass Spectrometry Reviews* 2009, *28,* 254–272.

15 Structural Characterization of Oligomer-Aggregates of β-Amyloid Polypeptide Using Ion Mobility Mass Spectrometry

Marius-Ionuţ Iuraşcu, Claudia Cozma,
James Langridge, Nick Tomczyk,
Michael Desor, and Michael Przybylski

CONTENTS

15.1 INTRODUCTION

A significant proportion of the elderly population is affected by progressive neurodegenerative diseases, characterized by the accumulation of insoluble fibrils and plaques in the brain. The formation and accumulation of fibrillar plaques and aggregates of β-amyloid peptide (Aβ) and α-synuclein (Syn) in the brain have been recognized as characteristics of Alzheimer's disease (AD) and Parkinson's disease (PD).[1–3] Aβ is a polypeptide containing 39–43 amino acid residues derived from proteolytic cleavage of the transmembrane Aβ precursor protein (APP). Recently the formation of Aβ-oligomers has become of particular interest, since oligomers have been suggested to be key neurotoxic species for progressive neurodegeneration[2,4,5]; however, molecular details of the pathophysiological degradation of APP and of Aβ-aggregation pathways are hitherto unclear.[6] Despite the lack of molecular mechanism(s), studies towards the development of immunotherapeutic approaches for AD[7] have shown initial success in the production of therapeutic antibodies that disaggregate Aβ-fibrils and improve the memory impairment in transgenic AD mice. [8–11] The identification of the epitopes recognized by both Aβ-plaque-specific and aggregation-preventing antibodies, and the discovery of their Aβ-oligomer specificity are presently causing enhanced interest in the elucidation of Aβ-oligomeric structures.[12–16]

Ion mobility mass spectrometry (IMS-MS) has received increasing attention as a tool for the characterization of molecular assemblies according to their conformation and/or topography.[17–20] The IMS-MS instrument employed in this study consists of two parts: (i) an ion mobility drift cell where ions are separated within an electric field according to their collisional cross section, and (ii) a quadrupole time-of-flight mass spectrometer (SYNAPT-QTOF-MS).[21] Thus the IMS-MS implements a new mode of separation that provides the differentiation of conformational states of polypeptides. In conjunction with spectroscopic, proteomic, and electron paramagnetic resonance (EPR) studies of spin-labeled Aβ-peptides in our laboratory aimed at the structural identification of Aβ-aggregates,[22–26] IMS-MS was applied in this study of Aβ-oligomer preparations of Aβ(1-40) in vitro. First applications revealed molecular details of Aβ-oligomerization such as the identification of oxidative structure modifications.

15.2 MATERIALS AND METHODS

15.2.1 SYNTHESIS AND PURIFICATION OF SPIN-LABELED Aβ-PEPTIDE DERIVATIVES

Aβ(1-40)(HO-DAEFRHDSGYEVHHQKLVFFAEDVGSNKGAIIGLMVGGVV-NH$_2$) and N-terminal Cys–Aβ(1-40) were synthesized by solid-phase peptide synthesis (SPPS) on a semiautomatic peptide synthesizer EPS 221 (Abimed/Intavis, Langenfeld, Germany) on a NovaSyn® TGR resin according to the Fmoc strategy. The spin label used was 3-(2-iodoacetamido)-2,2,5,5-tetramethyl-1-pyrrolidinyloxy (Iodo–Proxyl) (IPSL) (Sigma-Aldrich, München, Germany), and labeling reactions were performed using a 10-fold molar excess of IPSL as previously described.[27] The IPSL-labeled Aβ-peptides were purified by reverse-phase high-performance liquid chromatography (RP-HPLC) (Figure 15.1) on a semipreparative Vydac C4 column,

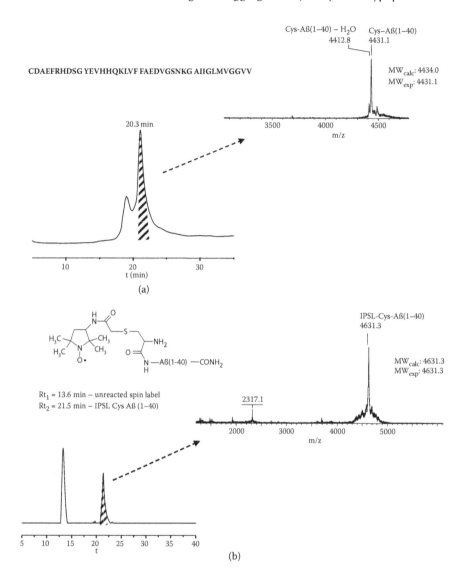

FIGURE 15.1 HPLC profiles (crude products) and MALDI-TOF mass spectra of HPLC purified (a) Cys-Aβ(1-40) and (b) IPSL-Cys-Aβ(1-40) peptides.

using a gradient of 0.1% TFA (trifluoroacetic acid) in milliQ-water (eluent A) and 80% acetonitrile (eluent B).

15.2.2 MASS SPECTROMETRY

Synthetic Aβ-peptides were separated by HPLC and analyzed by matrix-assisted laser desorption/ionization–time-of-flight mass spectrometry (MALDI-TOF MS) on a Bruker Daltonik Biflex ITM linear TOF mass spectrometer (Bruker Daltonik,

TABLE 15.1

ESI–MS Data of Synthetic Aβ-Peptide Derivatives

	MW Calculated	[M+6H]⁶⁺	[M+7H]⁷⁺	MW Found
Aβ(1–40)	4326.1644	722.1419	618.9992	4326.22
Cys-Aβ(1–40)	4429.1736	739.0073	—	4429.0
IPSL-Cys-Aβ(1–40)	4627.3104	772.1584	661.9733	4626.9

Bremen, Germany). Aliquots of 0.8 μL of peptide sample were mixed with 0.8 μL 4-hydroxy α–cyano cinnamic acid (HCCA) in AcCN: 0.1% TFA in water 2:1 (v:v). Upon lyophilization, peptide samples were redissolved in 0.2% formic acid at a concentration of 0.1 μg/μL. HPLC was performed on an Agilent 1100 instrument (Agilent Technologies, Waldbronn, Germany) equipped with a C8 column (Vydac 150 × 1 mm, 5 μm), using a flow rate of 50 μL/min. The following gradient was used: equilibration step, 2% B for 5 min, linear gradient from 2% B to 65% B in 63 min; from 65% B to 98% B in 10 min; 10 min washing step 98% B.

Electrospray (ESI) mass spectrometry was performed on an Esquire 3000+ ion trap spectrometer (Bruker Daltonik, Bremen, Germany) at the following conditions: capillary temperature 250°C, nebulizer gas 20 psi (Ar), drying gas 9 L/min (N₂), potential difference 4 kV (positive ion mode), end plate offset 500 V, skimmer 40 V, and capillary exit 136 V. The ion trap was locked on automatic gain control, and six microscans were collected for each full MS scan with a maximum accumulation time of 200 ms for each ion (Table 15.1).

15.2.3 PREPARATION OF Aβ-FIBRILS

For Aβ-oligomerization and fibril formation, Aβ(1-40) and PSL-Cys-Aβ(1-40) peptides were solubilized at a concentration of 1 μg/μL in a buffer containing 50 mM Na₃PO₄, 150 mM NaCl, 0.02% NaN₃ at pH 7.5.[27] Because of the low solubility of Aβ-fibrils, several sonication/vortex cycles were performed to ensure complete solubilization. Each solution was incubated for 5 days at 37°C, yielding a white precipitate at the end of the incubation period. The precipitate was briefly sonicated, centrifuged 15 min at 13,000 rpm, and the supernatant removed and replaced with an equal volume of deionized water (MilliQ). Both the supernatant and the resuspended precipitate were subjected to gel electrophoresis experiments.

15.2.4 GEL ELECTROPHORESIS

For *in gel* proteolytic digestion and mass spectrometry, the resuspended precipitate fraction was subjected to SDS-PAGE (sodium dodecyl sulfate polyacrylamide gel electrophoresis) separation. The fibril fraction (10 μL) was mixed with 10 μL running buffer, 4% SDS, 25% glycerol, and bromophenol blue staining reagent. After loading on a 15% gel, electrophoresis was developed for 15 min at 60 V, and subsequently for 1–2 h at 100 V. Gel electrophoresis of IPSL-Cys-Aβ(1-40) fibril preparations was carried out with a freshly prepared solution in fibril growth buffer, and a sample after incubation for 5 days. One-dimensional (1D)-Tris-Tricine poly-acrylamide gel electrophoresis (15%) was run as described above and visualized by Coomassie Blue staining.

15.2.5 In-Gel Tryptic Digestion and Peptide Extraction

Protein spots were excised from the gel (Figure 15.2a, bands 1–4), washed with MilliQ water, and the gel pieces shaken for 30 min at 25°C with 60% acetonitrile for dehydration, and dried in a Speed Vac centrifuge for 30 min. Spots were then destained with 50 mM NH_4HCO_3 (15 min), dehydrated with 60% acetonitrile solution (15 min) and dried in a Speed Vac centrifuge (30 min).

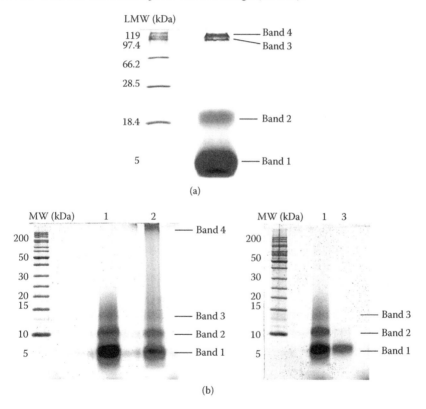

FIGURE 15.2 1D-gel electrophoresis (Coomassie blue staining) of Aβ-aggregates in vitro. (a) SDS-PAGE of Aβ(1-40) after 5 days of incubation at 37°C; band 1, Aβ(1-40) monomer; band 2, Aβ-oligomer; bands 3 and 4, high-molecular-weight aggregates ("soluble protofi-brils"). (b) Tris-Tricine PAGE of IPSL-Cys-Aβ(1-40) after 5 days of incubation at 37°C. Lane 1, a freshly prepared IPSL-Cys-Aβ(1-40) solution (in fibril growth buffer); lane 2, precipitate after 5 days of incubation—only the soluble part entered the separation gel, the rest remained in the stacking gel; lane 3, supernatant after 5 days of incubation. In band 1, the monomeric IPSL-Cys-Aβ(1-40) is present, in band 2 the dimer, in band 3 the trimer, and in band 4 high-molecular-weight aggregates ("soluble protofibrils").

The gel pieces were then swollen in digestion buffer (12.5 ng trypsin/μL 50 mM NH$_4$HCO$_3$) at 4°C (on ice) for 45 min, and incubated at 37°C overnight (12 h). After removal of supernatant, peptide extraction was performed at 25°C with a solution of 60% acetonitrile, 0.1% TFA in MilliQ water. The tryptic digestion mixtures were characterized by MALDI-TOF MS, and data were analyzed against the NCBInr protein database by means of the Mascot MS Ion search engine.

15.2.6 ION MOBILITY MASS SPECTROMETRY

Ion mobility mass spectrometry was performed with a Waters SYNAPT-G1 quadrupole time-of-flight (QTOF) mass spectrometer (Waters Corp., Milford, MA). A freshly prepared Aβ(1-40) peptide solution (1μg/μL) and the supernatant fraction from in vitro fibril preparation (5 days incubation at 37°C) were analyzed by injection of 5 μL (0.5 μg/μL) in a desalting cartridge and elution for 10 min at 20 μL/min with a gradient of acetonitrile from 10% to 90%. Ion mobility–MS was performed in the mass-to-charge ratio (m/z) range of 350–4000 at a pressure of 0.45 bar, cone voltage of 25 V, and a drift voltage (wave height) of 5–15 V.

15.3 RESULTS AND DISCUSSION

15.3.1 SYNTHESIS AND CHARACTERIZATION OF SPIN-LABELED Aβ-PEPTIDES

Aβ(1-40) and N-cysteinyl-Aβ(1-40) peptides were synthesized by SPPS/Fmoc protection strategy at a preparative scale (100 μM) and the products purified by HPLC (Figure 15.1). Aβ(1-40) and Cys-Aβ(1-40) were separated at HPLC retention times of 21.3 and 20.5 min, respectively, and thus the crude Cys-Aβ(1-40) could be used directly for introducing the IPSL- spin label by alkylation (Figure 15.1b). At these conditions, oxidation and disulfide formation at the N-terminal cysteine residue was effectively suppressed, and no dimeric Aβ-peptide derivatives were observed. The HPLC profile showed the presence of two peaks at 13.6 and 21.5 min, respectively (Figure 15.1b); however, MALDI-MS confirmed the presence of a single IPSL-Cys-Aβ(1-40), while the component at a retention time of 13.6 min was found to be unreacted spin-label reagent. The HPLC-purified peptides were also subjected to ESI-LC-MS analysis (Table 15.1), which confirmed the expected molecular masses. In all cases, the multiply charged molecular ions (6+ and 7+) were found to be the most abundant ions, with loss of one water molecule as the only detectable fragmentation.

15.3.2 GEL ELECTROPHORESIS AND MASS SPECTROMETRIC CHARACTERIZATION OF IN VITRO Aβ-OLIGOMER FORMATION

Due to the insolubility of Aβ-aggregates, fibril formation could not be analyzed directly by mass spectrometry. In order to determine the molecular composition of aggregates, *in gel* tryptic digestion and mass spectrometric analysis of the gel electrophoresis bands of Aβ-oligomers was performed. Aβ(1-40) was subjected to fibril growth at 1 μg/μL (220 μM) in buffer solution, pH 7.5, and incubated for 5 days at

37°C; at this time, high-molecular-weight fibrils were obtained as a white precipitate, sonicated and centrifuged for 15 min as described in Section 15.2, Materials and Methods, and the supernatant separated from the precipitate for ion mobility–MS. The precipitate was reconstituted in MilliQ water and subjected to polyacrylamide gel electrophoresis (Figure 15.2). Only the soluble fraction was found to enter the gel (Figure 15.2a) while the insoluble part remained at the top of the stacking gel (data not shown). MALDI-MS analysis of this soluble fraction revealed monomeric Aβ (Figure 15.2a, band 1) and Aβ-oligomers at approximately 20 kDa (pentamer, Figure 15.2a, band 2), while high-molecular-weight aggregates were found at >100 kDa (Figure 15.2a, bands 3 and 4). All gel bands were cut out, digested with trypsin, and analyzed by MALDI-MS. These results ascertained the presence of Aβ-peptides in all gel bands by identification of the corresponding peptide fragments (Table 15.2 and Figure 15.3).

The preparation of aggregates of the spin-labeled IPSL-Cys-Aβ(1-40) was performed at identical conditions, by incubation of a 200-μM solution for 5 days at 37°C. Following incubation, supernatant and precipitate were separated by centrifugation and the precipitate suspended in MilliQ water for gel electrophoresis. Tris-Tricine PAGE analysis was performed with a freshly prepared IPSL-Cys-Aβ(1-40) solution in fibril growth buffer, the supernatant, and the fibril suspension of the spin-labeled Aβ-aggregates (Figure 15.2b). The IPSL-Cys-Aβ(1-40) peptide started to aggregate immediately during the solubilization, as indicated by dimeric and trimeric gel bands (Figure 15.2b, lane 1). From the fibril precipitate, only the fraction soluble in running buffer entered the separation gel (Figure 15.2b), and the presence of monomer, dimer, and trimer and some high-molecular-weight aggregates with

---

TABLE 15.2

Identification of Aβ-Fragments in Oligomer Bands Separated by Gel Electrophoresis (cf. Figure 2a) by MALDI-TOF-MS

Gel Band No.	Tryptic Peptide Masses [M+H]$^+$	Aβ-Peptide Sequences
1	1084.6	[29–40]
	1326.0	[17–28]
	1337.0	[6–16]
	2393.7	[17–40]
	2643.6	[6–28]
2	1084.7	[29–40]
	1326.0	[6–16]
	1337.0	[17–40]
3	1085.8	[29–40]
	1336.3	[6–16]
	2392.6	[17–40]
4	1085.6	[29–40]
	1326.5	[17–28]
	1337.3	[6–16]
	2393.4	[17–40]
	3709.9	[6–40]

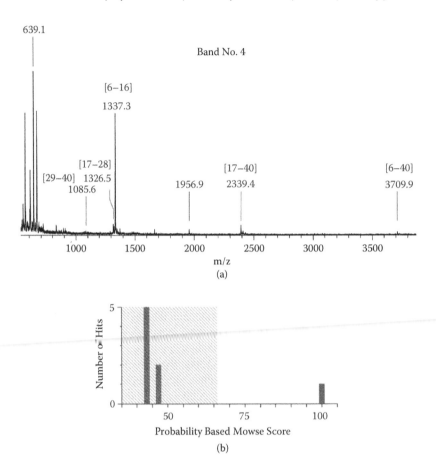

FIGURE 15.3   MALDI-MS of *in-gel* tryptic digestion mixture of Aβ(1-40), high-molecular-weight aggregates (band 4 from Figure 15.2a). (a) MALDI-TOF mass spectrum of tryptic digestion mixture; (b) database search result with identified peptide fragments.

molecular weights beyond 200 kDa were observed (Figure 15.2b, lane 2). In contrast, the supernatant showed only the band of the Aβ monomer (Figure 15.2b, lane 3).

15.3.3   ION MOBILITY MASS SPECTROMETRY OF Aβ-OLIGOMERS

The soluble fraction of the Aβ(1-40) fibril preparation obtained by incubation over 5 days at 37°C and a freshly prepared Aβ(1-40) peptide solution (220 μM) were subjected to comparative analysis by ion mobility–MS (Figure 15.4a and b). In the freshly prepared Aβ(1-40) solution, the [M+5H]$^{5+}$ ion was predominant, while in the fibril preparations higher charged ions ([M+6H]$^{6+}$, [M+7H]$^{7+}$ were most abundant. The signal-to-noise ratio of the [M+5H]$^{5+}$ and [M+6H]$^{6+}$ ions was lower in the fibril preparation sample than in the freshly prepared Aβ(1-40) sample, suggesting a lower amount of Aβ(1-40) monomer due to the aggregate formation. The extracted ion mobility profiles for the [M+5H]$^{5+}$ ion of the freshly prepared Aβ(1-40)

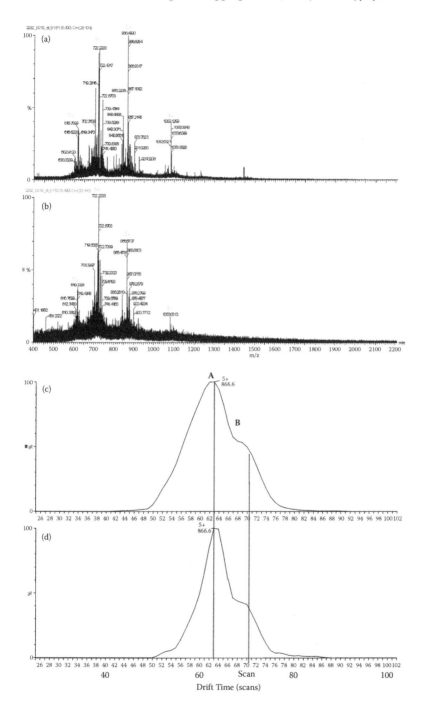

FIGURE 15.4 Ion mobility MS analysis of (a) freshly prepared Aβ(1-40); (b) supernatant after 5 days of incubation; (c) and (d) extracted ion mobility profiles for m/z 866.6, [M+5H]$^{5+}$ from freshly prepared Aβ(1-40) and supernatant.

(Figure 15.4c) and the fibril preparation (Figure 15.4d) indicate the presence of two conformational states (A and B) with different ion mobilities (cross section s), which may be indicative of the oligomerization process.

15.3.4 Ion Mobility-MS of Met^{35} Oxidized Aβ-Peptide

The extracted ion mobility plot for the [M+5H]$^{5+}$ ion of the fibril preparation is presented in Figure 15.5a and b together with the drift scope of the 5+, 6+, and 7+ charged ions. The resulting deconvoluted spectrum of the fibril preparation is shown in Figure 15.6. From the mass spectrometric data of the ion mobility profiles, modifications of Aβ by oxidation and loss of one water molecule could be identified, in addition to the intact Aβ(1-40) ion. The oxidation at the Met^{35} residue of Aβ was confirmed by MS/MS sequence determination (data not shown).

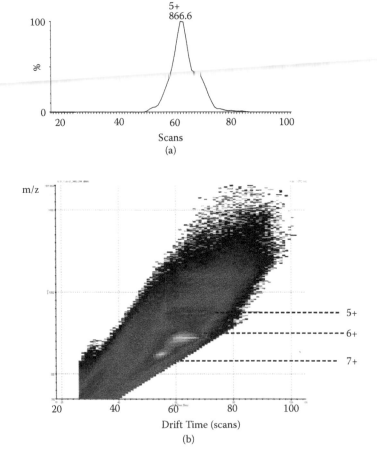

FIGURE 15.5 Drift time profile of the Aβ(1-40) supernatant after 5 days of incubation. (a) the ion mobility profile for the [M+5H]$^{5+}$ ion and (b) the drift scope.

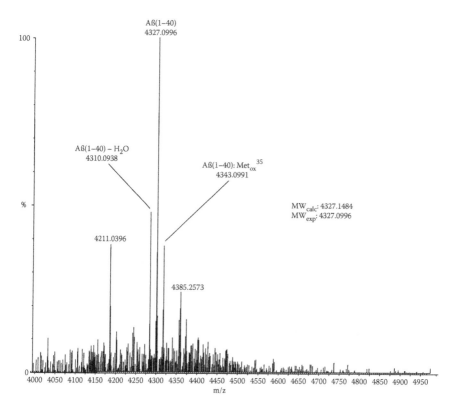

FIGURE 15.6   Deconvoluted spectrum of Aβ(1-40) supernatant after 5 days of incubation showing the presence of the Met[35]-sulfoxide oxidation product.

15.4   CONCLUSIONS

In the present study ion mobility–MS was explored for the characterization of intermediate products in the oligomerization and fibril formation reaction of β-amyloid in vitro. The results obtained here indicate that IMS-MS is an efficient tool for characterizing the composition of oligomerization products and their reaction intermediates, and to estimate the molecular dimensions of aggregate products. The gel electrophoresis data suggest that (i) Aβ-aggregation in vitro begins rapidly upon solubilization of the monomer peptide, and (ii) the insoluble fibril product contains large aggregates, while in the supernatant soluble Aβ-oligomers are present. The ion mobility–MS data indicate (i) the presence of at least two different conformational forms involved in Aβ-aggregation, and (ii) the formation of modifications such as Met[35] oxidation during the oligomerization. Hence the complementarity of both methods appears promising for the characterization of the Aβ-oligomerization and -aggregation pathway. The comparison of ion mobility–MS with EPR spectroscopic data of Aβ-oligomers showed good agreement with the gel electrophoresis study, as previously described.[27] Hence IMS-MS appears to be a promising tool for the characterization of oligomerization and aggregation pathways of β-amyloid and other "misfolding-aggregating" polypeptides.

ACKNOWLEDGMENTS

We thank Marilena Manea and Marcel Leist for expert help with the synthesis of spin-labeled Aβ-peptide derivatives, and for critical discussion of the manuscript. This work was supported by the International Research Center "Proteostasis" at the University of Konstanz, and by the Deutsche Forschungsgemeinschaft, Bonn, Germany.

REFERENCES

1. Jakobsen L.D., Jensen P.H. (2003) Parkinson's disease: alpha-synuclein and parkin in protein aggregation and the reversal of unfolded protein stress. *Methods Mol Biol*, 232: 57–66.
2. Morgan D. (2006) Immunotherapy for Alzheimer's disease. *J Alzheimers Dis*, 9: 425–32.
3. Uversky V.N. (2007) Neuropathology, biochemistry, and biophysics of alpha-synuclein aggregation. *J Neurochem*, 103: 17–37.
4. Salminen A., Ojala J., Kauppinen A., Kaarniranta K., Suuronen T. (2009) Inflammation in Alzheimer's disease: amyloid beta oligomers trigger innate immunity defence via pattern recognition receptors. *Prog Neurobiol*, 87: 181–94.
5. Crews L., Tsigelny I., Hashimoto M., Masliah E. (2009) Role of Synucleins in Alzheimer's Disease. *Neurotox Res.*, 16(3):306–17.
6. Hull M., Berger M., Heneka M. (2006) Disease-modifying therapies in Alzheimer's disease: how far have we come? *Drugs*, 66: 2075–93.
7. Gardberg A.S., Dice L.T., Ou S., Rich R.L., Helmbrecht E., Ko J., Wetzel R., Myszka D.G., Patterson P.H., Dealwis C. (2007) Molecular basis for passive immunotherapy of Alzheimer's disease. *Proc Natl Acad Sci U S A*, 104: 15659–64.
8. Luhrs T., Ritter C., Adrian M., Riek-Loher D., Bohrmann B., Dobeli H., Schubert D., Riek R. (2005) 3D structure of Alzheimer's amyloid-beta(1-42) fibrils. *Proc Natl Acad Sci U S A*, 102: 17342–7.
9. McLaurin J., Cecal R., Kierstead M.E., Tian X., Phinney A.L., Manea M., French J.E., Lambermon M.H., Darabie A.A., Brown M.E., Janus C., Chishti M.A., Horne P., Westaway D., Fraser P.E., Mount H.T., Przybylski M., St George-Hyslop P. (2002) Therapeutically effective antibodies against amyloid-beta peptide target amyloid-beta residues 4-10 and inhibit cytotoxicity and fibrillogenesis. *Nat Med*, 8: 1263–9.
10. Solomon B. (2007) Beta-amyloid based immunotherapy as a treatment of Alzheimers disease. *Drugs Today* (Barc), 43: 333–42.
11. Grau S., Baldi A., Bussani R., Tian X., Stefanescu R., Przybylski M., Richards P., Jones S.A., Shridhar V., Clausen T., Ehrmann M. (2005) Implications of the serine protease HtrA1 in amyloid precursor protein processing. *Proc Natl Acad Sci U S A*, 102: 6021–6.
12. Stefanescu R., Iacob R.E., Damoc E.N., Marquardt A., Amstalden E., Manea M., Perdivara I., Maftei M., Paraschiv G., Przybylski M. (2007) Mass spectrometric approaches for elucidation of antigenantibody recognition structures in molecular immunology. *Eur J Mass Spectrom* (Chichester, Eng), 13: 69–75.
13. Manea M., Hudecz F., Przybylski M., Mezo G. (2005) Synthesis, solution conformation, and antibody recognition of oligotuftsin-based conjugates containing a beta-amyloid(4-10) plaque-specific epitope. *Bioconjug Chem*, 16: 921–8.

14. Perdivara I., Deterding L.J., Cozma C., Tomer K.B., Przybylski M. (2009) Glycosylation profiles of epitope-specific anti-ss-amyloid antibodies revealed by liquid chromatography-mass spectrometry. *Glycobiology*,

15. Juszczyk P., Paraschiv G., Szymanska A., Kolodziejczyk A.S., Rodziewicz-Motowidlo S., Grzonka Z., Przybylski M. (2009) Binding epitopes and interaction structure of the neuroprotective protease inhibitor cystatin C with beta-amyloid revealed by proteolytic excision mass spectrometry and molecular docking simulation. *J Med Chem*, 52: 2420–8.

16. Dodel R., Hampel H., Depboylu C., Lin S., Gao F., Schock S., Jackel S., Wei X., Buerger K., Hoft C., Hemmer B., Moller H.J., Farlow M., Oertel W.H., Sommer N., Du Y. (2002) Human antibodies against amyloid beta peptide: a potential treatment for Alzheimer's disease. *Ann Neurol*, 52: 253–6.

17. Ruotolo B.T., Benesch J.L., Sandercock A.M., Hyung S.J., Robinson C.V. (2008) Ion mobility-mass spectrometry analysis of large protein complexes. *Nat Protoc*, 3: 1139–52.

18. Ruotolo B.T., Hyung S.J., Robinson P.M., Giles K., Bateman R.H., Robinson C.V. (2007) Ion mobility-mass spectrometry reveals long-lived, unfolded intermediates in the dissociation of protein complexes. *Angew Chem Int Ed Engl*, 46: 8001–4.

19. Kanu A.B., Dwivedi P., Tam M., Matz L., Hill H.H., Jr. (2008) Ion mobility-mass spectrometry. *J Mass Spectrom*, 43: 1–22.

20. Trimpin S., Clemmer D.E. (2008) Ion mobility spectrometry/mass spectrometry snapshots for assessing the molecular compositions of complex polymeric systems. *Anal Chem*, 80: 9073–83.

21. Zhou M., Sandercock A.M., Fraser C.S., Ridlova G., Stephens E., Schenauer M.R., Yokoi-Fong T., Barsky D., Leary J.A., Hershey J.W., Doudna J.A., Robinson C.V. (2008) Mass spectrometry reveals modularity and a complete subunit interaction map of the eukaryotic translation factor eIF3. *Proc Natl Acad Sci U S A*, 105: 18139–44.

22. Drescher M., Godschalk F., Veldhuis G., van Rooijen B.D., Subramaniam V., Huber M. (2008) Spin-label EPR on alpha-synuclein reveals differences in the membrane binding affinity of the two antiparallel helices. *Chembiochem*, 9: 2411–6.

23. Drescher M., Veldhuis G., van Rooijen B.D., Milikisyants S., Subramaniam V., Huber M. (2008) Antiparallel arrangement of the helices of vesicle-bound alpha-synuclein. *J Am Chem Soc*, 130: 7796–7.

24. Murakami K., Hara H., Masuda Y., Ohigashi H., Irie K. (2007) Distance measurement between Tyr10 and Met35 in amyloid beta by site-directed spin-labeling ESR spectroscopy: implications for the stronger neurotoxicity of Abeta42 than Abeta40. *Chembiochem*, 8: 2308–14.

25. Torok M., Milton S., Kayed R., Wu P., McIntire T., Glabe C.G., Langen R. (2002) Structural and dynamic features of Alzheimer's Abeta peptide in amyloid fibrils studied by site-directed spin labeling. *J Biol Chem*, 277: 40810–5.

26. LeVine H., 3rd (2005) Mechanism of A beta(1-40) fibril-induced fluorescence of (trans,trans)-1-bromo-2,5-bis(4-hydroxystyryl)benzene (K114). *Biochemistry*, 44: 15937–43.

27. Ionut Iurascu M., Cozma C., Tomczyk N., Rontree J., Desor M., Drescher M., Przybylski M. (2009) Structural characterization of ss-amyloid oligomer-aggregates by ion mobility mass spectrometry and electron spin resonance spectroscopy. *Anal Bioanal Chem*, 8: 2509–19.

16 The Conformational Landscape of Biomolecules in Ion Mobility–Mass Spectrometry

Jody C. May and John A. McLean

CONTENTS

16.1 INTRODUCTION

This chapter presents the application of ion mobility–mass spectrometry (IM-MS) to the separation of biomolecules as well as the elucidation of conformational information that can be derived from the IM-MS technique. Ion mobility has become a broadly encompassing term that now includes a family of gas-phase electrophoretic separation techniques. The realm of ion mobility analytical techniques can be broadly divided into two categories: broadband dispersive methods and narrow bandpass filtering methods, all of which exploit the fundamental differences in an ion's gas-phase mobility in order to effectuate a separation. While dispersive methods generate a complete snapshot of a sample during each measurement, filtering methods transmit ions possessing a single or narrow range of ion mobility value(s) and must be scanned in order to generate a mobility spectrum. The drift tube and traveling wave techniques are dispersive, while asymmetric/differential field and differential mobility methods are filtering. A particular ion mobility technique is chosen based on the kind of information that is sought: Dispersive mobility methods are useful for broadband signal interrogation while filtering mobility methods benefit from high sensitivity owing to efficient ion transmission and signal attenuation. In this chapter, we focus only on the dispersive techniques, and in particular the drift tube method from which exact collision cross section (CCS) values can be determined from the measurement. Although IM-MS has been used quite successfully for a wide variety of analytes, the emerging area of biomolecular structural interrogation is highlighted here in the context of the conformational landscape, which is how biochemical classes order themselves in the two-dimensional (2D) mobility-mass space.

16.2 THE EMERGING LANDSCAPE—A HISTORICAL PERSPECTIVE OF ION MOBILITY–MASS SPECTROMETRY

16.2.1 In the Beginning Was Ion Mobility

A concise definition of ion mobility might be "the movement of an ion through a gas" and while accurate, this definition offers no indication as to the principles governing the separation of different ions, i.e., the spectrometry method that makes the ion mobility phenomenon useful as an analytical technique. It is perhaps no surprise that ion mobility as a phenomenon was defined long before it was understood how to use it to distinguish different kinds of ions, just as the physical property of mass was known long before there was any sort of mass spectrometry. Early observations of fluorescence in gas discharge tubes were among the first accounts of artificially generated ions in air. Ions weren't explicitly "discovered" until the work of Goldstein identified that positively charged ions were formed in a gas exposed to a stream of electrons (cathode rays). His work with hydrogen gas is credited as the first observation of the proton.[1] A decade or so later in 1896, J. J. Thomson and Ernest Rutherford described their work on the effects of electric fields on ions in gases using Roentgen radiation (x-rays).[2] More careful measurements of ion mobilities were reported the following year by John Zeleny, a student of Thomson, using his novel counter-flow ion mobility device. Notable in Zeleny's work was the observation that two ion populations possessing different velocities were formed in air and that the measured differences in their mobilities were attributed to "...an inequality in the size of the two ions."[3] The first mass spectrometer was described a decade later by Thomson, and so ion mobility spectrometry preceded mass spectrometry as an analytical technique. While the next half-century of ion transport research was important in advancing our understanding of the ion mobility phenomenon, the instrumentation was relatively simple in design and myopic to determining the exact identity of ions being investigated.

16.2.2   THE EMERGENCE OF ION MOBILITY–MASS SPECTROMETRY

A golden era of ion mobility instrumentation began in the 1960s with the development of sophisticated drift tube spectrometers that could measure with precision the mobility of ions in pure gases.[4] Integral to the versatility of these instruments was the addition of a mass spectrometer to interrogate ion species by their mass-to-charge ratio (m/z). The first hybrid mobility-mass spectrometer instrument appearing in the literature was developed for studying ion-molecule reaction chemistry.[5] Other early experimental variations of the mobility-mass spectrometer were also described during this time, including a reverse configuration MS-IM and a drift tube bracketed between mass spectrometers (MS-IM-MS).[6,7] These instruments utilized magnetic sector mass analyzers that were commercially available from several companies following World War II. Another noteworthy ion mobility instrument configuration to come out of this era was the first drift tube coupled orthogonally to a time-of-flight mass spectrometer (Bendix), which is a common IM-MS configuration utilized today.[8] Identifying the potential utility of the ion mobility separation to analytical chemists, an ion mobility instrument was commercialized (Franklin GNO) and sold under the name Plasma Chromatograph, which was a somewhat misleading moniker as separations were, by definition, neither chromatographic nor conducted with a plasma. For many people, this instrument represents the first ion mobility spectrometer, as it brought a specialized technique into the realm of broad chemical analysis. Versions of the Plasma Chromatograph were configured as an IM-MS by combining the ambient pressure drift tube with a quadrupole mass spectrometer,[9,10] making it a sensitive instrument but prone to complicating ion chemistry as a result of the ambient chemical ionization source and drift chamber utilized in its design. The legacy of the Plasma Chromatograph continues today with countless stand-alone ion mobility spectrometers being utilized as chemical analyzers in security applications throughout the world.

16.2.3   THE BIOLOGICAL ION MOBILITY–MASS SPECTROMETRY ERA

The 1990s brought about the age of biological spectrometry research. Initiated by breakthroughs in soft ionization methods, biomolecular analysis by IM-MS has been the focus of considerable research efforts since the 1990s, and the technology has and continues to advance at a rapid pace in order to meet the analytical demands of high sensitivity, high resolution, and broad dynamic range. The groups of Bowers,[11] Jarrold,[12] and Clemmer[13] pioneered much of the structurally relevant biological IM-MS work in the 1990s. In their work, high precision, reduced pressure drift tubes were used to obtain some of our most accurate CCS measurements on peptides. More recently, Robinson and coworkers used IM-MS to elucidate quaternary structural information regarding macromolecular protein complexes.[14] In concert with these efforts has been a drive to assign some physical meaning to the measured cross sections. This is undertaken theoretically using sophisticated computational modeling, which generates a pool of candidate structures. These candidate structures are then linked to experimental CCS data through a calculation of the expected orientationally averaged cross section. Matched structures are used to infer the physical conformational properties of the analyte and are an approximation of what the actual structure could be, while ruling out structures that are inconsistent with the empirical results.

The modern era of IM-MS is in many ways defined by the emergence of another commercial technology, namely the traveling wave ion mobility (TWIM) technique, which was introduced into a high-end commercial (Waters) MS-IM-MS instrument in 2007. The TWIM technique is discussed in other contributions to this book and is briefly described in the following section. What merits mention of the TWIM technique in a historical context is the fact that until its introduction, biological IM-MS was still considered a specialized technique and limited in use to laboratories with access to one of the few existing drift tube instruments or with the resources to build their own. With the wide-scale production release of the Waters Synapt, as well as another recently introduced commercial ion mobility interface (FAIMS [high-field asymmetric waveform ion mobility spectrometry], Thermo Scientific), IM-MS technologies are now widely available analytical techniques.

16.3   ION MOBILITY–MASS SPECTROMETRY TECHNOLOGIES AS APPLIED TO BIOLOGICAL SEPARATIONS

16.3.1   THE DRIFT TUBE AND TRAVELING WAVE TECHNIQUES

It is useful here to provide a brief overview of the fundamental operational principles that govern separations in both the drift tube and traveling wave mobility techniques. Since there are several other excellent contributions in this book that cover the basis for the drift tube and traveling wave ion mobility experiments, only concepts important for understanding how the measurements relate to analyte structure are developed here.

The primary experimental components of an IM-MS instrument are depicted in Figure 16.1a. The separation parameter for all ion mobility methods is the ion-neutral collision cross section (CCS), or the integral of the analyte ion collisions (approximately 10^5–10^8) with the neutral gas (atoms or molecules) in which it drifts. For the drift tube technique, ions migrate under the influence of a uniform electric field against a background pressure of a neutral gas creating a net temporal displacement that is directly related to the ion's CCS or apparent surface area (Figure 16.1b). The measured temporal arrival of ions in the drift tube mobility technique is directly related to the CCS parameter due to the fact that all other contingent parameters (neutral gas pressure, electric field, and temperature) are kept constant during the separation.[15]

The TWIM device also consists of a gas-filled chamber partitioned by a series of electrodes, but this is where the similarities with drift tubes end. Instead of a

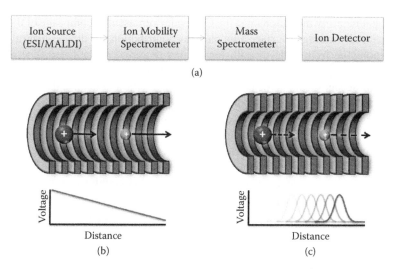

FIGURE 16.1 (a) A generalized schematic of the typical arrangement of IM-MS instrumentation used for biomolecular analysis. (b) Time-dispersive ion mobility utilizing a uniform electrostatic field. (c) Time-dispersive ion mobility utilizing an electrodynamic field, such as that used in traveling wave ion mobility. See text for a full description of the operating principles of both electrostatic and electrodynamic ion mobility.

uniform electric field, the TWIM method uses a dynamically pulsed potential that travels from one end of the chamber to the other at a fixed velocity (Figure 16.1c). This traveling wave pushes ions through the gas, while the drag force of the gas collisions impedes the ion's motion such that larger ions that experience more drag force tumble backwards over the wave. As the larger ions tumble backwards more frequently, they remain in the device longer, and separations are effectuated by the relative retention of different sized ions within the device.[16] The TWIM is a dispersive technology and so offers many of the same analytical benefits as the drift tube, with the significant technological advantage that ions enter and exit the TWIM device at the same potential, providing a convenient means to coupling with other technologies. Another important advantage of the TWIM technology is that a superimposed radio frequency (RF) potential is added to the ring electrodes to confine ions along the central axis of the device, which greatly improves instrument sensitivity. While the relationship between the measured drift time and the ion's CCS is complicated due to the nature of the TWIM separation, the instrument can be calibrated using drift tube data to obtain CCS values.[17,18] Observed resolving powers with the TWIM separation is not as high as with drift tubes, but this is expected to change in the coming years as the technology is further refined.

The precision at which the CCS can be determined in either ion mobility method is ultimately tied to how constant (or at least how well characterized) each experimental parameter is during the mobility measurement. So-called "end effects" must be accounted for, which is any time the ion spends outside of the defined drift region of static pressure, temperature, and field. Measurements of CCS within approximately 2% precision can be obtained with little effort from most research-grade drift tubes.

For all ion mobility measurements, the CCS value represents a composite of the ion's orientationally averaged surface area plus any contributions by the ion-neutral interaction potential from inelastic collisions, either reactive or interactive. The greater the polarizability of the drift gas and the closer in mass the ion and neutral drift gas are, the more significant is the contribution made by the van der Waals interactions. This is the motivation for using helium as the drift gas in high precision IM-MS experiments. It is generally accepted that the contribution of inelastic interactions is minimal for ions above approximately 500 Da, in which case the ion–gas interaction can be accurately depicted as elastic hard spheres.[19] Thus, for biomolecules, the measured CCS is an accurate depiction of the physical size of the analyte ion and can be utilized in making generalizations regarding the ion's conformation.

16.3.2 The Separation of Biomolecules in the IM-MS Experiment

Complex biological samples pose a considerable challenge for mass spectrometry as the diverse representation of endogenous molecular classes span as much as ten orders of magnitude in concentration, and oftentimes the highest abundant analytes represent a small fraction of the molecular diversity of the sample.[20] Commonly these samples are pretreated either through simple clean-up methods or with more sophisticated prefractionation such as multidimensional electrophoresis or chromatography. Liquid chromatography (LC) amends itself very well to direct coupling with mass spectrometers utilizing direct infusion sources (e.g., electrospray or chemical ionization) and LC-MS has become the de facto standard for addressing a wide range of biochemical analysis problems. IM-MS is a complimentary post-ionization separation that is extraordinarily fast (μs–ms vs. min–hrs for LC). Since separation occurs after ionization, there is no compromise made to the standard LC-MS setup, thus the practical benefit of current IM-MS technologies is that they are complementary to LC-MS separations. The compromise for the incredible speed of IM-MS separations is reduced peak capacity (approximately 10^3–10^4) as compared with more orthogonal multidimensional separation techniques. The peak capacity production rate for IM-MS is exceptional and can exceed 1 million peaks per second. The analytical advantages of IM-MS for biological analysis are fourfold: (i) simplify MS spectra for higher confidence-level interpretation, (ii) improve limits of detection for low abundant analytes through mobility dispersion of concomitant signal, (iii) improve identification/quantitation, and (iv) provide an added level of information (ion structure) for analyte identification and classification purposes. Figure 16.2 contains a typical example of a 2D mobility-mass spectrum for singly charged peptide ions derived from proteolytic digestion of bovine hemoglobin.[21] Several of the highest abundance ion signals are labeled in panel (i). Note that the 2D plot in panel (i) is a projection of 3D data in that relative abundance of the various signals is indicated through false coloring (heat mapping) in a way similar to spectra obtained in 2D nuclear magnetic resonance techniques. The total mass spectrum is also projected in panel (iii) to illustrate the typical one-dimensional spectrum that would be obtained in the absence of the IM separation. By integrating the mass data over small regions of ion mobility space (e.g., 1300–1400 μs), the resulting selected mass spectrum exhibits only those signals corresponding to the species present in a

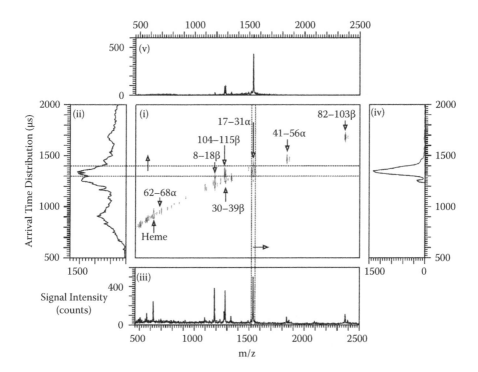

FIGURE 16.2 (i) A two-dimensional MALDI-IM-MS plot of conformation space illustrating the separation of peptides obtained from a tryptic digest of bovine hemoglobin (approximately 4 pmol deposited on the target). Several of the most intense peaks are assigned according to their position in the α- or β-chain of the protein, respectively. (ii) The ion mobility arrival time distribution integrated over all *m/z* space. (iii) The mass spectrum obtained by integrating over all ion mobility arrival time distribution space. (iv) The arrival time distribution integrated over the *m/z* range of 1525–1550 and (v) the mass spectrum obtained by integrating over the arrival time distribution of 1300–1400 μs (regions outlined by dashed lines). (Adapted from McLean, J. A., Russell, D. H. *J. Proteome Res.,* 2, 427–430, 2003. With permission. © 2003 American Chemical Society.)

particular arrival time window, as illustrated in panel (v). This selective interrogation of the data results in an attenuation of chemical noise that would otherwise obstruct the analyte signal of interest, as illustrated by comparing the spectra in panels (v) and (iii). Likewise the ion mobility profile can be deconvoluted by integrating over a narrow m/z range as depicted in panel (iv), which contains a baseline resolved mobility spectrum for the singly-protonated peptide.

16.4 THE CONFORMATIONAL LANDSCAPE OF BIOMOLECULES

16.4.1 THE EXPERIMENTAL MOBILITY-MASS CORRELATIONS FOR BIOMOLECULES

The spectrum of biological molecules represents a diverse set of structural classes with each class comprised of a family of subclasses, each expressing their own distributions of preferred structural ordering. Biomolecular structure is itself a dynamic process and the fact that there is any order at all in the conformational landscape is, at first glance, something of a surprise. As it would happen, there is a high degree of correlation between ion mobility and mass for a given analyte class. While this ordering tends to reduce the overall peak capacity of the IM-MS separation, it lends the technique a significant advantage in terms of predictive power and analyte characterization. In the research that was spawned following the release of the Plasma Chromatograph in the 1970s, there was considerable work undertaken with understanding the predictive nature of the mobility-mass correlation. Though initially it was thought that the mobility spectrum alone could be used to directly infer the ion mass without the need for MS, it was later established that correlations were chemical class specific rather than universal identifiers.[22] This holds true for both low- and high-molecular-weight analytes, and is especially useful for biomolecules that exhibit ordering into defined biomolecular classes (e.g., proteins, lipids, carbohydrates, and nucleic acids). Until recently, however, few researchers utilized the mobility-mass correlation for characterization, since for small molecules the correlations are very general and were often poorly resolved for all but the most extreme cases of disparate chemical classes.

For complex samples containing a mixture of analyte molecular classes (e.g., serum or whole cell lysates), it is empirically known that different molecular classes exhibit significant differences in their gas-phase packing efficiencies. Particularly for biomolecules, the relative packing efficiency follows the general trend in order of increasing density: lipids < peptides < carbohydrates < oligonucleotides. This ordering is depicted conceptually in Figure 16.3a for IM-MS analysis space. This trend represents the inherent degrees of freedom that different biopolymers possess and ultimately is a reflection of the averaged preferred gas-phase conformation. Experimental data for a mixture of biomolecules is contained in Figure 16.3b and demonstrates how the mobility-mass correlation can be a useful diagnostic tool for assigning an unknown analyte to a specific biomolecular class. Dashed lines are shown as visual aids to help locate those signals that correspond to a particular chemical class. An example of the practical utility of mobility-mass correlation information can be illustrated by considering a proteomics analysis experiment, which is commonly undertaken with samples containing a mixture of biomolecules in addition to the peptides of interest. By selecting only those signals that correspond to the peptide correlation, interfering signals can be removed prior to spectral interrogation, which lends itself as a significant advantage for improving confidence levels in protein identifications obtained by database searching. Additionally, separating the signals of interest from the chemical noise results in improved dynamic range for low-abundance proteins, which oftentimes are the proteins of biological interest (e.g., post-translationally modified and signaling mitogen-activated protein [MAP] kinases).

Figure 16.3c contains a combined map of mass versus collision cross section and represents a diverse sampling of biomolecules (n = 951). Note that the mobility data in Figure 16.3c is depicted as CCS as opposed to measured drift times (Figure 16.3b). Because the transformation of drift time data to CCS is nonlinear due to the inclusion

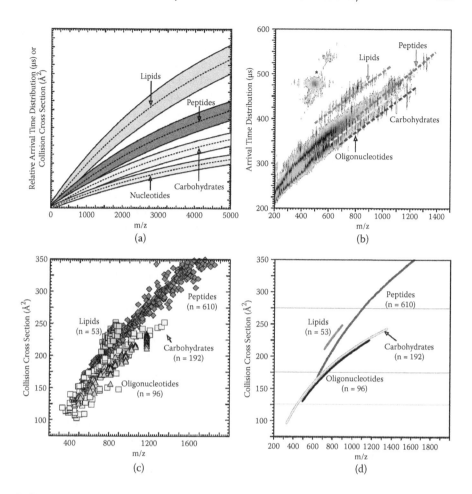

FIGURE 16.3 (a) A hypothetical plot illustrating the differences in IM-MS conformation space for different molecular classes based on different gas-phase packing efficiencies. (b) A plot of MALDI-IM-MS conformation space obtained for a mixture of model species representing each molecular class (ranging from 7 to 17 model species for each class, spanning a range of masses up to 1500 Da). Dashed lines are for visualization purposes of where each molecular class occurs in conformation space. Signals in the vicinity of the asterisk arise from limited post-IM fragmentation of the parent ion species. (c) A plot of collision cross section as a function of *m/z* for different biologically relevant molecular classes, including oligonucleotides (n = 96), carbohydrates (n = 192), peptides (n = 610), and lipids (n = 53). All species correspond to singly charged ions generated by using MALDI, where error ±1σ is generally within the data point. (d) A plot of the average collision cross section versus *m/z* fitted to logarithmic regressions for the data corresponding to each molecular class. (Part (a) is adapted from Fenn, L. S., and McLean, J. A., *Anal. Bioanal. Chem.*, 391, 906, 2008, Fig. 2(a), with kind permission from Springer Science+Business Media. Parts (b)–(d) are adapted from Fenn, L. S., Kliman, M., Mahsutt, A., Zhao, S. R., and McLean, J. A., *Anal. Bioanal. Chem.*, 394, 235, 2009, Fig. 1(a–c), with kind permission from Springer Science+Business Media.)

of the mass term within the Mason-Schamp relationship, mobility data projected as arrival (drift) time will exhibit a better separation than when projected in CCS. This can be readily seen by comparing Figure 16.3b with Figure 16.3c. Logarithmic regression fits to each class data set is contained in Figure 16.3d and represents an averaged CCS distribution across mass space. While a linear fit is adequate for quantifying the mobility-mass correlation across a narrow mass range, the logarithmic regression analysis is more appropriate over the broad mass ranges investigated here. Several conclusions can be drawn from the analysis. The relative deviation of CCS values from the logarithmic fit differs across different biomolecular classes, which observes the following trend: lipids < oligonucleotides < peptides < carbohydrates. This reflects the relative ordering present within each biomolecular class. For example, carbohydrates are structurally comprised of large numbers of small, monomeric units that exhibit a high degree of branching and are thus expected to represent a relatively broad distribution of conformations. In contrast, lipids and oligonucleotides can be characterized by relatively fewer possible orderings within their structures, though these latter conclusions are highly general given the small sample sets measured (i.e., lipids and oligonucleotides represent <20% of the total sample size). Of practical interest is the observation that resolution between biomolecular class distributions improves with an increase in mass, which intuitively results from a greater contribution of relative packing efficiencies as the number of atoms increases. Since biomolecules are composed of a relatively small representation of atoms (C, H, O, N, P, and S), the mass and volume of biological analytes are expected to be represented by a narrow range of densities. A summary of these results is depicted conceptually in Figure 16.4 for the four biomolecular classes investigated thus far. Figure 16.4 brings to light some predictions regarding the expected ordering of subclasses. Oligonucleotides are expected to exhibit a narrow distribution of structures as compared with other biomolecular classes. For lipids, there is predicted to be two disparate subclass distributions between the compact phosphatidyl compounds and elongated spingomyelins and cerebrosides. With a larger sampling of data, these fine structure predictions might be better resolved in the conformational landscape.

16.4.2 Using Computational Methods to Assign Structural Meaning to Empirical Data

Empirically measured collision cross sections are typically supported by complementary computational studies in order to infer structurally relevant details from the IM-MS experiment.[11,12,23] Although the specific procedural details for comparing experimental and computational results vary across different research laboratories, the general theoretical framework for comparisons consists of five steps: (i) generation of in silico model structures, (ii) exploration of the conformational space (e.g., by simultaneous molecular mechanics/molecular dynamics (MM/MD) and simulated annealing protocols), (iii) determination of expected collision cross sections for modeled structures (typically via MOBCAL developed by Jarrold and coworkers[19,24]), (iv) generation of scatter plots of relative energy versus collision cross section, and finally (v) interrogation of cluster plots for the lowest energy structures that

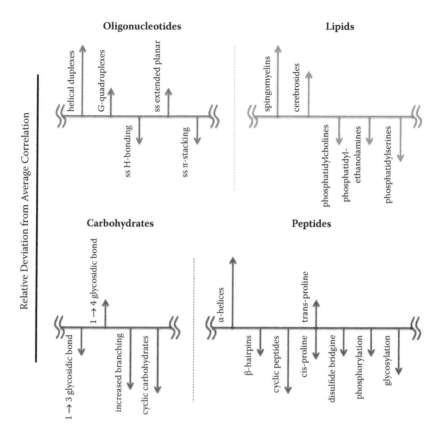

FIGURE 16.4   A plot depicting structural motifs and subclasses of molecules that result in deviations from the average ion mobility-*m/z* correlation for each of four different classes of biomolecules, including oligonucleotides, lipids, carbohydrates, and peptides. The relative length and direction of individual arrows corresponds to the relative degree of deviation and whether those species are greater than or less than the predicted correlation for the average of molecules of that class. (Adapted from McLean, J. A., *J. Am. Soc. Mass Spectrom.*, 20, 1775–1781, © 2009 with permission from The American Society for Mass Spectrometry.)

correlate to the empirically measured CCS. The reader is directed to an outstanding tutorial by Bowers and coworkers that describes the specifics of each step in this procedure as well as discussing the fundamental considerations behind the methods. [11] For biomolecular structural studies, MM/MD approaches are typically used due to the large size of the molecules, but these techniques require that suitably parameterized force fields exist, and there is some difficulty for these approaches to sample compact conformations. It is important to recognize that, ultimately, these methods do not result in a high-resolution structure, but rather a general representation of what is likely to be the actual conformation of the analyte under study. The correlated experimental and theoretical results are thus useful for interpreting structural motifs that may exist as well as narrowing down those structural conformations that are inconsistent with the experimental results.

16.5   NAVIGATING THE CONFORMATIONAL LANDSCAPE— TAILORING WITH CHEMICAL SHIFT STRATEGIES

The ability to construct an atlas of the conformation space in IM-MS measurements provides an opportunity to utilize the unoccupied regions where no signals are predicted to occur. This is a particularly appealing strategy in the analysis of extremely complex samples oftentimes encountered in the analysis of biological samples, for example, tissue homogenates, serum, and whole blood, where signals begin to overlap in conformation space. An illustrative example in glycoproteomics is shown in Figure 16.5a for the analysis of human glycoprotein, which is composed of over 50% carbohydrate. Through using a nonspecific protease to generate peptides, glycoconjugates, and glycans, a very complex IM-MS spectrum is obtained whereby signals higher in conformation space are identified as peptides and signals lower in the spectrum are identified as carbohydrate. However, in the intervening region between the two extremes it is challenging to characterize to what species these signals correspond. To obtain further chemical information in complex spectra such as those shown, two chemical labeling strategies have been developed to selectively shift signals in a predictive way based on the chemical moieties that are present within specific molecules, namely through the use of noncovalent and covalent ion mobility shift reagents.

Noncovalent labeling strategies involve chemical complexation of the label with specific functional moieties that are to be probed. For example, Creaser and colleagues initially described the utility of using crown ethers to complex primary amines.[25,26] Through this complexation reaction they demonstrated the ability to separate isomeric amines through the preferential coordination of crown ethers with primary amines, which in turn increased the CCS of the ether-amine coordinated species relative to isomeric amines that remained uncoordinated. This strategy was later extended to peptide analysis by Clemmer and coworkers, whereby complexation of crown ethers with lysine residues provides separation selectivity by shifting lysine-containing peptides to longer than predicted drift times.[27] Interestingly, through activating the noncovalent complex after the ion mobility separation causes the complex to fall apart, yielding the exact mass of the lysine-containing peptide without the attendant crown ether. This is illustrated in Figure 16.5b. Other noncovalent strategies have also been utilized such as the coordination of various alkali metals to distinguish cyclic versus linear peptides[28] and the use of metal and metal acetates to shift ion mobility signals of carbohydrates and glycans.[29,30] Although noncovalent labeling strategies have great utility and are relatively easy to perform, the degree of labeling is semiquantitative in that both labeled and unlabeled target species can often be observed.

Covalent labeling strategies are frequently utilized in MS-based quantitation of protein expression levels.[31] Current methods for MS-based protein quantitation typically utilize stable isotope labeling to compare protein expression levels from different experiments in the same analysis. Contemporary MS-based approaches for stable isotope labeling in protein quantitation were first introduced in 1999 by three independent labs[32–34] and is now implemented enzymatically, metabolically, or by chemical modification. Using similar strategies, chemical labeling can be used to change the apparent density of the target species to shift the signal to regions of conformation

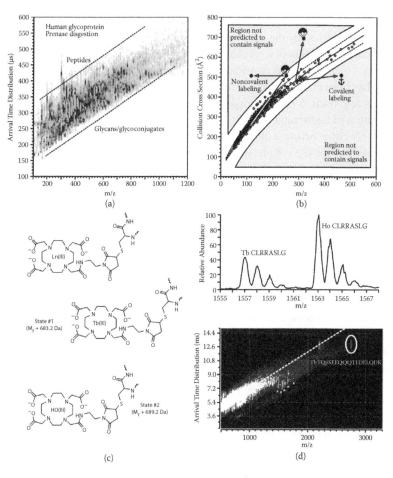

FIGURE 16.5 (a) A plot of IM-MS conformation space in the analysis of human glyco-protein following nonspecific proteolytic digestion with pronase. Dashed lines guide where peptide and glycan signals are observed. (b) A conceptual diagram of conformation space illustrating the modes in which ion mobility shift reagents can be used in noncovalent and covalent strategies. The data points are for a large suite of singly charged peptides where dashed lines represent ±7% deviation from the mobility-mass correlation, shown as a solid line. (c) Covalent labels selective for conjugation to cysteine residues incorporating lan-thanide metals. (d, top) A typical mass spectrum from a relative quantitation experiment using cysteine-selective labels to label the peptide CLRRASLG. Following labeling, a 2:1 mixture of Ho:Tb was made and analyzed. Relative quantitation information is obtained by ratioing the isotopic envelopes of the Ho and Tb species, respectively. (d, bottom) A typical IM-MS plot of conformation space utilizing phosphorylation-specific lanthanide labeling. The model phosphoprotein β-casein (bovine) was proteolytically digested with trypsin and the resulting peptides were subsequently covalently labeled incorporating Tb. The underiva-tized peptides are indicated by the white dashed line along the peptide mobility-mass cor-relation. The labeled phosphorylated peptide is shifted to a region of IM-MS conformation space that is not predicted to contain peptide signals in the absence of labeling. Peaks marked with an asterisk (*) are concomitant excess reagents in the labeling process. (Panel (d) is reproduced from Gant-Branum, R. L., Kerr, J. T., and McLean, J. A., *Analyst* 134, 1525–1530, 2009. With permission of The Royal Society of Chemistry.)

space not predicted to contain signals. For covalent labeling, this is conceptually illustrated in Figure 16.5b by the addition of a low mass/high surface area group, which is the molecular equivalent of a parachute, or a high mass/low surface area group, which is the molecular equivalent of an anchor. To accomplish the latter, lanthanide-based labeling has been utilized as illustrated in Figure 16.5c and d.[31,35–37]

In this approach, a label incorporating a macrocycle that has a high affinity for coordination of lanthanide 3+ [Ln(III)] ions is attached to peptides using chemistry selective for either primary amines (n-terminus and/or lysine)[35] or sulfhydryl groups (cysteine),[36] or for sites of phosphorylation (phosphoserine and phosphothreonine). [31,37] Importantly, the chelating moiety is relatively insensitive to which lanthanide is incorporated, because the ionic radii of all Ln(III) are nearly invariable. The mass differences between differentially labeled samples can then be tuned by selection of the Ln(III) that is loaded into the macrocycle (e.g., Lanthanum/Lutetium [La/Lu] result in a mass difference of 36 Da). The ability to selectively choose the mass difference and the broad mass difference that can be selected allows the analysis of approximately 15 different samples simultaneously, or to extend relative quantification to much larger peptides owing to the large mass differences available. Thus the label rapidly provides qualitative identification of the species that contain specific functional moieties on the basis of IM shift and quantitation capabilities on the basis of stable isotope labeling provided by the MS. Similar covalent strategies for IM shift reagents have also been described for labeling of carbohydrates using boronic acid derivatization.[38]

16.6   CONCLUSIONS

Further characterization of conformation space and devising new methodologies to use conformation space in rational ways should open new directions in how the IM-MS technique is utilized. Applications ranging from natural product discovery and nanomaterials characterization will fuel future studies aimed at understanding and exploring conformation space. Furthermore, instrumental advances to enhance IM resolution will lead to higher definition in the conformational landscape.

ACKNOWLEDGMENTS

Financial support for this work was provided by the Vanderbilt University College of Arts and Science, the Vanderbilt Institute of Chemical Biology, the Vanderbilt Institute of Integrative Biosystems Research and Education, the U.S. Defense Threat Reduction Agency (HDTRA-09-1-0013), and the NIH National Institute on Drug Abuse (RC2DA028981).

REFERENCES

1. Hedenus, M., "Eugen Goldstein and his laboratory work at Berlin Observatory", *Astronomische Nachrichten* 2002, *323*, 567–569.

2. Thomson, J. J.; Rutherford, E., "On the passage of electricity through gases exposed to Roentgen rays", *Philosophical Magazine Series 5* 1896, *42*, 392–407.
3. Zeleny, J., "Mobilities of the ions in gases at low pressures", *Philosophical Magazine* 1898, *46*, 120–154.
4. McDaniel, E. W.; Crane, H. R., "Measurements of the mobilities of the negative ions in oxygen and in mixtures of oxygen with the noble gases, hydrogen, nitrogen and carbon dioxide", *Review of Scientific Instruments* 1957, *28*, 684–690.
5. McDaniel, E. W.; Martin, D. W.; Barnes, W. S., "Drift tube-mass spectrometer for studies of low-energy ion-molecule reactions", *Review of Scientific Instruments* 1962, *33*, 2–7.
6. Bloomfield, C. H.; Hasted, J. B., "New technique for the study of ion-atom interchange", *Discussions of the Faraday Society* 1964, *37*, 176–184.
7. Kaneko, Y.; Megill, L. R.; Hasted, J. B., "Study of inelastic collisions by drifting ions", *Journal of Chemical Physics* 1966, *45*, 3741–3751.
8. McKnight, L. G.; McAfee, K. B.; Sipler, D. P., "Low-field drift velocities and reactions of nitrogen ions in nitrogen", *Physical Review* 1967, *164*, 62.
9. Cohen, M. J.; Karasek, F. W., "Plasma chromatography—New dimension for gas chromatography and mass spectrometry", *Journal of Chromatographic Science* 1970, *8*, 330–337.
10. Carr, T. W. *Plasma chromatography*; Plenum Press: New York, 1984.
11. Wyttenbach, T.; Bowers, M. T., "Gas-phase conformations: The ion mobility/ion chromatography method", *Topics in Current Chemistry* 2003, *225*, 207–232.
12. Jarrold, M. F., "Peptides and proteins in the vapor phase", *Annual Review of Physical Chemistry* 2000, *51*, 179–207.
13. Bohrer, B. C.; Merenbloom, S. I.; Koeniger, S. L.; Hilderbrand, A. E.; Clemmer, D. E., "Biomolecule analysis by ion mobility spectrometry", *Annual Review of Analytical Chemistry* 2008, *1*, 293–327.
14. Ruotolo, B. T.; Giles, K.; Campuzano, I.; Sandercock, A. M.; Bateman, R. H.; Robinson, C. V., "Evidence for macromolecular protein rings in the absence of bulk water", *Science* 2005, *310*, 1658–1661.
15. McLean, J. A.; Ruotolo, B. T.; Gillig, K. J.; Russell, D. H., "Ion mobility-mass spectrometry: A new paradigm for proteomics", *International Journal of Mass Spectrometry* 2005, *240*, 301–315.
16. Giles, K.; Pringle, S. D.; Worthington, K. R.; Little, D.; L., W. J.; Bateman, R. H., "Applications of a traveling wave-based radio-frequency-only stacked ring ion guide", *Rapid Communications in Mass Spectrometry* 2004, *18*, 2401–2414.
17. Williams, J. P.; Scrivens, J. H., "Coupling desorption electrospray ionisation and neutral desorption/extractive electrospray ionisation with a travelling-wave based ion mobility mass spectrometer for the analysis of drugs", *Rapid Communications in Mass Spectrometry* 2008, *22*, 187–196.
18. Ruotolo, B. T.; Benesch, J. L. P.; Sandercock, A. M.; Hyung, S.-J.; Robinson, C. V., "Ion mobility-mass spectrometry analysis of large protein complexes", *Nature Protocols* 2008, *3*, 1139–1152.
19. Mesleh, M. F.; Hunter, J. M.; Shvartsburg, A. A.; Schatz, G. C.; Jarrold, M. F., "Structural information from ion mobility measurements: Effects of the long-range potential", *Journal of Physical Chemistry* 1996, *100*, 16082–16086.
20. Liebler, D. *Introduction to proteomics: Tools for the new biology*; Humana Press: Totowa, NJ, 2002.
21. McLean, J. A.; Russell, D. H., "Sub-femtomole peptide detection in ion mobility-time-of-flight mass spectrometry measurements", *Journal of Proteome Research* 2003, *2*, 427–430.

22. Griffin, G. W.; Dzidic, I.; Carroll, D. I.; Stillwell, R. N.; Horning, E. C., "Ion mass assignments based on mobility measuremets. Validity of plasma chromatographic mass mobility correlations", *Analytical Chemistry* 1973, *45*, 1204–1209.

23. Hoaglund-Hyzer, C. S.; Counterman, A. E.; Clemmer, D. E., "Anhydrous protein ions", *Chemical Reviews* 1999, *99*, 3037–3080.

24. Shvartsburg, A. A.; Jarrold, M. F., "An exact hard-spheres scattering model for the mobilities of polyatomic ions", *Chemical Physics Letters* 1996, *261*, 86–91.

25. Creaser, C. S.; Griffiths, J. R.; Stockton, B. M., "Gas-phase ion mobility studies of amines and polyether/amine complexes using tandem quadrupole ion trap/ion mobility spectrometry", *European Journal of Mass Spectrometry* 2000, *6*, 213–218.

26. Creaser, C. S.; Griffiths, J. R., "Atmospheric pressure ion mobility spectrometry studies of cyclic and acyclic polyethers", *Analytica Chimica Acta* 2001, *436*, 273–279.

27. Hilderbrand, A. E.; Myung, S.; Clemmer, D. E., "Exploring crown ethers as shift reagents for ion mobility spectrometry", *Analytical Chemistry* 2006, *78*, 6792–6800.

28. Ruotolo, B. T.; Tate, C. C.; Russell, D. H., "Ion mobility-mass spectrometry applied to cyclic peptide analysis: Conformational preferences of gramicidin S and linear analogs in the gas phase", *Journal of the American Society for Mass Spectrometry* 2004, *15*, 870–878.

29. Dwivedi, P.; Bendiak, B.; Clowers, B. H.; Hill Jr, H. H., "Rapid resolution of carbohydrate isomers by electrospray ionization ambient pressure ion mobility spectrometry-time-of-flight mass spectrometry (ESI-APIMS-TOFMS)", *Journal of the American Society for Mass Spectrometry* 2007, *18*, 1163–1175.

30. Fenn, L. S.; McLean, J. A., "Structural resolution of carbohydrate positional and structural isomers based on gas-phase ion mobility-mass spectrometry", *Physical Chemistry and Chemical Physics* 2010, submitted.

31. Gant-Branum, R. L.; Kerr, T. J.; McLean, J. A., "Labeling strategies in mass spectrometry-based protein quantitation", *Analyst* 2009, *134*, 1525–1530.

32. Gygi, S. P.; Rist, B.; Gerber, S. A.; Turecek, F.; Gelb, M. H.; Aebersold, R., "Quantitative analysis of complex protein mixtures using isotope-coded affinity tags", *Nature Biotechnology* 1999, *17*, 994–999.

33. Oda, Y.; Huang, K.; Cross, F. R.; Cowburn, D.; Chait, B. T., "Accurate quantitation of protein expression and site-specific phosphorylation", *Proceedings of the National Academy of Sciences of the United States of America* 1999, *96*, 6591–6596.

34. Pasa-Tolic, L.; Jensen, P. K.; Anderson, G. A.; Lipton, M. S.; Peden, K. K.; Martinovic, S.; Tolic, N.; Bruce, J. E.; Smith, R. D., "High throughput proteome-wide precision measurements of protein expression using mass spectrometry", *Journal of the American Chemical Society* 1999, *121*, 7949–7950.

35. Kerr, T. J.; McLean, J. A., "Peptide quantitation using primary amine selective metal chelation labels for mass spectrometry", *Chemical Communications* 2010, *46*, 5479–5481.

36. Kerr, T. J.; McLean, J. A., "High density shift reagents for multiplexed peptide quantitation by ion mobility-mass spectrometry", *Analytical Chemistry* 2010, in preparation.

37. Gant-Branum, R. L.; Kerr, T. J.; McLean, J. A., "Relative quantitation of phosphorylated peptides and proteions using phosphopeptide element-coded affinity tagging (PhECAT)", *Analytical Chemistry* 2010, in preparation.

38. Fenn, L. S.; McLean, J. A., "Enhanced carbohydrate structural selectivity in ion mobility-mass spectrometry analyses by boronic acid derivatization", *Chemical Communications* 38:5, issue ii, 2008, 5505–5507.

39. Fenn, L. S.; McLean, J. A. "Biomolecular structural separations by ion mobility-mass spectrometry", *Analytical and Bioanalytical Chemistry* 2008, *391*, 906.

40. Fenn, L. S.; Kliman, M.; Mahsutt, A.; Zhao, S. R.; McLean, J. A., "Characterizing ion mobility-mass spectrometry conformation space for the analysis of complex biological samples", *Analytical and Bioanalytical Chemistry* 2009, 394, 235.
41. McLean, J. A., "The mass-mobility correlation redux: The conformational landscape of anhydrous biomolecules", *Journal of the American Society for Mass Spectrometry* 20, 2009, *41*, 1775–1781.

Index

Printed and bound by CPI Group (UK) Ltd, Croydon, CR0 4YY

21/10/2024

01777105-0011